Studies in Neuroscience, Psychology and Behavioral Economics

Series Editors

Martin Reuter, Rheinische Friedrich-Wilhelms-Universität Bonn, Bonn, Germany

Christian Montag, Institut für Psychologie und Pädagogik, Universität Ulm, Ulm, Germany

This interdisciplinary book series addresses research findings and methods in the intersection of the fields of neuroscience, psychology and behavioral economics. At the core of the series is the biological basis of human decision making, including aspects of an economic, cognitive, affective and motivational nature. Contributions from fundamental and applied sciences are equally represented and cover a broad range of scientific work, reaching from mathematical modeling to scholarly introductions into new research disciplines and related methodologies and technologies (e.g. genetics, psychophysiology, brain imaging, computational neuroscience, animal modeling). The book series presents empirical and methodological scientific innovations bridging the gap from brain to behavior. It includes research works in neuroeconomics and neuro-information systems (e.g. human-machine interaction) as well as social, cognitive and affective neuroscience, which aim at enhancing the understanding of human decision making.

More information about this series at https://link.springer.com/bookseries/11218

Halley M. Pontes
Editor

Behavioral Addictions

Conceptual, Clinical, Assessment, and Treatment Approaches

Springer

Editor
Halley M. Pontes
Department of Organizational Psychology
Birkbeck, University of London
London, UK

ISSN 2196-6605 ISSN 2196-6613 (electronic)
Studies in Neuroscience, Psychology and Behavioral Economics
ISBN 978-3-031-04771-8 ISBN 978-3-031-04772-5 (eBook)
https://doi.org/10.1007/978-3-031-04772-5

This Springer imprint is published by the registered company Springer Nature Switzerland AG
The registered company address is: Gewerbestrasse 11, 6330 Cham, Switzerland

Series Editor's Foreword

The study of behavioral addictions presents as an important topic in the behavioral and psychological sciences. In particular, addictive behaviors within the cyberspace has become a major focus of research for scientists around the globe. With more than 65% of the world being connected to the internet and the rapid rise of smartphone and social media use, it is not surprising that scientists are actively seeking to better understand how the overuse of diverse online content can be best characterized. Addictive use of the internet has now been studied for over two decades and the question how (over-)use of different online content impacts onto the psyche of internet users is still growing in importance in a world that is becoming increasingly digital.

This excellent book—edited by a highly recognized researcher investigating both Internet Use Disorders (IUD) and in particular disordered gaming, Dr. Halley Pontes—provides interested scientists, psychology students, and mental health practitioners alike with a thorough overview on the rapidly evolving field of behavioral addictions. The book includes ten unique chapters that not only focus on online addictive behaviors, but also addictive behaviors, which happen to appear in the "offline" world. It is likely that the distinction between online and offline behaviors will not be of relevance in the near future, when human mankind navigates through a completely connected world, hence the 'Internet of Things'.

The present book starts with two chapters providing the reader with insights regarding two officially recognized additive behaviors (Chaps. 1 and 2): Both Gambling Disorder (6C50) and Gaming Disorder (6C51) are included in the 11th International Classification of Diseases (ICD-11), and in particular, the inclusion of Gaming Disorder marked an important point in the study of online addictive behaviors. Six years after the inclusion of Internet Gaming Disorder in the 5th Diagnostic and Statistical Manual of Mental Disorders (DSM-5) as a tentative diagnosis in 2013, the World Health Organization decided to include Gaming Disorder as a bona fide addictive disorder in the latest revision of the ICD, stirring up a heated debate among scientists discussing if this diagnosis came too early. It needs to be mentioned that not only the Gaming Disorder diagnosis itself is still a matter of ongoing debate among scientists, but also the nature of excessive online behavior in general. Please

Preface

This book represents one of my best attempts at understanding the nature of (non-substance) addictions. I have been interested in the study of behavioral of addiction for over a decade now, and I can confidently say that studying addictive behaviors and addiction has truly been the most fascinating and important aspect of my life as a psychologist, scientist, and educator.

Often when I introduce myself to my psychology students in the first lecture of the year, I tell them that as a psychologist, I have two life questions that explain and capture really well why I do what I do in my work. For the purposes of this Preface, I will only focus on the first question: *Why do we do something too much, to a point where we disregard everything else?* Hopefully after reading this book, you will see that this question clearly lies at its heart as it can be seen as the main driving force behind it.

Addiction as a scientific notion and social phenomenon has been extensively debated over the last decades. Initially, scientists believed that addictions would only emerge in response to the continued use of psychoactive substances. Nevertheless, this limited understanding started to radically change around the 1980s with the formal medical recognition and introduction of 'pathological gambling' as a psychiatric disorder in the third revision of the Diagnostic and Statistical Manual of Mental Disorders (DSM-III) by the American Psychiatric Association.

Personally, I believe this marked a significant milestone within psychiatry and psychology, leading to the revolution and modernization of the concept of addiction. This is because if we can accept that one behavior that does not involve the use of a psychoactive substance (i.e., gambling) can become an addiction, then logically speaking, what would preclude other potentially problematic behaviors to share the same fate, provided that there is sufficient empirical evidence supporting this recognition?

In fact, this gradual shift from 'addiction as being exclusively related to substance use' to 'addiction as also including behavioral phenomena not involving substance use' has arguably led to many important scientific developments. In my view, the first one relates to the official definition of addiction in this day and age. The American Society of Addiction Medicine (ASAM) has recently redefined addiction as follows:

> *[...] a treatable, chronic medical disease involving complex interactions among brain circuits, genetics, the environment, and an individual's life experiences. People with addiction use substances* ***or engage in behaviors*** *that become compulsive and often continue despite harmful consequences. Prevention efforts and treatment approaches for addiction are generally as successful as those for other chronic diseases* (see https://www.asam.org/quality-care/definition-of-addiction, accessed on March 14, 2022).

Fundamentally, what we see today is that addiction is understood from different standpoints (see the seminal work by West, 2001[1]), with the field having clearly moved beyond the mere use and abuse of substance and its nefarious effects on people's lives. The current scientific view on addiction has been further cemented with the publication of the DSM-5, where 'pathological gambling' (now known as Gambling Disorder) has been reclassified and added to the 'non-substance-related disorders' and 'Internet Gaming Disorder' made an appearance as a tentative disorder—later on in 2019 being fully recognized as a bona fide addictive disorder and the second behavioral addiction; see Pontes and Griffiths (2020)[2] for a brief overview.

Perhaps the second key scientific development stemming from such a conceptual shift is related to the exponential growth in prominence and legitimacy of this field. According to a recent study by Sixto-Costoya and colleagues (2022),[3] research on behavioral addiction has greatly increased over the last 25 years, with scientific production in the field doubling up every five years from 1995 to 2019. The scientific growth in behavioral addiction research was mostly observed within the subfields of Gambling Disorder (33.83%), information and technology communication-related disorders (31.73%), Gaming Disorder (11.69%), Food Addiction (3.81%), Shopping Addiction (7.39%), Sports Addiction (3%), and Sex Addiction (3.56%). This growth in prominence and legitimacy as a field of research has also been continuously supported over the years by major academic journals playing a significant role in promoting behavioral addiction research, such as the *Journal of Behavioral Addictions* (ISSN: 20625871), *Addiction* (ISSN: 09652140), *Addictive Behaviors* (ISSN: 03064603), *Addictive Behaviors Reports* (ISSN: 23528532), and *Psychology of Addictive Behaviors* (ISSN: 0893164X). Additionally, behavioral addiction experts from all over the world gather every year at major events such as the annual *International Conference on Behavioral Addictions* and the *Lisbon Addictions* conference. Indeed, the field of behavioral addiction has grown so much that experts have come together to create the *International Society for the Study of Behavioral Addictions* (ISSBA) (http://issba.elte.hu), which is a society focusing on the study of

[1] West, R. (2001). Theories of addiction. *Addiction, 96*(1), 3–13. https://doi.org/10.1046/j.1360-0443.2001.96131.x.

[2] Pontes, H. M., & Griffiths, M. D. (2020). A new era for gaming disorder research: Time to shift from consensus to consistency. *Addictive Behaviors, 103*, 106059. https://doi.org/10.1016/j.addbeh.2019.106059.

[3] Sixto-Costoya, A., Castelló-Cogollos, L., Aleixandre-Benavent, R., & Valderrama-Zurián, J. C. (2021). Global scientific production regarding behavioral addictions: An analysis of the literature from 1995 to 2019. *Addictive Behaviors Reports, 14*, 100371. https://doi.org/10.1016/j.abrep.2021.100371.

behavioral addictions. Additional, researchers and international experts in the field have come together to promote research in behavioral addiction through dedicated research groups such as the world-leading 'The International Gaming Research Unit (IGRU)' (see https://www.ntu.ac.uk/research/groups-and-centres/groups/the-international-gaming-research-unit) which is led by Distinguished Professor Mark Griffiths (Nottingham Trent University, United Kingdom).

Given the rapidly evolving nature of our current understanding of addiction as a bio-psycho-social phenomenon, the latest societal changes, and technological developments we have seen, it is paramount and timely to publish this book as it intends to provide an updated and focused starting point for some of the most fascinating behavioral addictions that researchers have studied over the last few years. This is the mission of this book and its ten unique chapters as it intends to cover a wide range of behavioral addictions related to gambling, gaming, social media, smartphone, internet, love, sex, exercise, work, and shopping. All ten chapters were written by leading international experts with an extensive track record in behavioral addiction research. The language adopted in this work aims to make the book accessible to all audiences so that the reader can grasp the key scientific principles underpinning each behavioral addiction discussed in all chapters. This book is designed to be flexible and used in different contexts, both as a concise handbook that can be read as a single piece or as a document that can be consulted at a later date according to the reader's needs. I sincerely hope that you enjoy this book as much as I did when editing it.

As a forever curious psychologist and scientist, I recognize that after reading all ten chapters of this book I was left with more exciting research ideas and further questions about the nature of behavioral addictions than answers. And as a result, I am thrilled to say that I shall introduce myself to my students next year with the same old and yet relevant life questions I had prior to devoting my time to this book.

To all my readers, thanks for your interest in this book. Please feel free to visit my website at https://www.halleypontes.com and to connect with me on Twitter (https://twitter.com/DrHalleyPontes). I thank all my dedicated and incredibly talented colleagues for helping me with this book as without them and their wealth of expertise, this work would not have existed at all. My thanks also goes to Professor Christian Montag for his continued inspiration and dedication to the field and his work. I also extend my gratitude to Distinguished Professor Mark Griffiths for nearly a decade of mentorship, for his inspiration, and for having taught me important life and academic skills. Last but not least, I would like to dedicate this book to the one and only, *Emilia*—thank you for everything.

London, UK Halley M. Pontes

Contents

About the Editor

Dr. Halley M. Pontes is a chartered psychologist of the British Psychological Society (CPsychol), a member of the Australian Psychological Society (MAPS), and a Chartered Scientist (CSci) of the Science Council. He currently works as a researcher and a lecturer at Birkbeck, University of London, in the United Kingdom, and his main research interests relate to the mental health and psychological health and well-being issues linked with emerging addictive behaviors such as gaming disorder, social media addiction, Internet addiction, among other problematic behaviors (e.g., work addiction). He has received his Ph.D. from the Nottingham Trent University in the United Kingdom and has published over 100 studies in several scientific journals and presented his work on numerous international conferences. He has been previously awarded with the Durand Jacobs Award from McGill University (Canada) and the Early Career Research Award from the Australian Psychological Society (Australia) for his important contributions to the psychology of addictive behaviors and cyberpsychology.

Chapter 1
Gambling Disorder

Mark D. Griffiths and Filipa Calado

Abstract Gambling as a leisure activity has now become widespread in many countries. While the majority of individuals who gamble experience no significant negative detrimental effects, research has consistently shown that a small minority develop problems, and that for an even smaller minority, the behavior appears to be an addiction just like other more traditional psychoactive substance-based addictive behaviors. This chapter briefly overviews problem gambling behaviors by examining the (i) demographics of gambling and problem gambling, (ii) prevalence of gambling and problem gambling, (iii) psychological theories of problem gambling, (iv) importance of structural and situational characteristics in problem gambling, (v) biological bases of problem gambling, (vi) specific comorbidities, (vii) problem gambling assessment approaches, and (viii) treatment approaches for problem gambling. Problem gambling, like other addictions, results from an interaction and interplay between many factors including the individual's biological and/or genetic predisposition, their psychological constitution, their social environment, and structural characteristics of the gambling activity itself.

1.1 Introduction

Gambling is a heterogeneous activity that cuts across barriers of culture, class, and race (Griffiths 1995). There are many different types of gambling including the playing of slot machines, betting on sports events, buying a lottery ticket, playing bingo, and gambling on the stock market. Some of these gambling activities rely purely on chance (e.g., playing roulette, buying a scratchcard) whereas other gambling activities involve some form of skill (e.g., playing poker or blackjack, sports betting). In essence, gambling is defined as the staking of money (or something of financial value) on a future event that the outcome is unknown at the time of staking the money (Griffiths 1995). Gambling often takes place in dedicated

M. D. Griffiths (✉) · F. Calado
International Gaming Research Unit, Department of Psychology, Nottingham Trent University, Nottingham, UK
e-mail: mark.griffiths@ntu.ac.uk

H. M. Pontes (ed.), *Behavioral Addictions*, Studies in Neuroscience, Psychology and Behavioral Economics, https://doi.org/10.1007/978-3-031-04772-5_1

gambling venues such as casinos, betting shops, amusement arcades, bingo halls, amusement and arcades but can now be engaged in remotely via the internet on laptops, tablets, and smartphones.

For a small minority of individuals, gambling can become a problematic. There are many terms that have been used to describe such as problematic gambling including (among others) 'addictive', 'disordered', 'pathological', 'dependent', 'impulsive' and 'compulsive', (Griffiths and Delfabbro 2001). Although there is no total agreement among scholars, 'problem gambling' is arguably the most commonly used descriptor, and is a general term used to indicate varying patterns of damaging or disrupting gambling behavior. Some individuals can experience problems associated with their gambling that do not meet the full criteria for 'gambling disorder' as defined in diagnostic manuals such as the latest (fifth) edition of the Diagnostic and Statistical Manual of Mental Disorders (DSM-5; American Psychiatric Association 2013) or the latest (eleventh) revision of the International Classification of Diseases (ICD-11, World Health Association, 2019). Most problem gamblers do not display all the core components of genuine addictions (i.e., conflict, salience, withdrawal symptoms, relapse, tolerance, and mood modification; Griffiths 2005). Put very simply, all gambling addicts are problem gamblers, but not all problem gamblers are gambling addicts (Griffiths 2016).

The term 'problem gambling' is typically used to describe gambling behavior that damages, disrupts or compromises personal and family relationships (e.g., relationship breakdowns, divorce), as well as occupational/educational activities (e.g., poor work performance, job loss) (Griffiths 2004, 2006). Unsurprisingly, problem gambling typically results in great financial distress such as unpaid creditors, substantial debts, insolvency, and bankruptcy (Griffiths 2006). In the most extreme cases, legal problems will occur as a consequence of criminal behavior (e.g., stealing, fraud, embezzlement, etc.) in attempts to obtain money to gamble or to pay off gambling debts (Dennison et al. 2020; Griffiths 2006). This can also lead to homelessness (Nower et al. 2015). Other criminal acts are associated with problem gambling such as perpetration of intimate partner violence (IPV). A meta-analytic review of IPV in 14 studies by Dowling et al. (2016) reported that 36.5% of problem gamblers were perpetrators of physical IPV (36.5%), 38.1% reported being the victims of physical IPV (38.1%), and that among IPV perpetrators, the prevalence of problem gambling was 11.3%.

There are numerous screening instruments that have been developed to assess problematic gambling behavior and most individuals working in the gambling studies field agree that problematic behavior exists on a continuum from relatively minor problems at one end to extreme and pathological behavior at the other. The behavior can move in and out of being within the problematic status, and for many individuals there is a natural remission as they 'mature' out of the behavior (Meyer et al. 2009). This conception allies with the emphasis on general public health, which focuses on the personal, social, and physical 'harms' that problem gambling can create across various sectors of the general population, rather than the historically narrow focus on the problems of a minority of 'addicted' individuals. For example, the families and close friends of problem gamblers typically experience substantial difficulties (e.g., psychological, social and/or financial problems) (Griffiths and Delfabbo 2001; Griffiths 2006).

1.2 The Demographics of Gambling and Problem Gambling

To further understand gambling and problem gambling, it is important to examine its demographics (Griffiths and Delfabbro 2001). In fact, there are several groups that have been reported to be more likely to gamble and to develop gambling-related problems (Calado and Griffiths 2016). For instance, men have been found to gamble more than women (Costes et al. 2015; Sassen et al. 2011; Spritzer et al. 2011), to spend more money on gambling (Raisamo et al. 2014), and to participate more in most gambling activities, with the exception of bingo and scratchcards, which are more played by women (Potenza et al. 2006; Wardle et al. 2011). Gender differences have been attributed to a number of factors, including variations in how men's and women's roles in society are viewed and their motivations (Hayer and Griffiths 2015). Historically, gambling has been viewed as a male-dominated activity (Merkouris et al. 2016), with men being found to be more competitive than women (Burger et al. 2006), and more likely to report gambling for enhancement (excitement or achievement) (Wardle et al. 2011). Therefore, this motivation means more males choose more games of skill as their favorite gambling activities, whereas females choose more passive games, such as bingo and scratchcards (Burger et al. 2006). With regard to other demographic characteristics, individuals from the oldest age groups (aged 60 or more years) are less likely to gamble, whereas individuals aged between 24 and 45 years are more likely to gamble (Wardle et al. 2011; Welte et al. 2015). Furthermore, individuals with lower educational qualifications are more likely to gamble than individuals with higher qualifications (Costes et al. 2015; Wardle et al. 2011).

Problem gambling is more likely to occur among males (e.g., Abbott et al. 2014; Bondolfi et al. 2000; Bonke and Borregaard 2006; Brodbeck et al. 2009; Druine et al. 2006; Olason et al. 2015; Potenza et al. 2019; Williams et al. 2021), those with low education (Potenza et al. 2019), single or divorced individuals (Bakken et al. 2009; Çakici 2012; Druine et al. 2006; Makarovič 2010), black individuals (Potenza et al. 2019), as well as individuals of a younger age (Abbott et al. 2014; Kun et al. 2012; Olason et al. 2015). Moreover, youth constitutes a vulnerable risk factor in the development of problem gambling behavior. In addition, individuals with a lower level of education are more likely to develop problem gambling (Costes et al. 2015; Meyer et al. 2015; Olason and Gretarsson 2009), as well as individuals that belong to an ethnic minority (Makarovič 2010; Seabury and Wardle 2014; Wardle et al. 2011) or who had been born in another country to where they currently reside (Abbott et al. 2014; Bakken et al. 2009; Meyer et al. 2015). Furthermore, individuals unemployed or with a low income have been reported as being more likely to be problem gamblers

(Castrén et al. 2013; Costes et al. 2015; Druine et al. 2006; Ilkas and Turja 2003; Kun et al. 2012; Meyer et al. 2015; Sassen et al. 2011; Williams et al. 2021).

As can be seen by the brief analysis of socio-demographics, problem gambling is a complex and multidimensional phenomenon. The literature shows that social, psychological, and biological factors are involved in the development and maintenance of the behavior. Therefore, several theoretical models have attempted to explain the mechanisms underlying both gambling and problem gambling. Some of the major theories are outlined later in the chapter.

1.3 Prevalence of Gambling and Problem Gambling

There are far more nationally representative surveys assessing the prevalence of problem gambling than there are for other types of behavioral addiction. Most national surveys conducted worldwide have reported that most individuals have engaged in gambling at some point during their lives, and there are more individuals that gamble annually than those who do not. In a systematic review of worldwide problem gambling prevalence (from 2000 to 2015), Calado and Griffiths (2016) reported that the lifetime prevalence of problem and gambling disorder/pathological gambling (combined) in 69 studies ranged from 0.7% (in Denmark) to 6.5% (in Estonia)—although the latter figure was from research carried out in 2004. In the same study, the past-year problem gambling prevalence ranged from 0.12% (in Switzerland) to 5.8% (in Hong Kong). However, the prevalence rate of problem gambling in more recent Hong Kong studies is now much lower. The review by Calado and Griffiths also included a detailed country-by-country analysis of problem gambling in Europe.

Despite the variations in the prevalence rates of problem gambling, Calado and Griffiths (2016) reported that the socio-demographic findings in most European countries were very similar. More specifically, problem gambling was more likely to occur among males (e.g., Belgium, Cyprus, Denmark, Finland, Germany, Great Britain, Hungary, Iceland, Norway, Switzerland, and Sweden), single or divorced individuals (e.g., Belgium, Cyprus, Denmark, Great Britain, Iceland, Norway, and Slovenia), individuals of a younger age (e.g., Denmark, Estonia, France, Hungary, Norway, Sweden, Switzerland, Hungary, and Iceland), individuals with a lower level of education (e.g., Frances, Germany, Iceland, Norway, and Sweden), individuals that belong to an ethnic minority (e.g., Great Britain, Netherlands, and Slovenia), or who had been born abroad (e.g., Germany, Norway, and Sweden), and individuals unemployed or with a low income (e.g., Belgium, Finland, France, Germany, Hungary, and Netherlands). It should also be noted that research has consistently shown that there are high prevalence estimates for gambling disorder among specific clinical populations and/or those with specific neurological and medical conditions such as those with problematic psychoactive substance use (4.3–6.9%) or Parkinson's disease (2.2–7%) (Potenza et al. 2019). It has also been estimated that among those with gambling disorder, the majority of individuals have one or more psychiatric disorder (96%) and that two-thirds have three or more such disorders (64%) (Potenza et al. 2019).

The most popular types of gambling activities across European countries were lotteries, scratchcards, sports betting, and slot machines (e.g., Cyprus, Finland, Great Britain, Hungary, Iceland, Norway, and Sweden). However, the types of gambling activities most associated with problem gambling were slot machines and online gambling games (e.g., Germany, Norway, and Portugal). It was also noted that the prevalence rates of problem gambling remained relatively stable among those countries that have carried out more than one nationally representative survey (e.g., Germany, Great Britain, Netherlands, and Germany) (Calado and Griffiths 2016; Potenza et al. 2019). However, it should be noted that all the studies utilized self-report data, which are subject to many well known methods biases and there were many different screening instruments used.

In a systematic review of worldwide adolescent problem gambling prevalence (from 2000 to 2015), Calado et al. (2017) reported that the lifetime prevalence of problem and gambling disorder/pathological gambling (combined) in 44 studies among adolescents ranged from 1.6% (in Denmark and Brazil) to 5.6% (in Albania). In the same study, the past-year problem gambling prevalence among adolescents ranged from 0.2% (in Norway and Australia) to 5.8% (in Spain) although this figure was from a regional study and has decreased in more recent Spanish studies. They also noted that one study carried out in Croatia reported a problem gambling prevalence rate of 12.3% but this was a study with a three-month timeframe that used a screening instrument not employed in any other study. The review by Calado et al. also included a detailed country-by-country analysis of problem gambling in Europe.

As with studies of problem gambling among adults, Calado et al. (2017) reported that the socio-demographic findings in most European countries were very similar. Adolescent problem gambling was more likely to occur among males (e.g., Croatia, Denmark, Finland, United Kingdom, Iceland, Italy, Lithuania, Norway, Spain, Sweden, and Switzerland); among ethnic minority individuals (e.g., Great Britain, Norway, and Sweden); among individuals whose parents gambled (e.g., Great Britain, Iceland, and Spain); among individuals who did not live with both of their parents (e.g., Norway); and among older-aged adolescents (e.g., Denmark, Lithuania, and Switzerland). Adolescent problem gamblers were also more likely to gamble online (e.g., Denmark and Iceland). The review by Calado et al. (2017) also noted that the most prevalent gambling activities among adolescents were slot machines, card games, and scratch-cards (Bosnia and Herzegovina, Great Britain, Iceland, Luxembourg, and Norway). It was also reported that the types of gambling mostly engaged by problem gamblers were sports betting, slot machines, and card games (i.e., games with high event frequencies). Most European research on adolescent gambling has reported a clear association between gambling and substance abuse (Great Britain, Italy, Lithuania, Norway, Romania, Spain, and Switzerland). However, all the aforementioned adolescent gambling studies relied on self-report survey data and were typically carried out in school-based settings. Additionally, some of the studies did not have a nationally representative sample and there were many different screening instruments used (almost all of which were based on instruments originally developed for use among adults and then adapted by using more child-friendly wordings).

Within in each country, findings generally demonstrate that the prevalence rate of problem gambling among adolescents tends to be higher than the prevalence of problem gambling among adults (Calado and Griffiths, 2016). There are many possible reasons for this phenomenon, including but not limited to developmental and generational factors. More specifically, (i) the transitional period of adolescence is typically characterized by an increase in risky behaviors more generally (e.g., cigarette smoking, alcohol consumption, illicit drug use, as well as gambling) which then 'mature out' into adulthood as other life events take precedence (e.g., full-time employment, marriage, parenthood; Griffiths 1995); and (ii) millennials have grown up in a period where gambling is socially acceptable, widely available, and widely marketed (Volberg et al. 2010).

1.4 Psychological Theories of Problem Gambling

Psychoanalytic theories: The first psychologists to offer an explanation for problem gambling were the psychoanalysts. However, there is no single cohesive psychodynamic or psychoanalytic theory that has been advanced in order to explain the development of problem gambling (Rickwood et al. 2010). Psychoanalytic theorists adopted the view that problem gambling was the consequence of an underlying neurosis related to a regression to pre-genital psychosexual stages of development (Rosenthal 2008). Von Hattingberg (1914, cited by Moreyra et al. 2000), who conducted the first study of problem gambling, proposed that problem gamblers had a fixation in the anal stage of development, which explained the compulsive and masochistic traits in their personality. It was theorized that problem gamblers eroticized the tension and fear involved in gambling. The work of Von Hattingberg was further elaborated by Freud (1928) and Bergler (1957) who have largely influenced later psychoanalytic work on this topic. Freud (1928) suggested that gamblers played to lose and that it constituted a form of punishment that secondarily becomes a pleasurable activity. For Bergler (1957), problem gambling was viewed as masochistic and related to the Oedipus complex. According to his theory, a gambler's unconscious motivations included the desire to lose and to be punished. For Bergler (1957), the problem gambler unconsciously resents their parents and other early authority figures who forced them, as a child, to give up the pleasure principle in order to adopt the reality principle. Given the reliance on clinical case material and absence of empirically testable hypotheses, psychoanalytic models are generally considered to be of limited utility in explaining either the onset and maintenance of problem gambling. However, the psychoanalytic model is the only theory which acknowledges unconscious determinants in gambling.

Cognitive theories: Some authors emphasize the role of erroneous beliefs and irrational thinking, in the etiology and maintenance of gambling and problem gambling behaviors (e.g., Griffiths 1990; Petry 2005; Toneatto et al. 1997; Walker 1992). In order to investigate gamblers' beliefs, researchers have asked gamblers to verbalize their thoughts during gambling sessions, a technique called the 'thinking aloud' method. Using this method, researchers identified many cognitive distortions and misconceptions about gambling (Griffiths 1994). Some relevant cognitive distortions include (i) magnification of gambling in which gambling skill is overestimated, (ii) predictive control where cognitions are related to the ability to predict gambling outcomes, (iii) illusion of control where gamblers overestimate their personal abilities and skills in influencing game outcome, (iv) superstitious beliefs where gamblers carry specific objects, or perform specific actions and rituals, thinking this can influence the probability of winning; (v) selective memory where individuals recall large wins but have difficulty in recalling losses; and (vi) illusory correlations where gamblers assign false causality to salient features in the environment that they think are related to winning (Griffiths 1994; Toneatto 1999). Despite these strengths, cognitive theories have some limitations because they do not explain the functional interaction between arousal, conditioning, and cognitive activity, or the transition from recreational to problem gambling (Rickwood et al. 2010).

Behavioral theories: Behavioral theories view problem gambling as a learned behavior acquired through a process of reinforcement operating under operant and classical conditioning paradigms (Rickwood et al. 2010; Delfabbro 2013). Operant conditioning is established through fixed and variable ratio schedules of reinforcement (e.g., sensory stimulation, short interval between stake and payout in some gambling games), in which the behavior is maintained by intermittent schedules of reinforcement, most likely a variable-ratio schedule (Polezzi et al. 2012). The behavioral theory also postulates that both positive and negative reinforcement increase the probability of a gambling response being elicited and also explain persistence in gambling (Rickwood et al. 2010). The experience and the excitement of winning alongside various concomitant stimuli, act as positive reinforcers that help to maintain the behavior over time (Coventry and Constable 1999), whereas negative reinforcement is thought to occur when individuals use gambling as a way to escape from unpleasant experiences including depression, anxiety, boredom, and other adverse life-events (Petry 2005). Early wins are particularly important at the initial stages of a gambling career, with almost half of the gamblers reporting a significant monetary gain in the initial phase of their disorder that could act as a trigger for it (Kassinove and Schare 2001; Braverman and Shaffer 2012). Classical conditioning takes place when individuals continue to gamble as a result of becoming conditioned to states such as the excitement or arousal associated with gambling, leading them to feel bored, unstimulated, and restless when they are unable to gamble (Polezzi et al. 2012). Despite the existing evidence supporting behavioral theories, they fail to acknowledge the importance of internal events, as they tend to underestimate the power of individual motivation, emotions, and perceptions to influence gambling-related outcomes (Raylu and Oei 2002).

Need-state theories and addiction: Need state theories postulate that individuals gamble to escape from unpleasant feeling states such as anxiety, depression, and boredom (Blaszczynski and Nower 2002). Therefore, it can be assumed that principles of reinforcement derived from behavior theory are also incorporated in addiction models. According to such theory, susceptibility to addiction arises as a result of personal vulnerability and to the extent to which the experience of gambling assists individuals in dealing with underlying psychological problems (Delfabbro 2013). Both Walker (1992) and Jacobs (1986) proposed that problem gambling could often be described as a form of 'psychological addiction' in that engagement in the behavior can be negatively reinforcing. In Jacobs' (1986) view, there are specific physiological and psychological characteristics or experiences that can make individuals prone to this form of vulnerability. At a physiological level, some individuals may be chronically under- or over-aroused, so that they need to engage in alternative activities in order to obtain an optimal level of arousal that is hedonistically comfortable. For some, this may mean engaging in risky activity to gain excitement and an increase in physiological arousal. For others, the activity may have a cathartic or calming effect that enables individuals to reduce their arousal. At a psychological level, problem gamblers have been found to have lower self-esteem, mood disturbances, and often report a history of negative life events, early childhood trauma and rejection, and are more likely to have histories of early trauma, abuse or stressful life events that preceded the gambling problems (Scherrer et al. 2007; McCormick et al. 2012). Therefore, gambling is used to provide an escape from these problems. These models suggest that some individuals have a greater need for arousal while others have a need for relaxation. Therefore, it is assumed that it is unlikely that avoidance of negative feeling states will be common to all activities or all gamblers. However, as Griffiths and Delfabbro (2001) noted, it is unclear why some individuals have a greater need for relaxation and arousal, and whether this need would be sufficient to explain normal and problem gambling.

1.5 The Importance of Structural and Situational Characteristics in Problem Gambling

Structural characteristics in gambling refer to specific characteristics that are inherent within the gambling game. Such features include win probability, stake size, sound and lighting effects of the game, event frequency (how many wagers an individual can place in a given period of time), payout interval (the time between starting the game and payout), jackpot size, prize structure, near-miss opportunities (the psychological bias of interpreting losses as near wins or anticipatory of a winning streak), and whether the game has any element of skill (Griffiths 1993; Griffiths and Auer 2013; Parke and Griffiths 2006, 2007). The gambling industry has employed numerous game design features affecting the structural and situational characteristics of the gambling experience in order to entice the start and maintenance of gambling behaviors.

Each gambling activity has different structural characteristics that provide insights into the 'addictiveness' of a specific type of gambling. For instance, slot machines are sometimes colloquially referred to as the 'crack cocaine of gambling' due to its high association with problem gambling (Williams et al. 2021). This phenomenon is related to both high event frequencies (i.e., a player can typically gamble 1000 times an hour on some slot games) and high accessibility (i.e., increased opportunities to engage in the behavior). The structural characteristics of a particular gambling activity may act as reinforcers for a gambling behavior, may satisfy gamblers' needs, and may actually facilitate excessive gambling (Dowling et al. 2005). For instance, a gambling activity characterized by a short time between the initiation of a betting event and the result of that game means that little time are given over to financial considerations, and, winnings can be re-gambled almost immediately (Parke and Griffiths 2006).

In fact, it has been demonstrated in numerous studies that gambling activities, such as slot machines, characterized by short-event frequencies are the gambling activities most played by problem gamblers (Lupu and Todirita 2013; Calado et al. 2017; Williams et al. 2021). Many slot machine players overestimate the amount of skill involved when playing such games (Griffiths 1994). When combined with motivational factors (e.g., gambling as a way of escape; Wood and Griffiths 2007), slot machine gamblers may become conditioned to the 'tranquilizing' effect of engaging in the activity itself rather than the simple desire to win money. High event frequency activities such as slot machines also mean that the speed of play in such games is very fast, and a systematic review examining all studies that had empirically examined speed of play found that it was associated with problem gambling (Harris and Griffiths 2018). Research has also indicated that continuous forms of gambling (e.g., casino games, in-play sports betting, slot machines) with rapid play-rate are more associated with gambling problems (Griffiths 1999; Killick and Griffiths 2019).

Moreover, other structural characteristics that have been found to promote a gambler's desire to continue playing include near misses (Parke and Griffiths 2004; Pisklak et al. 2020). For example, Dixon et al. (2011) demonstrated that skin conductance responses and heart rate deceleration was significantly larger for near misses than either wins or losses, and these arousal responses were not mediated by players' problem gambling status. The authors hypothesized that these arousal patterns and responses are due to the experience of frustration in almost winning. In fact, the psychology of near misses postulates that gamblers become physiologically aroused when they almost win, and this stimulates further gambling behaviors (Griffiths 1991). In short, gamblers do not constantly lose, gamblers constantly nearly win (Griffiths 1994).

Macro-situational characteristics are primarily features of the environment, typically accessibility factors, such as the location of the gambling venue, the number of gambling venues in a specified area and possible membership requirements, free travel to and/or accommodation at the gambling venue, as well as or advertising and marketing that can stimulate individuals to gamble (Griffiths and Parke 2003; Hayer and Griffiths 2015). Micro-situational characteristics comprise the internal features of the gambling venue itself (e.g., decor, heating, lighting, color, background music, floor layout, refreshment facilities) or facilitating factors that may influence gambling in the first place (e.g., free bets or gambles on particular games) or influence continued gambling (e.g., the placing of a cash dispenser on the casino floor, free food and/or alcoholic drinks while gambling). Research has shown that peripheral features such as sounds, and "warm" lights play a key role in the development of some gambling behaviors (Stark et al. 1982; Finlay, Marmurek et al. 2010). For instance, Dixon et al. (2007) reported that fast tempo music (e.g., > 94 beats per minute) had a significant effect on participant betting speed while gambling. Similarly, Spenwyn et al. (2010) found that the combined effects of both high tempo music and red light resulted in faster bets by participants while playing a computerized version of roulette.

1.6 Biological Bases of Problem Gambling

Psychological theories on their own cannot fully explain problematic gambling behavior given that all human behavior is biopsychosocial in nature. Research into the biological bases of problem gambling dates back over three decades to the 1980s with studies examining biological and genetic predispositions. More recently, large twin studies have estimated that genetic factors account for 50–60% of the vulnerability for developing problem gambling (Eisen et al. 2001; Slutske et al. 2010; Giddens et al. 2011). On the other hand, gene association studies primarily report the involvement of genes for the dopaminergic and serotonergic systems (Comings et al. 2001; Lobo et al. 2010; Wilson et al. 2013). In a systematic review on the genetic basis of problem gambling and gambling disorder, Gyollai et al. (2014) identified 21 studies utilizing empirical data from eight independent samples (gene association data for 13 studies mainly examining the involvement of the serotonergic and dopaminergic systems, and twin data examining the role of genetic and environmental factors for the other eight studies). They concluded that even though the number of studies was small, the data clearly indicated that there was a genetic vulnerability in relation to problem gambling.

Molecular genetic studies examining problem gambling have noted that both serotonergic genes (e.g., MAOA, MAOB, SLC6A4) and dopaminergic genes (e.g., DRD1, DRD2, DRD4) may increase the vulnerability in developing problem gambling (Potenza et al. 2019). However, Potenza et al. (2019) caution that findings from molecular genetic studies examining problem gambling should be considered preliminary given that some of the methodological weaknesses (e.g., small sample sizes) and the fact that other heritable traits (e.g., impulsivity) are implicated in problem gambling and it is unclear how such overlaps contribute to genetic susceptibility.

Research within the fields of neurobiology and neurochemistry has demonstrated increasing evidence that implicates multiple neurotransmitter systems (e.g., dopaminergic, serotonergic) in the pathophysiology of problem gambling. The dopaminergic system is associated with reward mechanisms and addictive behaviors (Nestler 2004) and it is hypothesized that changes in dopaminergic pathways might underlie the seeking of rewards (e.g., gambling), that trigger the release of dopamine and produce feelings of pleasure (Zack and Poulos 2009). Problem gamblers have been shown to have increased dopamine transmission in the dorsal striatum (Potenza et al. 2019).

Neuroimaging studies of problem gamblers have indicated diminished ventral striatum and ventromedial prefrontal cortex and ventrolateral prefrontal cortex activity during rewarding events, which is suggestive of a blunted neurophysiological response to rewards and losses (De Ruiter et al. 2009; Reuter et al. 2005). Neurobiological research examining problem gamblers has shown that compared to non-problem gamblers, they have differences in the limbic and frontostriatal brain regions, including the amygdala, striatum, anterior cingulate cortex, orbitofrontal cortex, hippocampus, and insula (Potenza et al. 2019). These brain regions are associated with various characteristics that are central to problem gambling (e.g., stress dysregulation, maladaptive decision-making, chasing losses, excitement and reward sensitivity, and emotional/social problems) (Potenza et al. 2019).

Traditionally, serotonin function has been considered to be of substantial importance in mediating impulse control. Human studies of problem gamblers have shown decreased concentrations of platelet monoamine oxidase B (a peripheral marker of serotonergic function) and low concentrations of serotonin metabolites in the cerebrospinal fluid (Potenza 2001). In addition, research has suggested that problem gambling shares genetic vulnerability factors with other addictions (Slutske et al. 2000; Lang et al. 2016).

Biological models suggest that the drive toward intense is biologically prescribed, but they are unable to account for the full diversity of gambling patterns and behavior. In addition, these theories fail to explain demographic differences in the preference for activities and variations in motivation. Neither can they explain why some activities are more 'addictive' than others and why the structural characteristics of specific activities can influence behavior (Griffiths and Delfabbro 2001).

1.7 Specific Comorbidities

It is important to note that problem gambling is often comorbid with other psychological and behavioral disorders (e.g., depression, anxiety, suicidal ideation, etc.) (Wardle et al. 2020). These comorbid disorders can also exacerbate, or be exacerbated by, problematic gambling behavior (Griffiths 2004). There are many studies over the past three decades demonstrating a strong association between problem gambling and psychiatric comorbidity, particularly depression and anxiety (e.g., Assanangkornchai et al. 2016; Black and Moyer 1998; Brandt and Fischer 2019; Ford and Håkansson 2020; Håkansson et al. 2018; Ibáñez et al. 2001; Kerber et al. 2008; Kessler et al. 2008; Lorains et al. 2011; McCormick et al. 2012; Quigley et al. 2015). A recent national study in Sweden found that reporting anxiety or depression prior to gambling onset was a risk factor for the development of problem gambling among females but not for males (Sundqvist and Rosendahl 2019). Additionally, a recent review by Hartmann and Blaszczynski (2018) examined the longitudinal relationships between psychiatric disorders and problem gambling among 35 studies. Based on the findings of longitudinal studies, they concluded that psychiatric disorders could be both a precursor and a consequence of problem gambling. They also noted that other underlying interactive factors (with impulsivity being one such trait) may predict and drive the relationship between problem gambling and psychiatric distress. A recent systematic review by Richard et al. (2020) examined studies that had investigated the relationship between gambling, depression, and conduct disorders (e.g., aggression, anti-social behaviors, delinquency, etc.) across 47 studies. While cross-sectional studies showed that problem gambling was consistently related to both depression and conduct disorders, longitudinal studies showed that conduct disorders were a risk factor for problem gambling.

Many problem gamblers also experience irrational cognitive distortions and misperceptions in their thinking (e.g., overconfidence, superstitions, denial) (Ciccarelli et al. 2021; Griffiths 1994). Research has consistently shown that problem gambling has a high association with impulsivity as demonstrated in a recent meta-analytic review of 52 studies (Ioannidis et al. 2019). Furthermore, impulsivity may also play a role in the relationship between problem gambling and psychoactive substance abuse. Among problem gamblers, studies have also reported increased rates of substance abuse or dependence, attention-deficit hyperactivity disorder (ADHD), and personality disorders (e.g., narcissistic personality disorder, borderline personality disorder, antisocial personality disorder) (Jacob et al. 2018; Sussman et al. 2011; Theule et al. 2019).

A recent systematic review by Marchetti et al. (2019) examined the relationship between problem gambling and alexithymia (a personality trait in which individuals have a deficit in the cognitive processing of emotion). The review identified 20 studies (14 with community samples and six with clinical samples). The results of this study suggested that alexithymia was significantly more prevalent among problem gamblers (among both clinical and community samples), and that gambling may be used as a coping strategy by individuals in order to avoid negative emotions and increase emotional arousal.

Depression and anxiety caused by problem gambling may result in increased use of alcohol and/or other drugs as a means of coping, and vice versa (Griffiths et al. 2002). Such co-morbidities can differ among different demographic cohorts and different gambling types. Moreover, for some individuals, problem gambling is simply symptomatic of a more global disturbance in their biopsychosocial functioning. More severe problem gamblers tend to be easily bored, restless, highly competitive, energetic, restless (Parke et al. 2004), as well as more prone to experience stress-related physical illnesses including migraines, heart disease, insomnia, peptic ulcer disease, and hypertension (Daghestani et al. 1996; Griffiths et al. 2001; Griffiths 2004). Health-related problems can also be a consequence of withdrawal symptoms. Rosenthal and Lesieur (1992) reported that two-thirds of problem gamblers experienced at least one physical side-effect while undergoing withdrawal including loss of appetite, headaches, muscle aches, insomnia, stomach upsets, heart palpitations, chills and/or breathing difficulties. Such symptoms have also been reported in other studies (e.g., Blaszczynski et al. 2008; Griffiths and Smeaton 2002).

As aforementioned, gambling and problem gambling often co-occur with other potentially addictive behaviors. Among large older teen samples of heavy gamblers, co-occurrence with heavy use of alcohol or marijuana/other illicit drugs has been found to be 36% (Barnes et al. 2009) and 59% (Westphal et al. 2000). Among large samples of adult problem gamblers, 41–75% reported being current cigarette smokers (Becona 1993; Desai et al. 2007; Potenza et al. 2006). In several small samples, about 4–11.4% of adult problem gamblers reported alcoholism (Black and Moyer 1998; Lesieur and Rosenthal 1991; Netemeyer et al. 1998). In one large sample of adult problem gamblers that had called a gambling helpline, 18% reported problems with alcohol use (Potenza et al. 2006), and in large samples of Spanish and Swiss adults, 14% and 36%, respectively, of probable adult problem gamblers reported alcohol abuse (Bondolfi et al. 2000). Sampled from large representative samples of U.S. adults, 25% and 33% of problem gamblers reported alcohol dependence (Desai et al. 2007; Welte et al. 2001). In fact, drinking alcohol is the most commonly used substance by people with gambling problems (Gordon 2008).

Research has consistently shown that there is a high co-morbidity between problem gambling and alcohol use disorders in both community and clinical samples (e.g., Bland et al. 1993; Daghestani et al. 1996; Elia and Jacobs 1993; Lesieur et al. 1986; Smart and Ferris 1996; see also the review by Crockford and el-Guebaly 1998). To summarize, more specifically, high co-morbidity rates of gambling and alcohol consumption has been found among:

- Problem gamblers (e.g., Lister et al. 2015; Dannon et al. 2006; Ladd and Petry 2005)
- Treatment-seeking problem gamblers (e.g., Ciarrocchi and Richardson 1989; Ibáñez et al. 2001; Kausch 2003; Ladd and Petry 2003; Lesieur et al. 1986; Ramirez et al. 1984; Rash et al. 2011; Suomi et al. 2014; Stinchfield et al. 2005; Toneatto et al. 2002)
- Alcohol-dependent patients (e.g., Kovács et al. 2020; Maccallum and Blaszczynski 2001; Romanczuk-Seiferth et al. 2015)
- Psychiatric outpatients (e.g., Zimmerman et al. 2006)
- Casino employees (Shaffer et al. 1999; Shaffer and Hall 2002)
- Military personnel (Cowlishaw et al. 2020; Gallaway et al. 2019)
- Adolescents (Barnes et al. 2005, 2009; Cicarelli et al. 2020; Sutherland and Griffiths 1998; Walker et al. 2010; Westphal et al. 2000)
- Young adults (e.g., Afifi et al. 2016; Edgerton et al. 2019; Emond et al. 2020; Oksanen et al. 2019; Slutske et al. 2005)
- Elderly gamblers (e.g., Vander Bilt et al. 2004)
- Native Americans (e.g., Ella and Jacobs 1993)
- Adults in regional populations (e.g., Momper et al. 2010)
- Adults in large national and/or epidemiological surveys (e.g., Barnes et al. 2015; Petry et al. 2005; Wardle et al. 2007, 2011).

Survey population research at a national level has consistently found an association between levels of self-reported gambling-related problems and at-risk levels of alcohol consumption in many countries, including but not limited to the United States (Barnes et al. 2015; Cunningham-Williams et al. 1998; Feigelman et al. 1998; Gerstein, et al 1999; Kessler et al. 2008; Petry et al. 2005; Welte et al. 2001), Canada (Afifi et al. 2010), New Zealand (Abbott and Volberg 1992), Switzerland (Bondolfi et al. 2000, 2008), Korea (Park et al. 2010), United Kingdom (Griffiths et al. 2010), Sweden (Statens Folkalsoinstitut 2012; Fröberg et al. 2012), Finland (Oksanen et al. 2019), Australia (Dickerson et al. 1996), and Spain (Becona 1993).

Lorains et al. (2011) carried out a meta-analytic review examining the prevalence of common co-morbid disorders (including alcohol use disorder) and problem gambling in nationally representative population samples. Their results from across the 11 studies indicated that problem gamblers had high rates of other co-morbid disorders. The highest mean prevalence was for nicotine dependence (60.1%), followed by a substance use disorder (including alcohol use disorder) (57.5%), any type of mood disorder (37.9%) and any type of anxiety disorder (37.4%). However, they noted that *"there was evidence of moderate heterogeneity across studies, suggesting that rate estimates do not necessarily converge around a single population figure, and that weighted means should be interpreted with caution"* (p.495). In a similar meta-analytic review, Cowlishaw et al. (2014) examined studies that had assessed problem gambling among those undergoing treatment for substance abuse. They reported that 14% of those undergoing treatment for substance abuse were also problem gamblers.

In another comprehensive review of 11 types of addiction, Sussman et al. (2011) estimated that 50%, 30%, and 20% of problem gamblers also are addicted to cigarette, alcohol, and illicit drugs, respectively. The alcohol and drug use co-occurrence estimates are a slightly lower than those suggested by Freimuth et al. (2008), Kausch (2003), but are based on a larger pool of studies (albeit not a large pool). Yakovenko and Hodgins (2018) systematically reviewed the literature on comorbidity in problem gambling in a total of 251 studies. They noted that very few of the studies reviewed examined the mechanisms or temporal sequencing of co-morbidity, and that most studies evaluated the comorbidity prevalence rates between gambling and other co-occurring disorders. The high co-occurrence of gambling disorder and psychoactive substance use (and vice versa) among the general population may be indicative of common vulnerability factors, which are consistent with the findings from genetic and neuroimaging research (Potenza et al. 2019).

1.8 Problem Gambling Assessment Approaches

Although problem gambling has been studied and researched for over a century, formal assessment did not occur into 'pathological gambling' was included in the DSM-III (American Psychiatric Association 1980) as a disorder of impulse control alongside behaviors such as pyromania and kleptomania. In 1987, the criteria for pathological gambling were completely modified and modeled on the criteria for psychoactive substance use in the DSM-III-R (American Psychiatric Association 1987). The 1987 criteria were changed, in part, because of the increasing acceptance of gambling as a potentially addictive behavior. The criteria were slightly revised in the both the DSM-IV (American Psychiatric Association 1994) and DSM-5 (American Psychiatric Association 2013). However, the latest DSM-5 re-categorized 'pathological gambling' as 'gambling disorder' to emphasize its nature as a behavioral addiction rather than a disorder of impulse control (see Appendix 1). The DSM criteria are meant to be used by qualified practitioners during diagnostic clinical interviews, but have often been used in nationally representative epidemiological studies due to its satisfactory psychometric properties.

Since the publication of the DSM-III, criteria over 20 different screening tools have been developed to assess problem gambling (Dowling et al. 2019). The two most used (in addition to the DSM criteria) are arguably the South Oaks Gambling Screen (SOGS; Lesieur and Blume 1987) and its derivatives and the Problem Gambling Severity Index (PGSI; Ferris and Wynne 2001). None of these tools assesses 'gambling addiction', and problem gambling is typically operationally defined based on the number of criteria endorsed on each of these tools. For instance, in the DSM-5, and individual endorsing four (or more) out of nine criteria in a 12-month period are classed as having gambling disorder (see Appendix 1).

Research has indicated that different problem gambling instruments produce different prevalence rates of problem gambling among the same populations. Some studies suggest that the SOGS produces higher rates of problem gambling due higher rates of false positives (e.g., Stucki and Rihs-Middel 2007). When the SOGS is used simultaneously alongside another instrument, problem gambling rates assessed using the SOGS have been higher (e.g., Bonke and Borregaard 2006; Orford et al. 2003). In addition, different timeframes used when assessing problem gambling (e.g., past-year vs. lifetime) also result in different prevalence rates of problem gambling (with past-year rates unsurprisingly producing lower prevalence rates than a lifetime time-frame). A study by Orford et al. (2010) which compared the PGSI with the DSM-IV in a large nationally representative British sample found that some of the PGSI items displayed extreme men to women endorsement ratios and that the PGSI was possibly under-estimating the prevalence rates of problem gambling among females.

It should also be noted that there are also many brief assessment tools that have been developed to assess and screen for problem gambling. These brief screening instruments have typically been used (a) in largescale epidemiological surveys alongside several types of behaviors are studied simultaneously, as a way to reduce survey fatigue, and (b) by general practitioners to screen for problem gambling in health-based settings (Potenza et al. 2019). Dowling et al. (2019) recently reviewed 20 brief screening instruments and concluded that five of them met the criteria for satisfactory diagnostic accuracy in detecting both problem and at-risk gambling (i.e., NODS-CLiP [National Opinion Research Center Diagnostic Screen for Gambling Disorders—Loss of Control, Lying and Preoccupation], NODS-CLiP2, NODS-PERC

[National Opinion Research Centre Diagnostic Screen for Gambling Disorders—Preoccupation, Escape, Chasing and Risked Relationships], Brief Problem Gambling Screen (BPGS-2), and PGSI-SF [Problem Gambling Severity Index-Short Form]).

1.9 Treatment Approaches for Problem Gambling

Studies relating to the treatment of problem gambling appeared in the literature long before the formal medical recognition of pathological gambling and gambling disorder. Almost all forms of gambling treatment are based on specific theoretical perspectives (e.g., psychoanalytic therapy, behavioral therapy, cognitive therapy, pharmacotherapy, etc.) as well as multi-modal treatment combinations of these (Griffiths and MacDonald 1999; Griffiths et al. 2001; Hayer et al. 2005). Cognitive-based therapies (including cognitive therapy, cognitive-behavioral therapy, and motivational interviewing) have received empirical support for treatment efficacy of problem gambling (Potenza et al. 2019). Other reviews have concluded that gambling treatment efficacy is increased when pharmacotherapy is simultaneously combined with psychotherapies (Ribeiro et al. 2021).

Cognitive-based therapies for problem gambling primarily focusing on restructuring gambler's cognitive distortions have been found to reduce gambling frequency, diagnostic criteria, monetary risk, and urges to gamble (Ladouceur et al. 2001; Toneatto and Ladouceur 2003; Sylvain et al. 1997). These findings add support to the idea that erroneous perceptions play an important role in the development and maintenance of problem gambling. Most randomized trials have included cognitive-based interventions in one or more of their design elements, and most have reported some benefits for problem gamblers. There are a number of ways that cognitive-based treatments can be delivered (e.g., individual or group, via a workbook or online) (Petry et al. 2017).

In a systematic review of cognitive-based treatments of problem gambling, Petry et al. (2017) identified 21 randomized trial studies (three for cognitive therapy, three for cognitive-behavioral therapy, five for motivational interviewing, seven for a cognitive-behavioral workbook, and three for a cognitive-behavioral workbook feedback). Most of the studies reviewed reported the benefits of cognitive-behavioral therapy, but very few of the studies reported on their long-term efficacy. Petry et al. also noted that some self-directed minimal interventions may benefit individuals with less severe problem gambling symptoms, but that those with severe problem gambling benefit more from direct contact with therapists. They also noted that motivational interviewing on its own provides little evidence that it is effective in reducing problem gambling symptoms when not combined with other cognitive-based interventions.

A number of behavioral therapies have been developed to treat problem gambling utilizing aversive conditioning (Barker and Miller 1968; Koller 1972), response prevention (Symes and Nicki 1997), and imaginary desensitization (McConaghy et al. 1983). Some studies have compared aversive therapy to other modalities of treatment (McConaghy et al. 1983; McConaghy et al. 1991) and failed to corroborate the initial enthusiasm prompted by earlier case reports (e.g., Goorney 1968). More recently, Smith et al. (2015) reported that a behavioral (exposure-based) therapy with 21 participants who completed the treatment was very effective in reducing problem gambling at 12 weeks and at a 6-month follow-up. However, Ribeiro et al. (2021) concluded that in relation to the non-pharmacological treatment of problem gambling, at present there was currently no 'gold standard' treatment.

In a recent review, Ribeiro et al. (2021) also systematically examined the non-pharmacological treatments for problem gambling among studies that had utilized randomized controlled trials. They reviewed 22 studies (1694 patients, mean age = 42.94 years, 62.3% males). The results showed at least some efficacy in improving problematic gambling behavior for cognitive behavior therapy (n = 7), cognitive therapy (n = 3), exposure therapy (n = 1), combined or separate motivational interviewing and imaginal desensitization (n = 4), node-link mapping therapy (n = 1), couples therapy (n = 1), and 12-step facilitated and personalized feedback intervention (n = 1). They also reported that physical exercise had promising efficacy but the results were not statistically significant. The review demonstrated the heterogeneity of the available non-pharmacological therapies and that most studies presented with several methodological weaknesses (poor control groups, small sample sizes, etc.).

Pharmacotherapy has become increasingly used in the treatment of problem gambling including lithium, neuroleptics, glutamatergic agents, serotonergic antidepressants, opioid-receptor antagonists, and dopamine receptor antagonists (Potenza et al. 2019). Given the important role of serotonin function in problem gambling and impulse control disorders, serotonergic drugs have been utilized in the treatment of problem gambling (Brewer and Potenza 2008). Randomized placebo-controlled trials have been employed to evaluate the efficacy of selective serotonin reuptake inhibitors (SSRIs), such as sertraline (Saiz-Ruiz et al. 2005), fluvoxamine (e.g., Hollander et al. 2000), paroxetine (e.g., Grant et al. 2003), and escitalopram (Myrseth et al. 2011). In a 16-week double-blind, crossover study Hollander el al. (2000) reported the superior effect of fluvoxamine compared with placebo. However, in a six-month, double-blind, placebo-controlled study, Blanco et al. (2002) found that fluvoxamine treatment did not result in a statistically significant improvement, as measured by reduction in money and time spent gambling. In another double-blind, six-month, placebo-controlled trial using sertraline, Saiz-Ruiz et al. (2005) reported no significant differences compared to a placebo group among 60 problem gamblers. Moreover, Grant et al. (2003) in a 16-week study of paroxetine found no significant differences between active drug and placebo.

Myrseth et al. (2011) conducted a randomized controlled trial pilot study, in which one treatment group received cognitive-behavioral therapy, whereas the other group only received eight-week treatment of escitalopram, and reported that both groups had significant improvements in past-week gambling urges, thoughts, and behaviors. After eight weeks of treatment, the escitalopram group also received cognitive-behavioral therapy, and at 16 weeks of treatment, the authors did not find a significant differential group effect on the outcome measures between the two groups (cognitive-behavioral therapy versus escitalopram and cognitive-behavioral therapy). While overall showing mixed success of pharmacotherapy, limitations of previous SSRI studies include high drop-out rates and variability in the magnitude of the placebo response observed in different trials.

In addition, pharmacological treatment has been developed based on the dopaminergic system (which influences reward, motivation, reinforcement of reward, and appetitive urges). Given their ability to modulate dopaminergic transmission in the mesolimbic pathway, opioid receptor antagonists (naltrexone and nalmefene) have been utilized in the treatment of problem gambling (Hodgins et al. 2011). Results from double-blind, placebo-controlled studies of naltrexone and multi-center double-blind, placebo controlled trials of nalmefene suggest the efficacy of opioid antagonists in reducing the intensity of urges to gamble, gambling thoughts, and gambling behavior (Grant et al. 2009, 2010). However, a randomized control trial with bupropion, a drug with dopaminergic properties, did not differ from placebo in the treatment of problem gambling (Black et al. 2007).

In one of the few meta-analyses in the literature, Pallessen et al. (2007) reported that mood stabilizers, opioid-receptor antagonists and anti-depressants had better efficacy in treating problem gambling compared to no treatment or a placebo. However, Potenza et al. (2019) recently concluded that no specific pharmacological medications have been approved with an expressed indication for the treatment of problem gambling. They also noted that no pharmacological therapy has a formal indication for the treatment of gambling disorder but some medications (e.g., opioid receptor antagonists), may be helpful in the treatment of problem gambling.

Arguably the most popular treatment for problem gambling is Gamblers Anonymous (GA), with thousands of established groups worldwide. GA treatment utilizes the 12-step Minnesota Model Gamblers modeled after Alcoholics Anonymous, and comprises weekly meetings alongside other problem gamblers (although the GA Fellowship tends to use the term 'compulsive gambling') (Griffiths 1995). Most members in treatment have an individual sponsor and the treatment programme provides social and pragmatic peer assistance (e.g., helping group members to manage their gambling-related financial problems (Potenza et al. 2019). GA views problem gambling as a disease and that total abstinence is the main goal in alleviating the problem. The efficacy of the treatment is unknown given the Fellowship is underpinned by total anonymity (Griffiths 1995; Potenza et al. 2019). Drop-out rates are high (over 90% within the first year) but there are many problem gamblers who benefit from GA and achieve long-term abstinence (Griffiths 1995). Furthermore, compared to individuals who do not attend GA, those who do attend GA meetings while receiving other professional treatments have higher rates of gambling abstinence (Potenza et al. 2019).

1.10 Conclusions

Gambling is a highly prevalent leisure activity worldwide and problem gambling prevalence rates are relatively low and stable in mature gambling markets. Gambling and problem gambling are highly complex behaviors, which cannot be explained by any single theory or perspective. In many studies that have examined problem gambling, different factors have often been investigated descriptively or in isolation, without a theoretical or conceptual framework integrating the different elements. Nonetheless, problem gambling, like other addictions, result from an interaction and interplay between many factors including the individual's biological and/ or genetic predisposition, their psychological constitution, their social environment, and structural characteristics of the gambling activity itself (Griffiths and Delfabbro 2001).

Psychology plays a major role in further understanding problem gambling. This chapter has highlighted the complexity of this phenomenon, and provided a brief overview of some of the major theoretical models and perspectives related to problem gambling. Most theories (at least in part) are useful in providing a theoretical rationale for future studies. However, these theories are not completely independent from each other, and the weakness of one theory can be complemented by the strength of another. Therefore, gambling research is best served by a biopsychosocial model (Griffiths and Delfabbro 2001; Griffiths 2005) that stresses the biological and cognitive factors of the individual, but at the same time emphasizes the role of interpersonal variables, such as attitudes and gambling behavior of parents, the quality of relationship with them, contextual variables, such as situational factors of the gambling environment.

Consequently, researchers should not assume that individual factors will explain the full complexity of gambling and problem gambling behavior, and it is useful to explore how individual factors interact with other factors, such as gambling opportunities, attitudes of others, gambling behavior of friends and family, and inherent structural characteristics of gambling activities. It should also be noted that problem gambling does not occur in a vacuum and that it often clusters with other problematic behaviors (e.g., alcohol and drug addictions) and disorders (e.g., mood disorders) that can exacerbate and/or be exacerbated by problem gambling. Vulnerable individuals may attempt to continuously manipulate their neurobiological circuitry in order to obtain a more comfortable subjective state. Further research is needed to assess prevalence, and all possible patterns of co-occurrence, of addictive behaviors within the large samples of individuals.

Appendix 1: DSM-5 Diagnostic Criteria for Gambling Disorder (American Psychiatric Association 2013)

A. Persistent and recurrent problematic gambling behavior leading to clinically significant impairment or distress, as indicated by the individual exhibiting four (or more) of the following in a 12-month period:

    1. Needs to gamble with increasing amounts of money in order to achieve the desired excitement.
    2. Is restless or irritable when attempting to cut down or stop gambling.

3. Has made repeated unsuccessful efforts to control, cut back, or stop gambling.
4. Is often preoccupied with gambling (e.g., having persistent thoughts of reliving past gambling experiences, handicapping or planning the next venture, thinking of ways to get money with which to gamble).
5. Often gambles when feeling distressed (e.g., helpless, guilty, anxious, depressed).
6. After losing money gambling, often returns another day to get even ("chasing" one's losses).
7. Lies to conceal the extent of involvement with gambling.
8. Has jeopardized or lost a significant relationship, job, or educational or career opportunity because of gambling.
9. Relies on others to provide money to relieve desperate financial situations caused by gambling.

B. The gambling behavior is not better explained by a manic episode.

Specifiers for the severity are as follows:

- Mild: four or five criteria met.
- Moderate: six or seven criteria met.
- Severe: eight or nine criteria met.

References

Abbott M, Volberg RA (1992) Frequent gamblers and problem gamblers in New Zealand, Research Series No. 14. Department of Internal Affairs, Wellington

Abbott MW, Romild U, Volberg RA (2014) Gambling and problem gambling in Sweden: changes between 1998 and 2009. J Gambl Stud 30:985–999

Afifi TO, Cox BJ, Martens PJ, Sareen J, Enns MW (2010) The relationship between problem gambling and mental and physical health correlates among a nationally representative sample of Canadian women. Can J Public Health 101:171–175

Afifi TO, Nicholson R, Martins SS, Sareen J (2016) A longitudinal study of the temporal relation between problem gambling and mental and substance use disorders among young adults. Can J Psychiat 61(2):102–111

American Psychiatric Association (1980) Diagnostic and statistical manual of mental disorders (3rd ed). American Psychiatric Association, Washington DC

American Psychiatric Association (1994) Diagnostic and statistical manual of mental disorders (4th ed). American Psychiatric Association, Washington DC

American Psychiatric Association (2013) Diagnostic and statistical manual of mental disorders, 5th ed., Washington, DC

Assanangkornchai S, McNeil EB, Tantirangsee N, Kittirattanapaiboon P, Thai National Mental Health Survey Team (2016) Gambling disorders, gambling type preferences, and psychiatric comorbidity among the Thai general population: results of the 2013 National Mental Health Survey. J Behav Addic 5(3):410–418

Bakken IJ, Gotestam KG, Grawe RW, Wenzel HG, Øren A (2009) Gambling behavior and gambling problems in Norway 2007. Scand J Psychol 50:333–339

Barker JC, Miller M (1968) Aversion therapy for compulsive gambling. J Nervous Mental Dis 146:285–302
Barnes GM, Welte JW, Hoffman JH, Dintcheff BA (2005) Shared predictors of youthful gambling, substance use, and delinquency. Psychol Addict Behav 19:165–174
Barnes GM, Welte JW, Tidwell MCO, Hoffman JH (2015) Gambling and substance use: co-occurrence among adults in a recent general population study in the United States. Int Gambl Stud 15(1):55–71
Becona E (1993) The prevalence of pathological gambling in Galicia (Spain). J Gambl Stud 9:353–369
Black DW, Moyer T (1998) Clinical features and psychiatric comorbidity of subjects with pathological gambling behavior. Psychiatr Serv 49:1434–1439
Black DW, Arndt S, Coryell WH, Argo T, Forbush KT, Shaw MC et al (2007) Bupropion in the treatment of pathological gambling: a randomized, double–blind, placebo–controlled, flexible–dose study. J Clin Psychopharm 27(2):143–150
Blanco C, Petkova E, Ibanez A et al (2002) A pilot placebo–controlled study of fluvoxamine for pathological gambling. Ann Clin Psychiat 14:9–15
Bland RC, Newman SC, Orn H, Stebelsky G (1993) Epidemiology of pathological gambling in Edmonton. Can J Psychiat 38(2):108–112
Blaszczynski A, Nower L (2002) A pathways model of problem and pathological gambling. Addiction 97:487–499
Blaszczynski A, Walker M, Sharpe L, Nower L (2008) Withdrawal and tolerance phenomenon in problem gambling. Int Gambl Stud 8(2):179–192
Bonke J, Borregaard K (2006) Ludomani i Danmark: Udbredelsen af pengespil og problem-spillere [Pathological gambling in Denmark: prevalence of gambling and pathological gamblers]. Socialforskningsinstituttet, Copenhagen
Bondolfi G, Jermann F, Ferrero F, Zullino D, Osiek CH (2008) Prevalence of pathological gambling in Switzerland after the opening of casinos and the introduction of new preventive legislation. Acta Psychiatr Scand 117:236–239
Bondolfi G, Osiek C, Ferrero F (2000) Prevalence estimates of pathological gambling in Switzerland. Acta Psychiatrica Scandinavica 101:473–475
Bergler E (1957) The psychology of gambling. Hill and Wang, New York
Braverman J, Shaffer HJ (2012) How do gamblers start gambling: identifying behavioural markers for high–risk internet gambling. European J Pub Health 22(2):273–278
Brewer JA, Potenza MN (2008) The neurobiology and genetics of impulse control disorders: relationships to drug addictions. Biochem Pharmacol 75:63–75
Brandt L, Fischer G (2019) Adult ADHD is associated with gambling severity and psychiatric comorbidity among treatment-seeking problem gamblers. J Atten Disord 23(12):1383–1395
Brodbeck J, Duerrenberger S, Znoj H (2009) Prevalence rates of at risk, problematic and pathological gambling in Switzerland. Euro J Psychiatry 23:67–75
Burger TD, Dahlgren D, MacDonald CD (2006) College students and gambling: an examination of gender differences in motivation for participation. Coll Stud J 40(3):704–715
Çakici M (2012) The prevalence and risk factors of gambling behavior in Turkish Republic of Northern Cyprus. Anatolian J Psychiatry 13:243–249
Calado F, Griffiths MD (2016) Problem gambling worldwide: an update of empirical research (2000–2015). J Behav Addict 5:592–613
Calado F, Alexandre J, Griffiths MD (2017) Prevalence of adolescent problem gambling: a systematic review of recent research. J Gambl Stud 33:397–424
Castrén S, Basnet S, Pankakoski M, Ronkainen JE, Helakorpi S, Uutela A, Alho H, Lahti T (2013) An analysis of problem gambling among the finnish working-age population: a population survey. BMC Publ Health 13:519
Ciarrocchi J, Richardson R (1989) Profile of compulsive gamblers in treatment: Update and comparisons. J Gambl Behav 5:53–65

Ciccarelli M, Cosenza M, D'Olimpio F, Griffiths MD, Giovanna Nigro G (2021) Mentalizing failures, cognitive distortions, and emotional dysregulation among adolescent problem gamblers. J Gambl Stud 37:283–298
Ciccarelli M, Nigro G, Griffiths MD, D'Olimpio F, Cosenza M (2020) The associations between maladaptive personality traits, craving, alcohol use, and adolescent problem gambling: an Italian survey study. J Gambl Stud 36(1):243–258
Comings DE, Gade-Andavolu R, Gonzalez N, Wu S, Muhleman D, Chen C et al (2001) The additive effect of neurotransmitter genes in pathological gambling. Clin Genet 60(2):107–116
Costes JM, Eroukmanoff V, Richard JB, Tovar ML (2015) Les jeux de hasard et d'argent en France en 2014 [Gambling and gambling for money in France in 2014]. Les Notes De L'observatoire Des Jeux 6:1–9
Cowlishaw S, Metcalf O, Lawrence-Wood E, Little J, Sbisa A, Deans C, O'Donnell M, Sadler N, Van Hooff M, Crozier M, Battersby M, McFarlane AC (2020) Gambling problems among military personnel after deployment. J Psychiatr Res 131:47–53
Cowlishaw S, Merkouris S, Chapman A, Radermacher H (2014) Pathological and problem gambling in substance use treatment: a systematic review and meta-analysis. J Subst Abuse Treat 46(2):98–105
Crockford DN, el-Guebaly N (1998) Psychiatric comorbidity in pathological gambling: a critical review. Can J Psychiatry 43:43–50
Cunningham-Williams RM, Cottler LB, Compton WM, Spitznagel EL (1998) Taking chances: problem gamblers and mental health disorders – results from the St. Louis Epidemiologic Catchment Area study. Am J Public Health 88:1093–1096
Daghestani AN, Elenz E, Crayton JW (1996) Pathological gambling in hospitalized substance abusing veterans. J Clin Psychiatr 57:360–363
Dannon PN, Lowengrub K, Shalgi B, Sasson M, Tuson L, Saphir Y et al (2006) Dual psychiatric diagnosis and substance abuse in pathological gamblers: a preliminary gender comparison study. J Addict Dis 25:49–54
Delfabbro P (2013) Problem and pathological gambling: a conceptual review. J Gambl Bus Econ 7(3):35–53
Dickerson MG, Baron E, Hong SM, Cottrell D (1996) Estimating the extent and degree of gambling related problems in the Australian population: a national survey. J Gambl Stud 12:161–178
Dennison CR, Finkeldey JG, Rocheleau GC (2020) Confounding bias in the relationship between problem gambling and crime. J Gambl Stud:1–18.https://doi.org/10.1007/s10899-020-09939-0
De Ruiter MB, Veltman DJ, Goudriaan AE, Oosterlaan J, Sjoerds Z, Van Den Brink W (2009) Response perseveration and ventral prefrontal sensitivity to reward and punishment in male problem gamblers and smokers. Neuropsychopharmacology 34(4):1027–1038
Desai RA, Desai MM, Potenza MN (2007) Gambling, health, and age: data from the national epidemiologic survey on alcohol and related conditions. Psychol Addict Behav 21:431–440
Dixon L, Trigg R, Griffiths MD (2007) An empirical investigation of music and gambling behaviour. Int Gambl Stud 7:297–308
Dixon MJ, Harrigan KA, Jarick M, MacLaren V, Fugelsang JA, Sheepy E (2011) Psychophysiological arousal signatures of near–misses in slot machine play. Int Gambl Stud 11(3):393–407
Dowling NA, Merkouris SS, Dias S, Rodda SN, Manning V, Youssef GJ, Lubman DI, Volberg RA (2019) The diagnostic accuracy of brief screening instruments for problem gambling: a systematic review and meta-analysis. Clin Psychol Rev 74:101784
Dowling N, Smith D, Thomas T (2005) Electronic gaming machines: are they the 'crack-cocaine' of gambling? Addiction 100(1):33–45
Dowling N, Suomi A, Jackson A, Lavis T, Patford J, Cockman S, Thomas S, Bellringer M, Koziol-Mclain J, Battersby M, Harvey P, Abbott M (2016) Problem gambling and intimate partner violence: a systematic review and meta-analysis. Trauma Violence Abuse 17(1):43–61
Druine C, Delmarcelle C, Dubois M, Joris L, Somers W (2006) Etude quantitative des habitudes de Jeux de hasard pour l'offre classique et un ligne en Belgique [Quantitative study on online and offline gambling behaviour in Belgium]. Foundation Rodin, Bruxelles

Edgerton JD, Keough MT, Roberts LW (2019) An exploratory study of alcohol dependence and problem gambling co-development trajectories in young adults. Int Gambl Stud 19(1):1–21
Eisen SA, Slutske WS, Lyons MJ, Lassman J, Xian H, Toomey REEA et al (2001) The genetics of pathological gambling. Seminars Clin Neuropsychiat 6(3):195–204
Elia C, Jacobs DF (1993) The incidence of pathological gambling among native Americans treated for alcohol dependence. Int J Adhes 28(7):659–666
Emond A, Griffiths MD, Hollén L (2020) Problem gambling in early adulthood: a population-based study. Int J Ment Heal Addict:1–17.https://doi.org/10.1007/s11469-020-00401-1
Feigelman W, Wallisch LS, Lesieur HR (1998) Problem gamblers, problem substance users, and dual–problem individuals: an epidemiological study. Am J Public Health 88:467–470
Ferris J, Wynne HJ.(2001) The Canadian problem gambling index final report. Canadian Centre on Substance Abuse, Ontario, Ottawa
Finlay K, Marmurek HH, Kanetkar V, Londerville J (2010) Casino décor effects on gambling emotions and intentions. Environment Behav 42(4):524–545
Ford M, Håkansson A (2020) Problem gambling, associations with comorbid health conditions, substance use, and behavioural addictions: opportunities for pathways to treatment. PLoS One 15(1):e0227644
Freimuth M, Waddell M, Stannard J, Kelley S, Kipper A, Richardson A, Szuromi I (2008) Expanding the scope of dual diagnosis and co-addictions: behavioral addictions. J Groups Addict Recover 3:137–160
Freud S (1928) Dostoevsky and parracide. The standard edition of the complete psychological works of Sigmund Freud (ed. and trans. J. Strachey), vol 21. Hogarth Press, London, pp 177–194
Fröberg F, Johan Hallqvist J, Tengstro A (2012) Psychosocial health and gambling problems among men and women aged 16–24 years in the Swedish National Public Health Survey. Eur J Pub Health. https://doi.org/10.1093/eurpub/cks129
Gallaway MS, Fink DS, Sampson L, Cohen GH, Tamburrino M, Liberzon I, Calabrese J, Galea S (2019) Prevalence and covariates of problematic gambling among a US military cohort. Addict Behav 95:166–171
Gerstein D, Hoffmann J, Larison C, Laszlo E, Murphy S, Palmer A et al (1999) Gambling impact and behavior study: report to the national gambling impact study commission. National Opinion Research Center at the University of Chicago, Chicago
Giddens JL, Xian H, Scherrer JF, Eisen SA, Potenza MN (2011) Shared genetic contributions to anxiety disorders and pathological gambling in a male population. J Affect Disorders 132(3):406–412
Goorney AB (1968) Treatment of a compulsive horse race gambler by aversion therapy. Br J Psychiat 114:329–333
Gordon A (2008) Comorbidity of mental disorders and substance use: a brief guide for the primary care clinician. Drug and Alcohol Services South Australia, Adelaide
Grant JE, Kim SW, Potenza MN, Blanco C, Ibanez A, Stevens L et al (2003) Paroxetine treatment of pathological gambling: a multi–centre randomized controlled trial. Int Clin Psychopharm 18(4):243–249
Grant JE, Donahue CB, Odlaug BL, Kim SW, Miller MJ, Petry NM (2009) Imaginal desensitisation plus motivational interviewing for pathological gambling: randomised controlled trial. Br J Psychiatry 195(3):266–267
Grant JE, Odlaug BL, Potenza MN, Hollander E, Kim SW (2010) Nalmefene in the treatment of pathological gambling: multicentre, double–blind, placebo–controlled study. Br J Psychiatry 197(4):330–331
Griffiths MD (1990) The cognitive psychology of gambling. J Gambl Stud 6:31–42
Griffiths MD (1991) The psychobiology of the near miss in fruit machine gambling. J Psychol 125:347–357
Griffiths MD (1993) Fruit machine gambling: the importance of structural characteristics. Journal Gambl Stud 9:101–120

Griffiths MD (1994) The role of cognitive bias and skill in fruit machine gambling. Br J Psychol 85:351–369
Griffiths MD (1995) Adolescent gambling. Routledge, London
Griffiths MD (1999) Gambling technologies: prospects for problem gambling. J Gambl Stud 15:265–283
Griffiths MD (2004) Betting your life on it: problem gambling has clear health related consequences. BMJ 329:1055–1056
Griffiths MD (2005) A "components" model of addiction within a biopsychosocial framework. J Substance Use 10:191–197
Griffiths MD (2006) An overview of pathological gambling. In: Plante T (ed) Mental disorders of the new millennium. vol I: behavioral issues. Greenwood, New York, pp 73–98
Griffiths MD (2007) Gambling addiction and its treatment within the NHS. British Medical Association, London
Griffiths MD (2016) Problem gambling and gambling addiction are not the same. J Addict Dependence 2(1):1–3
Griffiths MD, Auer M (2013) The irrelevancy of game-type in the acquisition, development and maintenance of problem gambling. Front Psychol 3:621. https://doi.org/10.3389/fpsyg.2012.00621
Griffiths MD, Delfabbro PH (2001) The biopsychosocial approach to the study of gambling: contextual factors in research and clinical interventions. J Gambling Issues 5:1–33
Griffiths MD, MacDonald HF (1999) Counselling in the treatment of pathological gambling: an overview. Br J Guidance Counselling 27:179–190
Griffiths MD, Parke J (2003) The environmental psychology of gambling. In Reith G (ed) Gambling: who wins? who loses? Prometheus Books, New York, pp 277–292
Griffiths MD, Parke J, Wood RTA (2002) Excessive gambling and substance abuse: is there a relationship? J Substance Use 7:187–190
Griffiths MD, Smeaton M (2002) Withdrawal in pathological gamblers: a small qualitative study. Soc Psychol Rev 4:4–13
Griffiths MD, Sutherland I (1998) Adolescent gambling and drug use. J Comm Appl Soc Psychol 8:423–427
Griffiths MD, Bellringer P, Farrell–Roberts K, Freestone F (2001) Treating problem gamblers: a residential therapy approach. J Gambl Stud 17:161–169
Griffiths MD, Wardle J, Orford J, Sproston K, Erens B (2010) Gambling, alcohol consumption, cigarette smoking and health: findings from the 2007 British gambling prevalence survey. Addict Res Theor 18:208–223
Gyollai Á, Griffiths MD, Barta C, Vereczkei A, Urbán R, Kun B, Kokonyei G, Székely A, Sasvári-Székely M, Blum K, Demetrovics Z, Demetrovics Z (2014) The genetics of problem and pathological gambling: a systematic review. Curr Pharm Des 20(25):3993–3999
Håkansson A, Karlsson A, Widinghoff C (2018) Primary and secondary diagnoses of gambling disorder and psychiatric comorbidity in the Swedish health care system—a nationwide register study. Front Psych 9:426
Harris A, Griffiths MD (2018) The impact of speed of play in gambling on psychological and behavioural factors: a critical review. J Gambl Stud 34:393–412
Hartmann M, Blaszczynski A (2018) The longitudinal relationships between psychiatric disorders and gambling disorders. Int J Ment Heal Addict 16(1):16–44
Hayer T, Griffiths MD (2015) The prevention and treatment of problem gambling in adolescence. In: Gullotta TP, Adams G (eds) Handbook of adolescent behavioural problems: evidence-based approaches to prevention and treatment, 2nd edn. Springer, New York, pp 467–486
Hayer, T, Griffiths MD, Meyer G (2005) The prevention and treatment of problem gambling in adolescence. In: Gullotta TP, Adams G (eds) Handbook of adolescent behavioral problems: evidence–based approaches to prevention and treatment. Springer, New York, pp 467–486
Hodgins DC, Stea JN, Grant JE (2011) Gambling disorders. Lancet 378(9806):1874–1884

Hollander E, DeCaria CM, Finkell JN, Begaz T, Wong CM, Cartwright C (2000) A randomized double–blind fluvoxamine/placebo crossover trial in pathologic gambling. Biol Psychiat 47(9):813–817

Ibáñez A, Blanco C, Donahue E, Lesieur HR, Pérez de Castro I, Fernández-Piqueras J, Sáiz-Ruiz J (2001) Psychiatric comorbidity in pathological gamblers seeking treatment. Am J Psychiatry 158(10):1733–1735

Ilkas H, Turja T (2003) Penningsspelsunderso¨kning [Gambling research]. Ministry of Social Affairs and Health, Helsinki

Ioannidis K, Hook R, Wickham K, Grant JE, Chamberlain SR (2019) Impulsivity in gambling disorder and problem gambling: a meta-analysis. Neuropsychopharmacology 44(8):1354–1361

Jacobs DF (1986) A general theory of addictions: a new theoretical model. J Gambl Behav 2:15–31

Jacob L, Haro JM, Koyanagi A (2018) Relationship between attention-deficit hyperactivity disorder symptoms and problem gambling: a mediation analysis of influential factors among 7,403 individuals from the UK. J Behav Addict 7(3):781–791

Kassinove JI, Schare ML (2001) Effects of the "near miss" and the "big win" on persistence at slot machine gambling. Psychol Addict Behav 15(2):155–158

Kausch O (2003) Patterns of substance abuse among treatment-seeking pathological gamblers. J Subst Abuse Treat 25:263–270

Kerber CS, Black DW, Buckwalter K (2008) Comorbid psychiatric disorders among older adult recovering pathological gamblers. Issues Ment Health Nurs 29(9):1018–1028

Kessler RC, Hwang I, Labrie R, Petukhova M, Sampson NA, Winters KC et al (2008) DSM-IV pathological gambling in the national comorbidity survey replication. Psychol Med 38:1351–1360

Killick EA, Griffiths MD (2019) In-play sports betting: a scoping study. Int J Mental Health Addict 17:1456–1495

Koller K (1972) The gambling addicts vs. the gambling professional. Int J Addict 7:387–393

Kovács I, Demeter I, Janka Z, Demetrovics Z, Maraz A, Andó B (2020) Different aspects of impulsivity in chronic alcohol use disorder with and without comorbid problem gambling. PLoS ONE 15(1):e0227645

Kun B, Balázs H, Arnold P, Paksi B, Demetrovics Z (2012) Gambling in western and eastern Europe: the example of hungary. J Gambl Stud 28:27–46. https://doi.org/10.1007/s10899-011-9242-4

Ladd GT, Petry NM (2003) A comparison of pathological gamblers with and without substance abuse treatment histories. Exp Clin Psychopharmacol 11:202–209

Ladouceur R, Sylvain C, Boutin C, Lachance S, Doucet C, Leblond J, Jacques C (2001) Cognitive treatment of pathological gambling. J Nervous Mental Dis 189(11):774–780

Lang M, Leménager T, Streit F, Fauth–Bühler M, Frank J, Juraeva D et al (2016) Genome–wide association study of pathological gambling. Eur Psychiat 36:38–46

Lesieur HR, Blume SB (1987) The South Oaks gambling screen (The SOGS): a new instrument for the identification of pathological gamblers. Am J Psychiatry 144:1184–1188

Lesieur HR, Blume SB, Zoppa RM (1986) Alcoholism, drug abuse, and gambling. Alcohol Clin Exper Res 10:33–38

Lesieur HR, Rosenthal RJ (1991) Pathological gambling: a review of the literature (prepared for the American psychiatric association task force on DSM-IV committee on disorders of impulse control not elsewhere classified). J Gambl Stud 7:5–39

Lister JJ, Milosevic A, Ledgerwood DM (2015) Personality traits of problem gamblers with and without alcohol dependence. Addict Behav 47:48–54

Lobo DS, Souza RP, Tong RP, Casey DM, Hodgins DC, Smith GJ et al (2010) Association of functional variants in the dopamine D2–like receptors with risk for gambling behaviour in healthy Caucasian subjects. Biol Psychol 85(1):33–37

Lorains FK, Cowlishaw S, Thomas SA (2011) Prevalence of comorbid disorders in problem and pathological gambling: systematic review and meta-analysis of population surveys. Addiction 106:490–498

Lupu V, Todirita IR (2013) Updates of the prevalence of problem gambling in Romanian teenagers. J Gambl Stud 29:29–36

Maccallum F, Blaszczynski A (2001) Pathological gambling and comorbid substance use. Aust NZ J Psychiat 36:411–415
Makarovič M (2010) Towards the patterns of problem gambling in Slovenia. Res Discuss 3:21–48
Marchetti D, Verrocchio MC, Porcelli P (2019) Gambling problems and alexithymia: a systematic review. Brain Sci 9(8):191
McConaghy N, Armstrong MS, Blaszczynski A, Allcock CC (1983) Controlled comparison of aversive therapy and imaginal desensitization in compulsive gambling. Br J Psychiat 142:366–372
McConaghy N, Blaszczynski A, Frankova A (1991) Comparison of Imaginal desensitisation with other behavioural treatments of pathological gambling a two–to nine–year follow–up. Br J Psychiat 159(3):390–393
McCormick J, Delfabbro P, Denson LA (2012) Psychological vulnerability and problem gambling: an application of Durand Jacobs' general theory of addictions to electronic gaming machine playing in Australia. J Gambl Stud 28(4):665–690
Merkouris SS, Thomas AC, Shandley KA, Rodda SN, Oldenhof E, Dowling NA (2016) An update on gender differences in the characteristics associated with problem gambling: a systematic review. Curr Addict Rep 3(3):254–267
Meyer G, Hayer T, Griffiths M (2009) Problem gambling in Europe: challenges, prevention, and interventions. Springer, New York
Meyer C, Bischof A, Westram A, Jeske C, de Brito S, Glorius S, Schön D, Porz S, Gürtler D, Kastirke N, Hayer T, Rumpf HJ (2015) The "pathological gambling and epidemiology" (PAGE) study program: design and fieldwork. Int J Methods Psychiatr Res 24:11–31
Momper SL, Delva J, Grogan-Kaylor A, Sanchez AN, Volberg RA (2010) The association of at-risk, problem, and pathological gambling with substance use, depression, and arrest history. J Gambl Issues 24:7–32
Moreyra P, Ibáñez AA, Saiz–Ruiz J, Nissenson K, Blanco C (2000) Review of the phenomenology, etiology and treatment of pathological gambling. German J Psychiat 3(2):37–52
Myrseth H, Molde H, Støylen IJ, Johnsen BH, Holsten F, Pallesen S (2011) A pilot study of CBT versus escitalopram combined with CBT in the treatment of pathological gamblers. Int Gambl Stud 11(1):121–141
Nestler EJ (2004) Historical review: molecular and cellular mechanisms of opiate and cocaine addiction. Trends Pharmacol Sci 25(4):210–218
Netemeyer RG, Burton S, Cole LK, Williamson DA, Zucker N, Bertman L, Diefenbach G (1998) Characteristics and beliefs associated with probable pathological gambling: a pilot study. J Public Policy Mark 17:147–160
Nower L, Eyrich-Garg KM, Pollio DE, North CS (2015) Problem gambling and homelessness: results from an epidemiologic study. J Gambl Stud 31(2):533–545
Oksanen A, Sirola A, Savolainen I, Kaakinen M (2019) Gambling patterns and associated risk and protective factors among finnish young people. Nordic Stud Alcohol Drugs 36(2):161–176
Olason DT, Gretarsson SJ (2009) Iceland. In: Meyer G, Hayer T, Griffiths M (eds) Problem gambling in Europe. Challenges, prevention, and intervention. Springer, New York, pp 137–151
Olason DT, Hayer T, Brosowski T, Meyer G (2015) Gambling in the mist of economic crisis: results from three national prevalence studies from Iceland. J Gambl Stud 31:759–774
Orford J, Sproston K, Erens B (2003) SOGS and DSM-IV in the British gambling prevalence survey: reliability and factor structure. Int Gambl Stud 3:53–65
Orford J, Wardle H, Griffiths MD, Sproston K, Erens B (2010) PGSI and DSM-IV in the 2007 British gambling prevalence survey: reliability, item response, factor structure and inter-scale agreement. Int Gambl Stud 10:31–44
Pallesen S, Molde H, Arnestad HM, Laberg JC, Skutle A, Iversen E, Støylen IJ, Kvale G, Holsten F (2007) Outcome of pharmacological treatments of pathological gambling: a review and meta-analysis. J Clin Psychopharmacol 27(4):357–364
Park S, Cho MJ, Jeon HJ, Lee HW, Bae JN, Park JI, Sohn JH, Lee YR, Lee JY, Hong JP (2010) Prevalence, clinical correlations, comorbidities, and suicidal tendencies in pathological Korean

gamblers: results from the Korean epidemiologic catchment area study. Soc Psychiatry Psychiatr Epidemiol 45:621–629. https://doi.org/10.1007/s00127-009-0102-9
Parke J, Griffiths MD (2004) Gambling addiction and the evolution of the 'near miss.' Addict Res Theor 12:407–411
Parke J, Griffiths MD (2006) The psychology of the fruit machine: the role of structural characteristics (revisited). Int J Ment Heal Addict 4:151–179
Parke J, Griffiths MD (2007) The role of structural characteristics in gambling. In: Smith G, Hodgins D, Williams R (eds) Research and measurement issues in gambling studies. Elsevier, New York, pp 211–243
Parke A, Griffiths MD, Irwing P (2004) Personality traits in pathological gambling: sensation seeking, deferment of gratification and competitiveness as risk factors. Addict Res Theory 12:201–212
Petry NM (2005) Pathological gambling: etiology, comorbidity and treatment. American Psychological Association, Washinton, D.C.
Petry NM, Stinson FS, Grant BF (2005) Comorbidity of DSM-IV pathological gambling and other psychiatric disorders: results from the national epidemiologic survey on alcohol and related conditions. J Clin Psychiatry 66:564–574
Petry NM, Ginley MK, Rash CJ (2017) A systematic review of treatments for problem gambling. Psychol Addict Behav 31(8):951–961
Pisklak JM, Yong JJ, Spetch ML (2020) The near-miss effect in slot machines: a review and experimental analysis over half a century later. J Gambl Stud 36:611–632
Polezzi D, Casiraghi E, Vidotto G (2012) The somatic marker hypothesis in pathological gambling. In: Cavana A (eds) Psychology of gambling: new research. Nova Science Publishers, New York, pp 9–22
Potenza MN (2001) The neurobiology of pathological gambling. Seminars Clin Neuropsychiat 6(3):217–226
Potenza MN, Balodis IM, Derevensky J, Grant JE, Petry NM, Verdejo-Garcia A, Yip SW (2019) Gambling disorder. Nat Rev Dis Primers 5(1):1–21
Potenza MN, Maciejewski PK, Mazure CM (2006) A gender-based examination of past-year recreational gamblers. J Gambl Stud 22(1):41–64
Quigley L, Yakovenko I, Hodgins DC, Dobson KS, El-Guebaly N, Casey DM, Currie SR, Smith GJ, Williams RJ, Schopflocher DP (2015) Comorbid problem gambling and major depression in a community sample. J Gambl Stud 31(4):1135–1152
Raisamo SU, Mäkelä P, Salonen AH, Lintonen TP (2014) The extent and distribution of gambling harm in Finland as assessed by the problem gambling severity index. Eur J Pub Health 25(4):716–722
Ramirez LF, McCormick RA, Russo AM, Taber JI (1984) Patterns of substance abuse in pathological gamblers undergoing treatment. Addict Behav 8:425–428
Rash CJ, Weinstock J, Petry NM (2011) Drinking patterns of pathological gamblers before, during, and after gambling treatment. Psychol Addict Behav 25:664–674
Raylu N, Oei TP (2002) Pathological gambling: a comprehensive review. Clin Psychol Rev 22(7):1009–1061
Reuter J, Raedler T, Rose M, Hand I, Gläscher J, Büchel C (2005) Pathological gambling is linked to reduced activation of the mesolimbic reward system. Nature Neurosci 8(2):147–148
Ribeiro EO, Afonso NH, Morgado P (2021) Non-pharmacological treatment of gambling disorder: a systematic review of randomized controlled trials. BMC Psychiatry 21:105
Richard J, Fletcher É, Boutin S, Derevensky J, Temcheff C (2020) Conduct problems and depressive symptoms in association with problem gambling and gaming: a systematic review. J Behav Addict Adv Online Publ. https://doi.org/10.1556/2006.2020.00045
Rickwood D, Blaszczynski A, Delfabbro P, Dowling N, Heading K (2010) The psychology of gambling. Inpsych 32(6):11–21

Romanczuk-Seiferth N, Koehler S, Dreesen C, Wüstenberg T, Heinz A (2015) Pathological gambling and alcohol dependence: neural disturbances in reward and loss avoidance processing. Addict Biol 20(3):557–569

Rosenthal RJ (2008) Psychodynamic psychotherapy and the treatment of pathological gambling. Brazilian J Psychiat 30:S41–S50

Rosenthal RJ, Lesieur HR (1992) Self–reported withdrawal symptoms and pathological gambling. Am J Addict 1(2):150–154

Saiz-Ruiz J, Blanco C, Ibáñez A, Masramon X, Gómez MM, Madrigal M, Díez T (2005) Sertraline treatment of pathological gambling: a pilot study. J Clin Psychiat 66(1):28–33

Sassen M, Kraus L, Bühringer G, Pabst A, Piontek D, Taqi Z (2011) Gambling among adults in Germany: prevalence, disorder and risk factors. Sucht 57:249–257

Seabury C, Wardle H (2014) Gambling behaviour in England and Scotland. Gambling Commission, Birmingham

Scherrer JF, Slutske WS, Xian H, Waterman B, Shah KR, Volberg R, Eisen SA (2007) Factors associated with pathological gambling at 10-year follow-up in a national sample of middle-aged men. Addiction 102(6):970–978

Shaffer HJ, Hall MN (2002) The natural history of gambling and drinking problems among casino employees. J Soc Psychol 142(4):405–424

Shaffer HJ, Hall MN, Vander Bilt J (1999) Estimating the prevalence of disordered gambling behavior in the United States and Canada: a research synthesis. Am J Public Health 89:1369–1376

Slutske WS, Caspi A, Moffitt TE, Poulton R (2005) Personality and problem gambling: a prospective study of a birth cohort of young adults. Arch Gen Psychiatry 62:769–775

Slutske WS, Eisen S, True WR, Lyons MJ, Goldberg J, Tsuang M (2000) Common genetic vulnerability for pathological gambling and alcohol dependence in men. Arch Gen Psychiatr 57(7):666–673

Slutske WS, Zhu G, Meier MH, Martin NG (2010) Genetic and environmental influences on disordered gambling in men and women. Arch Gen Psychiat 67(6):624–630

Smart RG, Ferris J (1996) Alcohol, drugs and gambling in the ontario adult population. Can J Psychiat 41:36–45

Smith DP, Battersby MW, Pols RG, Harvey PW, Oakes JE, Baigent MF (2015) Predictors of relapse in problem gambling: a prospective cohort study. J Gambl Stud 31(1):299–313

Spenwyn J, Barrett DKR, Griffiths MD (2010) The role of lights and music in gambling behavior: an empirical pilot study. Int J Mental Health Addict 8:107–118

Spritzer DT, Rohde LA, Benzano DB, Laranjeira RR, Pinsky I, Zaleski M, Tavares H (2011) Prevalence and correlates of gambling problems among a nationally representative sample of Brazilian adolescents. J Gambl Stud 27(4):649–661

Stark GM, Saunders DM, Wookey P (1982) Differential effects of red and blue lighting on gambling behaviour. Curr Psychol Res 2:95–100

Statens Folkalsoinstitut (2012) Spel om pengar och spelproblem I Sverige 2009/2010. Osterund: Statens Folkalso Institut. Located at: http://www.fhi.se/PageFiles/15707/R2012-04-Spel-om-pengar-och-spelproblem-i-Sverige-2009-2010.pdf

Stinchfield R, Kushner MG, Winters KC (2005) Alcohol use and prior substance abuse treatment in relation to gambling problem severity and gambling treatment outcome. J Gambl Stud 21:273–297

Stucki S, Rihs-Middel M (2007) Prevalence of adult problem and pathological gambling between 2000 and 2005: an update. J Gambl Stud 23:245–257

Suomi A, Dowling NA, Jackson AC (2014) Problem gambling subtypes based on psychological distress, alcohol abuse and impulsivity. Addict Behav 39(12):1741–1745

Sundqvist K, Rosendahl I (2019) Problem gambling and psychiatric comorbidity—risk and temporal sequencing among women and men: results from the Swelogs case–control study. J Gambl Stud 35(3):757–771

Sussman S, Lisha N, Griffiths MD (2011) Prevalence of the addictions: a problem of the majority or the minority? Eval Health Prof 34:3–56

Sylvain C, Ladouceur R, Boisvert JM (1997) Cognitive and behavioral treatment of pathological gambling: a controlled study. J Consult Clin Psychol 65(5):727–732
Symes BA, Nicki RM (1997) A preliminary consideration of cue–exposure, response–prevention treatment for pathological gambling behaviour: two case studies. J Gambl Stud 13(2):145–157
Theule J, Hurl KE, Cheung K, Ward M, Henrikson B (2019) Exploring the relationships between problem gambling and ADHD: a meta-analysis. J Atten Disord 23(12):1427–1437
Toneatto T (1999) Cognitive psychopathology of problem gambling. Subst Use & Misuse 34(11):1593–1604
Toneatto T, Ladouceur R (2003) Treatment of pathological gambling: a critical review of the literature. Psychol Addict Behav 17(4):284–292
Toneatto T, Blitz–Miller T, Calderwood K, Dragonetti R, Tsanos A (1997) Cognitive distortions in heavy gambling. J Gambl Stud 13(3):253–266
Toneatto T, Skinner W, Dragonetti R (2002) Patterns of substance use in treatment-seeking problem gamblers: impact on treatment outcomes. J Clin Psychol 58:853–859
Vander Bilt J, Dodge HH, Pandav R, Shaffer HJ, Ganguli M (2004) Gambling participation and social support among older adults. A longitudinal community study. J Gambl Stud 20:373–390
Volberg R, Gupta R, Griffiths MD, Olason D, Delfabbro PH (2010) An international perspective on youth gambling prevalence studies. Int J Adolesc Med Health 22:3–38
Von Hattingberg H (1914) III. Analerotik, angstlust und eigensinn. Internationale Zeitschrift für Psychoanalyse 2(3):244–258
Walker DM, Clark C, Folk J (2010) The relationship between gambling behavior and binge drinking, hard drug use, and paying for sex. UNLV Gaming Rev Res J 14:15–26
Walker MB (1992) The psychology of gambling. Pergamon Press, Oxford
Wardle H, Moody A, Spence S, Orford J, Volberg R, Jotangia D, Griffiths MD, Hussey D, Dobbie F (2011) British gambling prevalence survey 2010. The Stationery Office, London
Wardle H, Sproston K, Orford J, Erens B, Griffiths MD, Constantine R, Pigott S (2007) The British gambling prevalence survey 2007. National Centre for Social Research, London
Wardle H, John A, Dymond S, McManus S (2020) Problem gambling and suicidality in England: secondary analysis of a representative cross-sectional survey. Publ Health 184:11–16
Welte JW, Barnes GM, Tidwell M, Hoffman JH, Wieczorek WF (2015) Gambling and problem gambling in the United States: changes between 1999 and 2013. J Gambl Stud 31:695–715
Welte J, Barnes G, Wieczorek W, Tidwell M, Parker J (2001) Alcohol and gambling pathology among U.S. Adults: prevalence, demographic patterns and comorbidity. J Stud Alcohol 62:706–712
Westphal JR, Rush JA, Steven L, Johnson LJ (2000) Gambling behavior of Louisiana students in grades 6 through 12. Pediatric Serv 51:96–99
Williams RJ, Leonard CA, Belanger YD, Christensen DR, El-Guebaly N, Hodgins DC, McGrath DS, Nicoll F, Smith GJ, Stevens RM (2021) Predictors of gambling and problem gambling in Canada. Can J Publ Health Adv Online Publ. https://doi.org/10.17269/s41997-020-00443-x
Wilson D, da Silva Lobo DS, Tavares H, Gentil V, Vallada H (2013) Family–based association analysis of serotonin genes in pathological gambling disorder: evidence of vulnerability risk in the 5HT–2A receptor gene. J Molec Neurosci 49(3):550–553
Yakovenko I, Hodgins DC (2018) A scoping review of co-morbidity in individuals with disordered gambling. Int Gambl Stud 18(1):143–172
Zack M, Poulos CX (2009) Parallel roles for dopamine in pathological gambling and psychostimulant addiction. Curr Drug Abuse Rev 2(1):11–25
Zimmerman M, Chelminski I, Young D (2006) Prevalence and diagnostic correlates of DSM-IV pathological gambling in psychiatric outpatients. J Gambl Stud 22:255–262

Chapter 2
Gaming Disorder

Cuneyt Evren

Abstract Internet gaming as a leisure activity is not uncommon, it is entirely legitimate and for most gamers unproblematic. Nevertheless, a minority of players may suffer from "disordered gaming" when excessive usage of video games becomes pathologic and dysfunctional. As with other addictions, diminished control in disordered gaming may lead to the continuation of detrimental behavior even in the face of negative consequences including distress and functional impairment. In the 1990s, disordered gaming was assessed according to the criteria for pathological gambling in the 4th revision of the Diagnostic and Statistical Manual of Mental Disorders (DSM-IV). When the DSM-5 was published in 2013, it included the tentative term "Internet Gaming Disorder" as a condition requiring further research. In 2016, the World Health Organization (WHO) included the diagnosis of "Gaming Disorder" as a "behavioral addiction" to the 11th International Classification of Diseases (ICD-11), and in 2019 the WHO officially recognized disordered gaming as a "mental health disorder". When investigating the effects of playing video games, a number of personality traits such as neuroticism, introversion, impulsivity, and aggressiveness need to be taken into consideration. Other important variables include the genre of games preferred by individual players, and it is important to investigate what psychological motives underlie people's engagement with this activity. When disordered gaming is suspected, individuals should be evaluated for common comorbidities such as depression, anxiety, attention deficit hyperactivity disorder, and obsessive–compulsive disorder. Neurobiological studies found similarities between disordered gaming and substance use disorder. The treatment of disordered gaming is still in the early stages of research, and it is thus premature to attempt reaching strong conclusions. Nevertheless, it has been suggested that a combination of psychological and pharmacological treatment, together with treatments that are effective for other addictions (e.g., family involvement) may show positive results.

Keywords ADHD · Comorbidity · Gaming disorder · Internet · Personality · Video games

---

C. Evren (✉)
Department of Psychology, Istanbul Gelisim University, Faculty of Economics, Administrative and Social Science, Istanbul, Turkey
e-mail: cuneytevren@yahoo.com

H. M. Pontes (ed.), *Behavioral Addictions*, Studies in Neuroscience, Psychology and Behavioral Economics, https://doi.org/10.1007/978-3-031-04772-5_2

2.1 Introduction

Internet gaming is a legitimate and commonly practiced leisure activity around the globe. Previous research has consistently demonstrated positive effects of healthy gaming (Connolly et al. 2012; Griffiths 2019); however, for a minority of players gaming may become dysfunctional, leading to functional impairments that are harmful to their social, occupational, familial, educational, and psychological functioning (Gentile et al. 2011). Researchers and society have gradually become more interested in the problems caused by the excessive use of video games (Zajac et al. 2017). Indeed, for some gamers, video game playing may become a repetitive and rewarding behavior, and coupled with the negative consequences of excessive gaming, this leisure activity may become what has been termed as disordered gaming.

As with other addictions, diminished control in disordered gaming affecting participation in the gaming may lead gamers to continue this behavior despite the experience of negative consequences including distress and functional impairments in their personal, relational, occupational, educational, or other life domains (Antons et al. 2020). Similarly to other addictions, cue-reactivity and craving may be increased and inhibitory control lowered, representing key mechanisms in disordered gaming, particularly in the presence of specific gaming-related cues (Antons et al. 2020).

Investing excessive amounts of time in gaming activities (often 8–12 h per day) is a primary way in which disordered gamers are negatively affected by this condition (Baggio et al. 2016). King and Delfabbro (2019a) reported that the time spent by disordered gamers causes individuals to miss opportunities and interferes with or even replaces normal routine activities and diminishes social functioning, including basic activities such as sleeping, eating, personal hygiene, real-world social interaction (e.g., talking to people face-to-face, meetings with friends, and/or visiting family), and fulfilling important responsibilities in school, at work, or caring for pets or children. Users' priorities are markedly changed when they experience disordered gaming, leading to a significant reduction in the interest shown for non-gaming-related information, people, events, and the capacity to process information outside the game. Persons suffering from disordered gaming are often absorbed by thoughts related to gaming even when not actually playing, pondering ideas and opportunities for future play. At the same time, positive reinforcement and success in non-gaming areas are lacking, which further encourages the user to seek refuge in gaming activities. When the person has less or no opportunity to play, negative mood states such as irritability, sadness, and boredom may manifest (King and Delfabbro 2019a).

This chapter will focus on providing a review of disordered gaming research based on recent findings published in the field. Moreover, the chapter will provide a review based on the following lines of disordered gaming research: (i) introduction (definition and related subjects), (ii) neurobiological findings, (iii) associated comorbidities, (iv) prevalence rates, (v) diagnostic and assessment approaches, and (vi) treatment approaches.

2.1.1 *History of Video games*

The emergence of computer games accompanied the appearance of the first computers in the 1950s. The proliferation of computer games began in the 1980s when video game consoles and computers (popular devices of that period such as Atari, Spectrum and Comodore 64) became more common. In this period, called the "Golden Age" of computer games, many companies producing computers and computer games emerged (Şengül and Büber 2016). The first publication about disordered gaming appeared in this period (Soper and Miller 1983). The rapid spread of personal computers in the 1990s and the development of related games triggered the growth of the games market all over the world. The introduction of the internet in this period allowed computer gamers to play together with others and further expanded the gaming industry (Şengül and Büber 2016).

Later, with the emergence of a new generation of consoles (e.g., PlayStation, Nintendo Wii, and Xbox) and mobile phones becoming suitable for playing games, people began to carry a game machine at all times. Nowadays, the game market has become such that companies producing smartphones and tablets have started to get more revenue from the sales of applications and games than they earn from the sale of the devices (Şengül and Büber 2016).

The population of online gaming experienced exponential growth in the early 2000s when the first massively multiplayer online role-playing games (MMORPGs) such as *World of Warcraft* were released, providing millions of players worldwide simultaneously playing with other gamers great stimulation in a dynamic social context.

2.1.2 *The State of Video games Today*

While computer games began with quite simple foundations, they have vastly diversified today, and millions of dollars are spent on the development and updating of these games. Thus, the computer and video game industry enjoyed a record-breaking year in 2018 when video game sales exceeded a total of $43.4 billion (Entertainment Software Association 2019). Gaming is now one of the most common pastimes in highly developed societies for all social groups. The Entertainment Software Association (ESA) recently published data indicating that approximately 65% of all adults in the United States of America (USA) regularly play video games, while about 75% of all households in the USA include at least one active gamer (ESA 2019). Gamers in the USA are using a variety of electronic devices for gaming, primarily smartphones (60%), personal computers (52%), and dedicated game consoles (49%) (ESA 2019). A similar pattern of gaming has been observed in other highly developed countries. As the Interactive Games and Entertainment Association (IGEA) reported for Australia, around 67% of all inhabitants play video games and 97% of all households with children own computer games (IGEA 2017).

2.1.3 Definition of Disordered Gaming

The negative effects and consequences of disordered gaming have been investigated since the early 1980s (Soper and Miller 1983), when researchers administering self-report measures observed that some children and young adults were unable to stop playing video games (Shotton 1989). Interestingly, the results of studies conducted in this period with unstandardized scales were later confirmed with standardized scales. The 1990s was the decade when the criteria for disordered gaming were determined, based on the definition of pathological gambling in the Diagnostic and Statistical Manual of Mental Disorders, 4th edition (DSM-IV) (American Psychiatric Association [APA] 2000). Evidently, pathological gambling and disordered gaming are not equivalent, but this definition was an important step towards the development of specific criteria for disordered gaming. Thus, the definition and six criteria for pathological gambling from the DSM-IV (APA 2000) were adapted by studies on "gaming addiction" or similar constructs (e.g., internet addiction), and many researchers adopted the term "pathological gaming" to describe this type of behavior accordingly (King et al. 2013). With the spread of the internet and the great increase in the number of online games, the number of publications in this area grew in the 2000s (King et al. 2013). The inconsistency of terminologies and the application of non-standardized assessment tools in previous research into disordered gaming (King et al. 2013) triggered numerous debates about the nature of this phenomenon as a unique clinical entity that should officially be recognized as a behavioral addiction (Griffiths and Pontes 2014; Pontes and Griffiths 2014, 2015).

The latest (fifth) edition of the DSM (DSM-5) also includes the six criteria (i.e., preoccupation, tolerance, withdrawal, persistence, escape, and problems) among the proposed criteria for "Internet Gaming Disorder" (IGD) (APA 2013), adding deception, displacement, and conflict as diagnostic indicators of dysfunction (Lemmens et al. 2015). After consulting a multidisciplinary expert working group, the APA decided to include IGD as a tentative condition in the DSM-5 (Petry and O'Brien 2013), labeling it as a "condition that needs further research before being fully recognized and accepted as an independent disorder in subsequent revisions of the DSM" (Petry and O'Brien 2013). According to Pontes and Griffiths (2020), this was the first major milestone for the study of disordered gaming.

According to Pontes and Griffiths (2020), the second major milestone for disordered gaming research was marked, when in 2016 "Gaming Disorder" (GD) was included as a "behavioral addiction" in the beta draft of the 11th edition of the International Classification of Diseases (ICD-11) (World Health Organization 2018), based on the realization that by then research had demonstrated the clinical significance of this phenomenon alongside its related health burden, and neurobiological similarities with other substance use disorders (Saunders et al. 2017). The third key milestone, according to Pontes and Griffiths (2020), was passed on May 25, 2019 at the 72nd World Health Assembly, when the WHO, in the a historic and long-awaited decision taken after an extensive and iterative review process, officially recognized disordered gaming as a mental health disorder (WHO 2019).

While some believe online disordered gaming to be a specific type of video game addiction or a variant of internet addiction in general, others consider it as an independent condition. Nevertheless, studies aiming to explore whether disordered gaming should be considered a subtype of generalized internet addiction suggest that these entities appear to be conceptually distinct from each other and that disordered gaming should not be regarded as a subtype of generalized internet addiction (Balhara et al. 2020; Griffiths and Pontes 2014).

2.1.4 *Time Spent Gaming*

Excessive involvement in gaming (i.e., 8–12 h per day) is arguably the main reason for individuals with disordered gaming to be negatively impacted (Baggio et al. 2016). While there may be excessive use of video games without it being regarded as an addiction (Van Rooij et al. 2011), excessive gaming is usually related to disordered gaming (e.g., addictive MMORPG use positively correlated with the weekly time spent playing MMORPGs, Zanetta Dauriat et al. 2011). A systematic review suggested an association between spending increased amounts of time playing games more frequently and over a greater number of years and IGD (Mihara and Higuchi 2017). For a recent discussion on the issue of time spent gaming and disordered gaming, see the work by Pontes et al. (2022).

In large samples of adult (Elliot et al. 2012a, b) and adolescent gamers (Rehbein et al. 2010), it was found that time spent gaming differed significantly between different genres of games, with 'real-time strategy' (RTS) and 'role playing games' [RPG; in particular MMORPGs (Rehbein et al. 2010)] being played for the longest periods. RPGs are intended to immerse the player in a rich and complex narrative connected to the player's character or avatar. The creation of avatars has in fact been described as a long and persistent process (Billieux et al. 2013). Finally, while time spent playing games both online and offline is associated with disordered gaming, these associations are far stronger in online games (Lemmens and Hendriks 2016; Mihara and Higuchi 2017). Online connectivity provides not only the ability to play with or against other players, but also allows game developers to introduce new features, updates, and contents, ensuring that the game can offer novel experiences to players (King and Delfabbro 2019a).

2.1.5 *Game Genres*

For half a century, video games have changed continuously, from early standalone games such as *Space Marines* (1962) and *Pong* (1972) to collaborative and competitive games located in massively multiplayer online environments, with millions of players simultaneously playing against non-player enemies in the game or against

Table 2.1 Game genres

1. Social network games (e.g. Farm Ville, Mafia Wars)
2. Puzzles or other small games (e.g. Scrabble, Gravity Pops, Candy crush, Solitaire, Tetris, Angry Birds)
3. Social games (e.g. Second Life, Habbo Hotel, goSupermodel)
4. The Sim's games (e.g. The Sims, Second Life)
5. Music games (e.g. Guitar Hero, Singstar)
6. RPG—role-playing game (e.g. Skyrim, Zelda, Mass Effect vs.)
7. MMORPG—Massively multiplayer online role-playing game (e.g. World of Warcraft, Lord of the Rings Online, The Secret World)
8. RTS—Real time strategy games (e.g. Age of Empires, Civilization, StarCraft)
9. MOBA—Multiplayer online battle arena (e.g. LoL, Dota 2)
10. Platform games (e.g. Super Mario, Ratchet and Clank, Skylander)
11. Action games/fighting games (e.g. Mount and Blade, Assassin's Creed, Uncharted, Tekken vs.)
12. Sports games/car games (e.g. FIFA, NHL, Gran Turismo, Need for Speed)
13. Simulator games (e.g. Flight simulator, Silent Hunter, Gran Turismo)
14. Horror-themed/survival games (Resident Evil, Outlast)
15. FPS—First person shooter games (e.g. Call of Duty, Battlefield, Counterstrike)
16. TPS—Third person shooting games (e.g. God of War, Metal Gear Solid, Just Cause)
17. Battle Royale (e.g. Fortnite, PUBG, Apex)

other players (Bányai et al. 2019). Common genres of video game are listed in Table 2.1.

According to the ESA, the most popular gaming genres in the USA are casual (71%), action (53%), and shooter games (47%), while the best-selling video game genres are action (26.9%), shooter (20.9%), RPGs (11.3%), and sports games (11.1%) (ESA 2019). Types of video games played most often by gamers in the USA in 2019 were casual single-player games such as Candy Crush and Angry Birds, casual multiplayer, first-person shooter games, single-player role-plays, battle royale, MMORPGs, and multiplayer online battle arena (MOBA) games, respectively (Statista 2020).

Different game genres may be related with specific needs of individual gamers. For example, MOBA games might be played in an attempt to regulate emotions, while negative affective disturbances may lead subjects to play MMORPG as a maladaptive coping strategy (Bonnaire and Baptista 2019). Certain video game genres (e.g., MMORPGs) are more strongly related to disordered gaming (Bertran and Chamarro 2016; Kuss and Griffiths 2012). Users of RPGs (Lee et al. 2007) and MMORPGs (Stavropoulos et al. 2017) showed significantly higher internet addiction scores, and these game genres have the highest enjoyment ratings (Chang et al. 2016). Studies evaluating MMORPGs suggested their addictive potential (Hussain et al. 2012; Müller et al. 2015; Stetina et al. 2011) in relation to their specific structural features (Billieux et al. 2015). Indeed, playing MMORPGs can be particularly problematic as these games require a high degree of dedication and time investment from the players to the detriment of employment, social and other recreational activities, and relationships (Kuss et al. 2012).

According to a systematic review, persons with disordered gaming preferred certain types of game genres (i.e., MMORPG, first person shooter, fighting games and RTS games) (Mihara and Higuchi 2017). In another review, King et al. (2019) indicated that playing MMORPGs had the strongest positive association with disordered gaming. Finally, a recent study found that the risk of disordered gaming was 1.52 times higher with the genres of RPGs, 1.59 times higher with simulation games, and 1.51 times higher with MOBA among middle school students (Han et al. 2020a, b). However, not only the type of video game played is related with disordered gaming, but also participating in a variety of different of video game genres (versatility) (Donati et al. 2015).

2.1.6 Esports

Gaming brings many players together in the real world. Today, gaming has become a major industry with large congresses and tournaments attended by thousands of people (King and Delfabbro 2019a). Gaming has become professionalized, and for a minority of players, competitive gaming has become a career option (Bányai et al. 2019; Griffiths 2017). This novel type of professional video gaming has been termed as "esports" (electronic sports). Esports includes professional or amateur competitions organized in coordination with leagues or tournaments supported by commercial organizations (Hamari and Sjöblom 2017). Esports is a new area in gaming culture gaining prominence and popularity in parts of the video game communities, especially among adolescents and young adults (Bányai et al. 2019). Some of these games played individually or by teams are displayed on large screens watched by thousands of people. For example, in the game titled "League of Legends", teams consisting of five people are struggling with the features they choose and a set of more than 100 heroes.

Currently the new MOBA games are the most popular esports genre, while the FPS and the RTS genres have kept their popularity. A recent report found a growth of the global esports economy in 2017 by 41.3% (up to $696 million), expecting esports brand investment to double by 2020 (Newzoo 2017). Esports spectators, an important element for esports competitions, are defined as individuals watching, supporting, and following professional esports contents (Smith and Stewart 2010). Online streaming platforms such as *Twitch* and *YouTube* also enable a large following for esports (Bányai et al. 2019). The esports audience is estimated to have reached 385 million globally, with 45% of them being active players, 23% viewing streams of esports, and 32% both playing games and viewing streams (Newzoo 2017).

A number of studies have included esports in the framework of traditional sports (Hallman and Giel 2018). To gain the status of sport, esports need to be accepted as such worldwide (van Hilvoorde and Pot 2016). At a time when the popularity and attraction of esports increase, concerns are expressed regarding not only the psychology of video gaming, but also the lack of physical activity and the sedentary nature of esports (van Hilvoorde 2016; van Hilvoorde and Pot 2016) or the intensive and excessive participation in gaming (Griffiths 2017). Emerging esports research suggests that esports players may be at greater risk for experiencing disordered gaming when compared to non-professional gamers due to their increased exposure and engagement with the activity (see Maldonado-Murciano et al., 2022), a finding that is supported by Montag et. al (2022) showing that esports players differ from non-espors players in regards to their basic gaming motives.

2.1.7 Gambling and Disordered Gaming

With growing interest in esports, operators for betting in online gambling and opportunities for esports players to earn money have emerged. In addition to the existing online gambling operators adding esports betting options, new specialized esports betting sites have been launched. Bets can be placed on the team that is thought to win, as in professional sports, or on additional predictions (e.g., who will win the first part played with the pistol in the *Counter-Strike: Global Offensive* [CS:GO] game). Bets can be made for real money or with in-game items such as skins (King 2018).

One of the features of the video games believed to be related with both disordered gaming and gambling is loot-box engagement (Raneri et al., 2022; Li et al. 2019). Loot boxes are virtual goods containing randomized rewards that can be purchased in many video games; thus, players “gamble” for chance items (Kristiansen and Severin 2020; Li et al. 2019; Zendle and Cairns 2019). In some games, loot boxes are provided to players as a reward for completing a designated stage, level, or any other specified in-game achievement; other games allow loot boxes to be purchased at will, some games award loot boxes in what seems to be random fashion (Macey and Hamari 2019). In previous studies, nearly half of the participants reported that they were engaged in loot box activities in the past year (Kristiansen and Severin 2020; Li et al. 2019). These individuals find rarer game items hedonically rewarding and motivating (Larche et al. 2019). Given the similarities found between certain loot boxes and gambling, a vivid debate investigates whether these items actually constitute a form of gambling (Drummond et al. 2020; Larche et al. 2019). Recent studies that evaluated this association of loot box engagement and problematic gambling among adolescents (Kristiansen and Severin 2020; Zendle 2019; Zendle et al. 2019; Zendle and Cairns 2019) and adults (Brooks and Clark 2019; Drummond et al. 2020; Larche et al. 2019; Li et al. 2019) suggest a relationship between loot boxes and gambling. Accordingly, loot box purchasing may facilitate the transition from recreational video gaming and online gambling to problematic engagement in video gaming and/or gambling (Li et al. 2019). The legal status of this feature has even been questioned, indicating a potential requirement to regulate their use as gambling (Zendle and Cairns 2019), and researchers have discussed ways to reduce the impacts of spending on loot boxes (Drummond et al. 2020; Kristiansen and Severin 2020). In 2016, the Chinese government passed legislation that required game developers to disclose the odds of receiving certain items from loot boxes. In 2018, the Belgium Gambling Commission declared that loot boxes are in violation of gambling legislation (King and Delfabbro 2018). While some researchers suggest that setting limits to the sale of loot boxes might protect gamers at risk of gambling problems (Drummond et al. 2019), others suggest that this may not be sufficient on its own to stop players overspending (King and Delfabbro 2019b). A cautious approach to public policy may be needed to set up adequate regulation for loot boxes (McCaffrey 2020).

2.1.8 Negative Consequences of Video Gaming

Although research has been concerned with the potential negative effects of widespread gaming, evaluating negative physical and psychosocial effects in their studies, video gaming cannot be considered universally harmful and disruptive to all users. According to the ESA, American adults playing video games believe that gaming positively impacts their lives by means of providing mental stimulation (79%) or relaxation and a way for relieving stress (78%) (ESA 2019). Similarly, gamers in Australia often report that their thinking skills (84%), dexterity (78%), and pain management (59%) are improved by gaming (IGEA 2017). However, gaming-related organizations are not alone in reporting favorably about these activities; they are supported by scholarly research indicating a range of positive and beneficial outcomes for the majority of gamers, including among others psychomotor, therapeutic, and educational benefits (Granic et al. 2013; Griffiths 2019). It has even been reported by some researchers that playing games has benefits such as increasing focus, improving processing memory and multitasking, increasing hand–eye coordination, and improving strategic skills (Mihara and Higuchi 2017).

Although a large majority of gamers generally experience positive outcomes of their pursuit, there are consistent reports in the scientific literature about video games having negative behavioral and psychological effects including aggressive (Ferguson 2018) and addictive (Pontes 2018) behaviors.

2.1.8.1 Behavioral and Psychosocial Problems

Arguably, another important factor supporting the recent formal recognition of disordered gaming as a mental health condition was related to the evidence of its associated behavioral problems and psychosocial functional impairments as shown by numerous studies linking disordered gaming to many problematic outcomes (Kircaburun et al. 2020). Disordered gaming explains the unique variance in psychological distress (1%) and disability (3%) after controlling for common comorbid conditions such as depression, obsessive compulsive disorder (OCD), attention deficit hyperactivity disorder (ADHD), and anxiety (Pearcy et al. 2017).

Šincek et al. (2017) found that committing violence against peers (both in traditional form and cyber aggression) significantly predicts disordered gaming, which supports the premise that children and adolescents at risk are prone to exhibiting different forms of risky behavior in different settings. According to a systematic review by Mihara and Higuchi (2017), people with disordered gaming often display aggression and violence, expressed in three different forms: (i) general aggressive tendencies; (ii) accepting aggression and violence psychologically; and (iii) actually carrying out aggressive behavior, such as breaking rules, fighting physically, and carrying weapons. In line with these manifestations, aggression, and physical aggression in particular, is related with the severity of disordered gaming (Evren et al. 2019a, b, c, d). In addition, aggression is an important variable for treatment as it contributes to a less favorable prognosis (Lee et al. 2018a, b). Therefore, as previous studies suggest, youths diagnosed with disordered gaming should be screened for aggression. Nevertheless, contrary to these data, a study examining whether violent games cause aggression in the long term suggested that the negative effect of the games is of short duration if any (Kühn et al. 2019). A study examining one group playing violent video games, one playing non-violent video games, and a control group not playing any video games, reported no evidence of negative effects such as aggression and hostility after two months in any of the groups (Kühn et al. 2019).

According to a recent rapid scoping review by Darvesh et al. (2020), the five most frequently reported health-related variables significantly associated with disordered gaming were depression (67 times), internet addiction (54 times), anxiety (48 times), impulsiveness (37 times), and ADHD (24 times). A recent meta-analysis by Männikkö et al. (2020) reviewing a total of 50 empirical studies on disordered gaming found the condition to be significantly associated with a wide range of outcomes related to health and well-being, including higher levels of depression, anxiety, loneliness, ADHD, and OCD, as well as a decreased levels of quality of life, life satisfaction, self-esteem, and success in school. In addition to being able to identify a link between disordered gaming with greater general psychiatric distress, the meta-analysis by Männikkö et al. (2020) also provided additional data on the individual differences and behavioral outcomes associated with disordered gaming such as increased impulsivity, poorer self-control, greater attention deficit, hyperactivity, and decreased levels of physical activity.

Cognitive functions in relation with disordered gaming constitute another area of interest for researchers. In a meta-analysis across 40 studies conducted by Ioannidis et al. (2019), problematic internet use, even with gaming being the predominant type of online activity, was found to be associated with significant impairment in inhibitory control (Stroop task, stop-signal task, go/no-go task), decision-making, and working memory. In a recent review, King et al. (2019) reported differences between problem gamers and controls in neurocognitive tasks measuring executive functions, attention, processing, and decision-making. According to King et al. (2019), problematic players also tend towards making poorer decisions under risky conditions, fail to take objective probabilities into account, demonstrate prepotent deficits in response inhibition, and are less able to delay gratification for a larger reward.

A systematic review by Mihara and Higuchi (2017) suggested that IGD was associated with problematic peer relations, a higher prevalence of being bullied as well as bullying others, and having friends with disordered gaming. The same review found many studies reporting lower educational and career attainment, including lower school grades, skipping classes, and truancy, indicating lower social skills, competence, and integration, which the authors suggested might largely be related to effects of disordered gaming (Mihara and Higuchi 2017).

2.1.8.2 Physical Negative Consequences

Sleep Disorder: Results of the systematic review suggest that disordered gaming, particularly using MMORPGs, might be associated with poorer quality of sleep and subjective insomnia (Lam 2014). Time spent playing games on the internet has also been associated with a gradually increasing delay in sleep time and decrease in time and quality of sleep, especially regarding children and adolescents (Mihara and Higuchi 2017). It has been reported that games played before bedtime can be responsible for these problems by causing increased activation in the central and autonomic nervous system (Mihara and Higuchi 2017).

Vision Problems: The widespread use of digital screens in devices used throughout the day has led to the emergence of "digital eye strain" as a new clinical syndrome that affects every individual who spends a large period of time fixated on multiple screens for work or leisure. A new subcategory, "video game vision" has been proposed specifically to address vision issues related to large periods of continuous use of screen-enabled devices in order to play video games (Mylona et al. 2020).

The majority of vision problems reported by players relate to blurred vision, difficulty in focusing and headache after screen usage, or discomfort in the form of a stinging/burning sensation due to dry eye syndrome, or photosensitivity (Jaiswal et al. 2019). Eye strain (67%), eye burn (61.9%), blurred vision (51.6%), dry eyes (26.2%), and double vision (28.9%) were experienced among gamers (Mowatt et al. 2018).

Musculoskeletal System Problems: A previous study reported that 75.1% of the players complained about neck pain related with the time spent at the screen and 65.5% complained about shoulder pain (Mowatt et al. 2018). A systematic review found a number of studies reporting physical pain as a consequence of disordered gaming (Mihara and Higuchi 2017).

General Physical Health Problems: As the time spent playing games increases, daily life activities such as sleep, eating and personal hygiene are neglected in disordered gaming. In order not to stop playing, eating times are shortened, weight loss can be seen in some individuals, while others may gain weight due to unhealthy foods being prepared and eaten quickly. In regards to game genre, RTS and RPG games demonstrate the poorest physical activity and screen-time behaviors, which may be characteristic of a more sedentary behavior profile (Thorne et al. 2014). Because disordered gaming is associated with both sedentary behavior and metabolic risk, one can assume that it can also be associated with higher reported body mass index (BMI) to the level of obesity (Weaver et al. 2009). Consistent with this observation, studies found that patients with disordered gaming display higher BMI than patients with gambling disorder (Mallorquí-Bagué et al. 2017) and severity of disordered gaming symptoms predicts BMI of 25 or more even when controlling for depression and anxiety (Evren et al. 2018a). In addition, the incidence of general health problems such as tension, headaches, and fatigue increase (King and Delfabbro 2019a).

Another important complication related with gaming is seizure. Most of gaming-related seizures are generalized tonic–clonic, myoclonic, and absence type, which can be categorized into idiopathic generalized epilepsies. Photosensitivity, behavioral and higher mental activities are seen as significant seizure precipitants (Chuang 2006).

2.1.9 Gaming Motivations

While most studies emphasize the risks and downsides of online gaming, there are suggestions that online games may also help address human needs emerging from modern culture in novel ways, indicated by the popularity of the games (Demetrovics et al. 2011). If gaming satisfies people's basic needs, it cannot be categorically evaluated as bad or good (Demetrovics et al. 2011). Rather, the features of individual games should be assessed from a motivational view, exploring gamers' reasons and motives for playing them, without judging them as helpful or dangerous. Motives have been defined as sets of information representing emotional preferences expressed by thoughts and ideas that refer to specific desirable or undesirable goals and categories of goals (McClelland 1985). These motives can be considered as factors stimulating and shaping behavior (Demetrovics et al. 2011).

Motives have been categorized using different approaches. For instance, Bartle (1996) identified four types of multiplayer computer gamers (i.e., killers, achievers, socializers, and explorers). Further testing of Bartle's model led to the development of a five-factor model of gaming motives (i.e., with the factors relationship, manipulation, immersion, escapism, and achievement) (Yee 2006a), and a ten-factor model (including three higher-level factors: achievement [advancement, mechanics, competition], immersion [discovery, role-playing, customization, escapism], and social [socialization, relationship, teamwork]) (Yee 2006b). The distinct motivational categories were found not to be mutually exclusive as players can be included in more than one category simultaneously (Demetrovics et al. 2011).

Measures validating the motivation of gaming have been developed by a number of researchers. Thus, the Gaming Motivation Scale (GAMS) was designed by Lafrenière et al. (2012) using the self-determination theory to assess factors such as intrinsic motivation, integrated, identified, introjected, external regulation, and amotivation. The Electronic Gaming Motives Questionnaire (EGMQ), adapted from the four-dimensional Gambling Motives Questionnaire-Revised (GMQ-R) was validated by Myrseth et al. (2017) to measure four gaming motives, namely, enhancement, coping, social, and self-gratification.

The Motives for Online Gaming Questionnaire (MOGQ) was created by Demetrovics et al. (2011) using information from a literature review and survey data to measure seven dimensions of motives for gaming: The social motive regards the desire to play and make friends; escape indicates gaming to avoid dealing with real-life problems; competition is aimed at defeating others; skill development enhances the player's coordination, concentration, and other abilities; coping is an attempt to reduce stress, tension, or aggression and improving one's mood by gaming. Accordingly, some studies suggested that gaming was both an inadequate coping behavior and a source of self-efficacy (Rehbein et al. 2010). Fantasy allows gamers to try out new identities and/or activities that are not available in the everyday life. The last dimension, recreation, means having fun gaming (Demetrovics et al. 2011; Király et al. 2015; Wu et al. 2017). With the MOGQ, for the first time motives of internet gamers from a broader age range can be assessed. With high internal consistency, the instrument encompasses the previously established main motives of gaming in line with earlier studies (Wu et al. 2017). Going beyond former approaches (Bartle 2003; Frostling-Henningsson 2009; Yee 2006b), the MOGQ can distinguish between seven distinct motives for internet gaming The MOGQ has been validated for different languages such as Chinese (Wu et al. 2017) and Turkish (Evren et al. 2020a).

A review by King et al. (2019) suggests that stronger gaming motivations (e.g., escapism, achievement) are among the significant vulnerabilities predisposing players for disordered gaming. When studying the behavioral effects of playing video games, examining the underlying psychological motives that may lead a person to engage with the activity is paramount (Montag et al. 2019). Researching the motivational basis of gaming behaviors is an important contribution to furthering an understanding of the general nature of gaming activities while illuminating the specific factors that determine if gamers engage with the activity in healthy, excessive, or disordered ways (Pontes et al. 2019). For a more recent review on the intersection between gaming motives and disordered gaming, please see the work by Wang and Cheng (2022).

2.1.10 Personality in Relation with Disordered Gaming

Studies have evaluated individual differences such as personality factors to provide insights concerning the etiology of disordered gaming. Examining the Big Five personality traits in a sample of 2.553 people aged 12–25 years, Walther et al. (2012) reported that disordered gaming was specifically associated with the characteristics of high impulsivity, irritability/aggression, social anxiety, attention deficit issues, and low self-esteem. In the area of personality and psychological problems, reported risk factors for disordered gaming included loneliness, impulsivity, neuroticism, and conduct problems (Mihara and Higuchi 2017). Other studies have identified personality factors and psychological characteristics such as impulsivity, neuroticism, introversion, lower agreeableness, lower conscientiousness, low openness to experience, aggressiveness, rule-breaking, trait anxiety, and narcissism (King and Delfabbro

2019a; King et al. 2019). In the systematic review by Schimmenti et al. (2017), it was suggested that, while many studies quite consistently mentioned certain personality factors (e.g., high neuroticism, impulsivity, and aggressiveness) as significant predictors of disordered gaming, a key role in the acquisition, development, and maintenance of disordered gaming might be played by different personality traits, quite commonly in combination. Therefore, establishing whether there are specific patterns of personality traits predisposing people to gaming disorder will require further research.

A recent review identified 21 studies suggesting a negative correlation between disordered gaming and extraversion, conscientiousness, agreeableness, histrionic traits, openness to experience, grit, oppositional traits, and self-demeaning traits (Şalvarlı and Griffiths 2019). The same review found a positive correlation between disordered gaming and negative valence, neuroticism, sensation seeking, inhibition, introversion, egotism, narcissism, sadism, Type D personality traits, negative affectivity, detachment, antagonism, disinhibition, psychoticism, novelty-seeking, harm avoidance, and schizotypal traits (Şalvarlı and Griffiths 2019).

Among the psychological tendencies, other reports include loneliness, low self-esteem, low self-efficacy, and low life-satisfaction (Mihara and Higuchi 2017). In contrast, some personality traits, such as high extraversion, high agreeableness, low impulsivity, high conscientiousness, and greater emotional regulation, have been identified as protective factors (King and Delfabbro 2019a; Mihara and Higuchi 2017). Also, higher-level social competence and self-esteem at baseline (Lemmens et al. 2011), social integration and school-related well-being (Rehbein et al. 2013) were among the factors found to be protective against disordered gaming.

2.1.11 Gender and Age

Previous studies suggest that the risk of disordered gaming is higher among males than females (Dong et al. 2018; López-Fernández et al. 2020). Gender differences were found in frequency of video game play (Huang et al. 2017), severity (Ko et al. 2005) and prevalence of disordered gaming (Mihara and Higuchi 2017), platform and genre preference (Chang et al. 2016; Huang et al. 2017; Zanetta Dauriat et al. 2011), motives for playing (Ko et al. 2005; Király et al. 2015) and engaging with loot boxes (Kristiansen and Severin 2020).

According to King and Delfabbro (2019a), one reason for this gender difference may be that many games have been developed primarily by males for males, with game modes and features (e.g., competition, territorial invasion, violence, and domination) and aesthetics (e.g., sexualized images of women, crude humor) that are generally more appealing to males, particularly among adolescents. There are also many games that appeal more strongly to female players, including social games, story-driven, and casual games. However, these types of games are often less time-consuming and risky (King and Delfabbro 2019a).

A systematic review reported an association between younger age and a greater prevalence of disordered (Mihara and Higuchi 2017). Adolescence, usually defined for convenience as the period between the ages of 12 and 18 years, is the most vulnerable time for the acquisition of addictive disorders, including disordered gaming (King and Delfabbro 2019a). Gaming may be used as a mechanism for coping with various physical and psychological changes and new role expectations during adolescence (King and Delfabbro 2019a). Given that age appears to have an inverse u-shaped relationship with disordered gaming, the risk of developing this behavior will usually decrease as individuals enter their late 20 s (Kuss and Griffiths 2012). According to longitudinal studies, the course of disordered gaming in adolescents tends to be more stable than in adult populations (Mihara and Higuchi 2017). Also, game genre preferences may change according to age (i.e., older adults may be more interested in puzzles and intellectually stimulating games, whereas their interest in popular games like MMORPGs and FPSs may be very low) (Blocker et al. 2014). Thus, when evaluating a variable related with disordered gaming, it is important to take gender and age into consideration (Ko et al. 2005; López-Fernández et al. 2020; Mihara and Higuchi 2017).

2.2 Neurobiological Bases of Disordered Gaming

Although numerous neurobiological studies, particularly Magnetic Resonance Imaging (MRI)/Diffusion Tensor Imaging studies, have been carried out to characterize disordered gaming, this research is in its early stage and it is premature to reach strong conclusions. Nevertheless, neurobiological studies demonstrating similarities between disordered gaming, substance abuse (Weinstein and Lejoyeux 2015), and pathological gambling (Fauth-Bühler and Mann 2017) led to the official inclusion of gaming disorder in the beta draft of the ICD-11 as a "behavioral addiction" (Saunders et al. 2017; WHO 2018). Patients with gaming disorder as well as persons engaged in pathological gambling exhibited a decreased sensitivity to loss, and greater reactivity to cues of gaming and gambling, respectively, as well as enhanced impulsive choice behavior, aberrant reward-based learning, but no changes in cognitive flexibility (Fauth-Bühler and Mann 2017).

Kuss et al. (2018), selecting 27 out of 853 studies on the neurobiological correlates involved in disordered gaming, reached the conclusion that in disordered gamers, response-inhibition and emotion regulation are lower than in healthy controls, prefrontal cortex functioning and cognitive control are impaired, working memory and decision-making capabilities are poorer, visual and auditory functioning decreased, and their neuronal reward system displays a deficiency, similar to those found in individuals with substance-related addictions (Kuss et al. 2018). A number of neurobiological factors similar to those found in substance use disorders may be involved in the development and maintenance of addictive behaviors, including brain networks related to reward processing, executive functioning, salience attribution, and habit formation being altered as well as neurochemical pathways being modified, possibly involving dopamine, serotonin, opioids, and other neurotransmitters (Antons et al. 2020). Consequently, as substance-related addictions share common predisposing factors with behavioral addictions, the latter may also be part of an addiction syndrome (Kuss et al. 2018).

This subsection will provide an overview of currently available information about the neuroscientific basis of disordered gaming. In particular, this subsection draws on meta-analyses (Meng et al. 2015; Yao et al. 2017; Zheng et al. 2019) that summarize the existing findings in the field. Out of 61 relevant articles, Meng et al. (2015) selected 10 qualified studies applying voxel-wise whole-brain analysis and undertook a comprehensive series of meta-analyses. In subjects with disordered gaming, the bilateral medial frontal gyrus (MFG) and the left cingulate gyrus, as well as the left medial temporal gyrus and fusiform gyrus showed a significantly higher level of activation compared to controls. The time subjects with disordered gaming spent online positively correlated with activations in the left MFG and the right cingulate gyrus. According to Meng et al. (2015), these findings indicate the important role of prefrontal lobe dysfunction in the neuropathological mechanism of disordered gaming.

Yao et al. (2017) conducted a meta-analytic study with the primary aim to evaluate structural neural alterations in disordered gaming, finding that patients with disordered gaming show gray-matter volume reductions in several brain regions including the anterior cingulate, dorsolateral prefrontal, orbitofrontal, and premotor cortices when compared to healthy controls. Consistent with this observation, a longitudinal study by Zhou et al. (2019a) showed a decrease in lower right orbitofrontal gray matter volume after six weeks of daily gaming, suggesting that the orbitofrontal cortex may play an important role in the development of disordered gaming, demonstrating excessive engagement in online gaming being directly associated with structural deficits in this particular brain region. Interestingly, though, the often observed association between lower striatal volume and higher addictive tendencies was not reported in the meta-analysis by Yao et al. (2017), although new evidence points towards the possibility that a similar association may exist in disordered gaming (Zhou et al. 2019b).

Functional neural alterations in disordered gaming are also evaluated in meta-analytic studies (Yao et al. 2017; Zheng et al. 2019). Functional MRI studies investigate whether certain brain areas are differently activated when a gamer (compared to a control subject) is confronted with gaming-related stimuli or follows another task while lying in the brain scanner. Conducting a meta-analysis to evaluate functional neural alterations in disordered gaming, Yao et al. (2017) found that among disordered gamers, (1) the anterior and posterior cingulate cortices, caudate, and posterior inferior frontal gyrus (IFG), areas predominantly associated with studies measuring reward and cold-executive functions, were hyperactivated and (2) the anterior IFG (related to hot-executive function), the posterior insula, and somatomotor and somatosensory cortices (related to reward function) showed hypoactivation compared to controls.

Zheng et al. (2019) selected 40 studies for a comprehensive series of meta-analyses. In cue-reactivity tasks, the authors observed hyperactivation in gamers compared to healthy controls in the area of the bilateral precuneus and the cingulate area and hypoactivation in the insula, but there were no differences in the striatum. Indeed, the lack of striatal hyperactivation when processing gaming-related cues could mark a difference between disordered gaming and other forms of addictive behaviors, which is an assumption recently backed up in a longitudinal study by Yu et al. (2020). In executive control tasks, disordered gamers showed significant hyperactivation in the right superior temporal gyrus, bilateral precuneus, bilateral cingulate, and insula along with hypoactivation in the left inferior frontal gyrus. In risky decision-making paradigms, the left striatum, right inferior frontal gyrus, and insula exhibited significant hyperactivation in disordered gamers while the left superior frontal gyrus, left inferior frontal gyrus, and right precentral gyrus showed hypoactivation.

The existing neurobiological evidence in disordered gaming suggests an association with both functional and structural neural brain alterations, partly overlapping with those observed in other addictive behaviors, such as substance use disorders. While neurobiological research of disordered gaming has made significant progress, it is still early for understanding what may be the neurobiological mechanisms of disordered gaming (Antons et al. 2020). The reported findings should be replicated in different cultural settings to support a neurobiological basis for the classification of disordered gaming and related disorders (Kuss et al. 2018). Finally, it will be of greatest importance to understand how the gamers' age affects neural processes linked to disordered gaming, as different developmental stages may present with greater or lesser vulnerability to addictive disorders such as disordered gaming (Sugaya et al. 2019).

2.3 Specific Comorbidities

Many studies reported psychiatric comorbidities associated with disordered gaming, including ADHD, depression, anxiety, and substance use disorder (Mihara and Higuchi 2017). A review of 24 studies by González-Bueso et al. (2018) indicated significant levels of correlation between disordered gaming and anxiety (92%), depression (89%), ADHD (85%), as well as social phobia/anxiety and symptoms of OCD (75%). However, the direction of these associations remains unclear (González-Bueso et al. 2018). Persons with disordered gaming may easily use their escape into online games as a maladaptive strategy for coping with adverse emotional experiences (Blasi et al. 2019). Thus, the risk of disordered gaming may be increased by a comorbid psychiatric disorder. On the other hand, psychiatric disorders including depression and phobia also occurred as a consequence of disordered gaming rather than being a risk factor (Mihara and Higuchi 2017). In line with these findings, Kuss and Pontes (2019) suggested that there may be a vicious circle in the relationship between disordered gaming and depression, as depressed individuals are likely to play excessively to be able to cope with daily stressors, such as low and negative mood states (also known as the mood-enhancing hypothesis). Conversely, negative outcomes associated with disordered gaming (e.g., social isolation and loneliness) can further intensify the experience of depressive symptoms (also known as the social displacement hypothesis) (Kuss and Pontes 2019). These hypotheses are supported by multiple longitudinal studies showing that depression can predict subsequent onset of disordered gaming and that disordered gaming can also predict later onset of depression (Kuss and Pontes 2019).

Among other comorbidities, ADHD is of special importance as it has consistently been associated with disordered gaming (Bioulac et al. 2008; Evren et al. 2019c; Ha et al. 2006; Han et al. 2017). In addition, recent reviews suggested high correlations between disordered gaming and ADHD (Mihara and Higuchi 2017; González-Bueso et al. 2018). In a study finding a large effect size for the association between disordered gaming and ADHD, it was suggested that young adults suffering from both disordered gaming and ADHD had higher impulsivity and hostility, traits that may mediate the association between ADHD and disordered gaming (Yen et al. 2017). Evren et al. (2019d) suggested a relation between the presence of probable ADHD and severity of disordered gaming symptoms as well as aggression and depression. In another study by Evren et al. (2020b), probable ADHD and alexithymia in university students were related with the severity of disordered gaming symptoms, even after controlling for the effects of depression and anxiety. A recent study suggests a mediating role for current ADHD symptoms, especially inattention, on the effect of immersive tendency on disordered gaming (Jung et al. 2020). A study conducted in Australia and USA found participants with stronger ADHD symptoms exhibiting disordered gaming behaviors at higher levels (Stavropoulos et al. 2019). A study using the game and player statistics as well as the self-esteem measure of a *PlayerUnknown's Battlegrounds* (PUBG), a MOBA game claims to be able to predict ADHD with an accuracy of 81.8% (Aggarwal et al. 2020). In several community-based longitudinal studies, ADHD symptoms were found to predict the development of disordered gaming (Gentile et al. 2011; Wartberg et al. 2019). One longitudinal clinical cohort study that investigated 755 disordered gamers over a 5-year period found comorbid ADHD to be one of the key predictive factors for long term recovery of disordered gaming (Han et al. 2018). Finally, Chang et al. (2020) suggested that disordered gaming may be a strong trigger for emotional dysregulation; therefore, future DSM criteria could consider adolescents with disordered gaming as a specific sub-class of ADHD.

A large amount of data provides empirical evidence for overlaps of different types of addictive behaviors in their etiology, phenomenology, and in the underlying psychological and biological mechanisms (Kotyuk et al. 2020). A review of 20 studies by Burleigh et al. (2019) reported the co-occurrence of disordered gaming with other potential behavioral addictions (e.g., social media addiction, internet addiction, and gambling disorder) mainly among adolescents, while potential substance use disorders (e.g., alcohol, nicotine, and caffeine use disorder) were more likely to occur among adult gamers. The co-occurrence of different addictions might be related to (i) the use of maladaptive coping strategies (e.g., emotional avoidance) in order to avoid unpleasant affective states and associated mental disorders, and (ii) diminished emotional regulation that may encourage engaging in risky behaviors including elevated substance use (Burleigh et al. 2019). Evren et al. (2019b) found the severity of disordered gaming symptoms to be related with alcohol and/or cannabis use over the past year, with the impact of alcohol and cannabis use interacting. A more recent epidemiological analysis of a large sample corroborated the association of disordered gaming with alcohol consumption and cannabis use (Kotyuk et al. 2020). When disordered gaming is comorbid with alcohol use disorder, psychopathological impairments such as higher impulsivity, impaired self-control, and depressive symptoms are more severe, and the affected persons spend more money on gaming than with either of the diagnoses alone (Na et al. 2017).

It is well known that treatment of individuals with disordered gaming is more difficult in case of comorbidity with any other psychiatric disorder. In a study conducted among 186 adult treatment-seeking gamers with gaming-related problems, participants with mood symptoms (40% of the total) reported significantly more symptoms of disordered gaming, stronger maladaptive gaming cognitions (e.g., overvaluing game rewards), more previous occurrences of gaming problems, and poorer quality of life (King et al. 2018). Abnormal psychological status, especially, depressive mood and attention deficit, may lead individuals with disordered gaming to seek treatment at the hospital (Seong et al. 2019). Thus, individuals seeking treatment for symptoms of anxiety, depression, ADHD or OCD should also be considered for disordered gaming.

2.4 Prevalence Rates of Disordered Gaming

The prevalence of gaming disorder reported in different studies vary according to the definitions, measures, and/or the use of self-selected samples (Pontes and Griffiths 2015). However, the introduction of IGD in DSM-5 has led to significant progress (King et al. 2013; Kircaburun et al. 2020). Considering robust studies with large and representative samples, prevalence rates of disordered gaming according to the APA framework among the general population ranged from 0.03% in Finland (Karhulahti and Koskimaa 2019) to 5.8% in Netherlands (Lemmens and Hendriks 2016). Figures for adolescents were 1.3% in the Netherlands (Carras et al. 2017) and 9.3% in Lithuania (Ustinavičienė et al. 2016), respectively. In young adults, rates varied from 2.2% in Switzerland (Henchoz et al. 2016) to 9.1% in Finland (Männikkö et al. 2015). Thus, prevalence rates may change according to the population that is included in the study and even depending on the characteristics of individual participants in the same sample. A study conducted among both university students and gamers in Turkey found a prevalence of disordered gaming according to the APA framework of 0.96% for the whole sample and 2.57% among esports players (Evren et al. 2018b). Another large-scale epidemiological study carried out in Turkey in 2018 with 24.494 participants aged 18–81 years reported a prevalence rate of 1.6% (Ünübol et al. 2020).

Several systematic reviews and meta-analyses evaluated the prevalence rates of disordered gaming. A systematic review conducted by Mihara and Higuchi (2017) found a range of .7%–27.5% for prevalence in total samples, with non-adolescent groups (prevalence between .3 and 27.5%) being similar to the adolescent group (between 1.3% and 19.9%). The same review found higher prevalence among males compared females in most studies, while in some studies, the rates tended to be higher among younger rather than older people. A review by Feng et al. (2017) analyzing studies published between 1998 and 2016 reported a range of disordered gaming prevalence between .70% and 15.6%, whereas another systematic review by Paulus et al. (2018), examining studies conducted in children and adolescents, found a wider range .6–50%. A meta-analysis conducted by Fam (2018) found a pooled prevalence of disordered gaming among adolescents of 4.6% (6.8% for males, 1.3% for females). Subgroup analyses found the highest prevalence estimates for studies (i) conducted in the 1990s, (ii) using DSM criteria for pathological gambling, (iii) examining gaming disorder, (iv) been carried out in Asia; and (v) using small samples (< 1.000). Chia et al. (2020) conducted a meta-analysis of studies from Southeast Asia and found the pooled prevalence rate for gaming disorder of 10.1%. Finally, in a most recent rapid scoping review evaluating 160 studies that used 35 different methods, Darvesh et al. (2020) reported a prevalence of disordered gaming ranging from .21% to 57.50% in general populations, 3.20%–91% in clinical populations, and 50.42%–79.25% in populations undergoing an intervention (severe cases).

2.5 Assessment Approaches to Disordered Gaming

Although the disorder is labeled "Internet" Gaming Disorder, the DSM-5 states that "Internet Gaming Disorder most often involves specific internet games, but it could involve non-internet computerized games as well, although these have been less researched" (APA 2013). Of the nine criteria used in the DSM-5 (i.e., preoccupation with internet games, withdrawal symptoms, tolerance, unsuccessful attempts to control participation in internet games, loss of interest in previous hobbies, continued excessive use of internet games, deceiving family members, the use of internet games to escape, and loss of a significant relationship, job or education, or career opportunity), seven are identical to those of gambling disorder and five to substance use disorder (Petry et al. 2014). More severe degrees of IGD involving problematic behavior displacing usual and expected social, work and/or educational, relationship and family activities may result in academic failure, job loss or marriage breakdown (Montag et al. 2019). For a positive diagnosis, five or more out of the nine criteria need to be endorsed over a period of 12 months (APA 2013). In the DSM-5, IGD is characterized clinically by a "persistent and recurrent use of the internet to engage in games, often with other players, leading to clinically significant impairment or distress" (APA 2013).

Psychometric tools have been employed before IGD was introduced in the DSM-5, but they had a number of shortcomings (King et al. 2013; Kircaburun et al. 2020), including (i) inconsistent core criteria used to define disordered gaming, (ii) lack of a temporal dimension for assessing disordered gaming, (iii) variety in the cutoff scores set to identify disordered gaming, (iv) insufficient inter-rater reliability and predictive validity, and (v) low consistency in the dimensionality of those tools. Since the APA initially included IGD in the DSM-5, seven clinical psychometric tools have been developed covering the nine IGD criteria (APA 2013; Kuss and Pontes 2019). These psychometric tests include the Internet Gaming Disorder Test (IGD-20 Test) (Pontes et al. 2014), the Internet Gaming Disorder Scale (IGDS; Lemmens et al. 2015), the nine-item Internet Gaming Disorder Scale–Short-Form (IGDS9-SF; Pontes and Griffiths 2015), the 10-item Internet Gaming Disorder Test (IGDT-10; Király et al. 2017), the Clinical Assessment Tool (C-VAT 2.0) (Van Rooij et al. 2017), the Personal Internet Gaming Disorder Evaluation (PIE-9) (Pearcy et al. 2016), and the Chinese Internet Gaming Disorder Scale (C-IGDS) (Sigerson et al. 2017). Clear superiority for any one of these tools could not be established, but according to a recent comprehensive systematic review, evidential support for their psychometric properties was better for the IGDS, IGDS9-SF, and the IGDT-10 compared to the other instruments (King et al. 2020). These measures have received particular clinical attention and have been validated for cross-cultural use (Evren et al. 2017, 2018b, 2020c).

Another milestone for the study of disordered gaming was its inclusion as a "behavioral addiction" in the beta draft of the ICD-11 (WHO 2018). More specifically, the beta draft of the ICD-11 defined disordered gaming as a pattern of persistent or recurrent online and/or offline gaming behavior manifested by three core diagnostic criteria: (1) impaired control over gaming (e.g., onset, frequency, intensity, duration, termination, context); (2) increasing priority given to gaming to the extent that gaming takes precedence over other life interests and daily activities; and (3) continuation or escalation of gaming despite the occurrence of negative consequences (WHO 2018). The recurrent gaming behavior involved in gaming disorder may happen online or offline, and the clinical assessment covers a 12-month period (which may be reduced if all diagnostic requirements are met and symptoms are severe), determining if the severity of the condition significantly affects personal, family, social, educational, and occupational life domains and/or other broad areas of functioning (WHO 2019). The WHO proposes a number of exclusion criteria for a differential diagnosis of gaming disorder, screening the patient for hazardous gaming and bipolar type I and type II disorders (WHO 2019).

Pontes et al. (2019) developed the Gaming Disorder Test (GDT) to promote research and improve the understanding of disordered gaming within the WHO's new diagnostic framework, assessing gaming disorder using a brief four-item measure with a single-factor structure. The first three items of the GDT were developed to map on to the following clinical criteria: (i) impaired control over gaming (i.e., *"I have had difficulties controlling my gaming activity"*), (ii) increased priority given to gaming (*"I have given increasing priority to gaming over other life interests and daily activities"*), and (iii) continuation despite negative consequences (*"I have continued gaming despite the occurrence of negative consequences"*). The fourth item of the GDT investigates the experience of major life problems occurring at extreme levels of gaming disorder in the light of the potential functional impairments caused by severe gaming disorder (***"I have experienced significant problems*** *in life [e.g., personal, family, social, education, occupational] due to the severity of my gaming behavior"*). Pontes et al. (2019) demonstrated for both the Chinese and English versions that the four items of the GDT are valid and reliable, and the instrument proved to be highly suitable for measuring gaming disorder in a cross-cultural context. Other studies, like the German (Montag et al. 2019) and Turkish (Evren et al. 2020d) versions of the GDT further supported the WHO's diagnostic framework for gaming disorder and its measurement using this tool.

The diagnostic frameworks for disordered gaming according to the APA and WHO have been compared in subsequent studies. A recent large-scale study used both diagnostic frameworks in a comparison of the co-occurrence of psychiatric symptoms in gamers assessed with both the APA and WHO diagnostic frameworks for disordered gaming and found them to be relatively consistent in their predictions of potential psychopathological symptoms associated with disordered gaming, indicating that both the APA and WHO diagnostic frameworks are appropriate for assessing disordered gaming and its accompanying comorbidities (Montag et al. 2019). Establishing consistency in the assessment and measurement between the criteria in DSM-5 (APA 2013) and ICD-11 (WHO 2019) is crucial to avoid the pitfalls in earlier disordered gaming research and to facilitate future research examining its detrimental psychological and social effects on society (Kircaburun et al. 2020).

2.6 Treatment Approaches to Disordered Gaming

Not much knowledge is available about first-line pharmacological and psychological treatment for disordered gaming and its effectiveness (Stevens et al. 2019). For almost all treatment studies, methodological shortcomings have been reported and information about efficacy and treatment effectiveness of the interventions is still largely inconsistent (Kircaburun et al. 2020).

Some previous meta-analyses and reviews have evaluated treatment approaches to disordered gaming. Winkler et al. (2013) analyzed 16 studies regarding the treatment of internet addiction (including studies on disordered gaming), suggesting that both pharmacological and psychological treatment approaches were deemed as effective in treating symptoms of internet addiction like time spent online and symptoms of common comorbidities such as anxiety and depression. Similarly, Przepiórka et al. (2014) suggested that a combination of psychological treatment (e.g., cognitive behavioral therapy [CBT]) and pharmacological treatment (e.g., opioid receptor antagonists such as naltrexone combined with sertraline), antidepressants (e.g., bupropion and escitalopram), antipsychotics (e.g., quetiapine and olanzapine), glutamate receptor antagonists (e.g. memantine), mood stabilizers (e.g., valproate), and psychostimulants (e.g., methylphenidate) would yield the best results in treating disordered gaming symptoms. A systematic review of forty-six studies found both psychological and pharmacological treatments to be effective in the treatment of disordered gaming (Kuss and Lopez-Fernandez 2016), while psychiatric comorbidities may pose important obstacles to the treatment of disordered gaming (King and Delfabbro 2019a). Similarly, Kuss and Lopez-Fernandez (2016) reported that with (primary or secondary) comorbidities such as OCD and ADHD, drugs commonly used for these disorders can help reduce disordered gaming problems. Therefore, results from studies including pure disordered gamers cases only in the evaluation of gaming disorder treatment cannot be generalized to cases with comorbidities (King and Delfabbro 2019a).

Recent reviews of disordered gaming treatment (Gioia and Boursier 2019; King and Delfabbro 2019a; Nakayama et al. 2017; Zajac et al. 2020) analyzed 28 studies, 7 of which evaluated pharmacological treatments (e.g., administering bupropion, methylphenidate, escitalopram, and atomoxetine), 13 assessed a CBT-based treatment, and 8 investigated other approaches (e.g., a speaking and writing course, family therapy, eclectic psychotherapy, self-discovery camp). Unfortunately, despite the relatively rapid expansion in research on disordered gaming treatment, none of the studies evaluating treatment approaches have been meticulously designed to address efficacy (Zajac et al. 2020). Almost all studies included adolescent or young adult populations. Two of the 7 drug studies, 5 of the 13 CBT studies and one of 8 studies evaluating other approaches were designed as randomized controlled trials (Zajac et al. 2020).

2.6.1 Pharmacological Treatments

Pharmacological studies, all of which so far have been carried out in South Korea, have typically evaluated drugs used to treat either depression (e.g., bupropion, escitalopram) or ADHD (e.g., methylphenidate, atomoxetine). In 3 studies using a pre-test post-test design, there was a decrease in disordered gaming symptoms after treatment with bupropion for 6 weeks (Han et al. 2010) or 12 weeks (Bae et al. 2018) or an 8-week treatment with methylphenidate (Han et al. 2009), respectively.

Two studies compared two different drugs; (i) 12-week treatment of bupropion and escitalopram (Nam et al. 2017) and (ii) 12-week atomoxetine and methylphenidate (Park et al. 2016). There was no placebo control group in either study. In both studies, disordered gaming symptoms decreased in response to the respective drugs, with no significant difference being found in the effectiveness of the two drugs.

Two pharmacological studies used randomized designs and had a control group. In the first study, an 8-week bupropion treatment was found to be superior to placebo regarding playing time and reduction of disordered gaming symptoms (Han and Renshaw 2012). In the second study, 6 weeks of treatment with bupropion or escitalopram were found to be superior to placebo in the treatment of disordered gaming symptoms, and the decrease of symptoms was greater with bupropion than escitalopram (Song et al. 2016).

Finally a study that evaluated the co-administration of CBT and bupropion suggested that the combination was superior to bupropion alone (without psychotherapy), and this difference persisted for 4 weeks of follow-up (Kim et al. 2012).

2.6.2 CBT Based Psychotherapy

Recent reviews suggest that CBT is one of the most widely adopted psychological methods for the treatment of disordered gaming symptoms (Gioia and Boursier 2019; King and Delfabbro 2019a; Nakayama et al. 2017; Stevens et al. 2019; Zajac et al. 2020).

In a recent study, Han et al. (2020a, b) examined the effectiveness of CBT in patients with disordered gaming, comparing patients who completed the CBT program (n = 101) and those who completed supportive therapy (n = 104). In the CBT program one therapist and four to five patients met once or twice per week for a total of 14 sessions lasting 90 min each, whereas the supportive therapy group attended a psychiatric outpatient department once or twice a week until they had completed 14 visits. Compared with the supportive therapy group, the CBT group showed greater improvement in disordered gaming (66.3%) and greater decrease in internet addiction, anxiety, impulsivity, and social avoidance. The authors reported that in the CBT group, the greatest improvements among the patients who improved were in the areas of internet addiction, attention, depression, anxiety, impulsivity, social avoidance, and family cohesion.

A recent meta-analysis of 12 independent CBT studies suggested that CBT can improve disordered gaming symptoms and comorbid depression and anxiety; although treatment gains tend to be short-term and their effect in reducing time spent gaming is unclear (Stevens et al. 2019). The authors also expressed the need for more rigorous studies to determine the potential long-term benefits of CBT for disordered gaming.

2.6.3 Other Treatment Approaches

The final seven studies examined an array of intervention approaches that did not fit into either the medication or CBT categories (Zajac et al. 2020). The first five were pretest–posttest studies. The first study suggested that family therapy (five sessions over three weeks) was related to a decrease in gaming time and disordered gaming symptoms (Han et al. 2012). The second study evaluated the effects of brief (i.e., 84 hours) voluntary abstinence, and although no significant improvement in gaming time was found for the entire sample (n = 24), those who met IGD criteria (n = 9) showed a significant decrease (King et al. 2017). The third study found that transcranial direct current stimulation (12 sessions over 4 weeks) was related to significant decreases in time spent gaming and symptoms of disordered gaming among young adults (Lee et al. 2018a, b). The fourth study reported that with an eclectic treatment approach for adolescents (CBT, family therapy, motivational interviewing, and solution-focused therapy), there was a significant improvement in parent-reported but not adolescent-reported disordered gaming symptoms (Palleson et al. 2015). The final pretest–posttest study found that attending a 9-day self-discovery camp was associated with a decrease in two of three measures of time spent gaming (i.e., hours per day and hours per week but not days per week) in an adolescent sample (Sakuma et al. 2017).

Two studies of other treatment approaches included control groups. The first used a non-randomized design where participants chose one of four treatment methods. Among these methods, a 7-day residential camp, 8 sessions of parent management and residential camp together with parent management were superior to the psychoeducation for improving disordered gaming symptoms at post-treatment and 6-month follow-up (Pornnoppadol et al. 2018). Finally, a randomized controlled trial found that an 8-week speaking and writing course focused on MMORPGs was not superior to a general education control in reducing time spent gaming among adolescents (Kim et al. 2013).

In general, strong conclusions about the efficacy of treatments for disordered gaming are limited by weaknesses in the design of the studies reviewed, including the lack of appropriate control groups, non-random assignment to treatment conditions, and small sample sizes (Zajac et al. 2020). Over half of the studies examining disordered gaming treatment were conducted in South Korea and thus a much larger number of treatment studies using cross-culturally validated tools would need to be conducted across countries in order to examine cross-cultural differences (Zajac et al. 2020). Researchers should consider adapting treatments known to be effective for other addictive behaviors, including substance use (i.e., encourage family involvement in treatments) (Zajac et al. 2020).

2.7 Conclusion

Disordered gaming has been steadily developing as a consequence of the modern way of life. While video gaming remains as a very popular activity and healthy gaming may be possible for most of the gamers, there are growing concerns among researchers and society due to problems arising from excessive use of video games, even though very few individuals will ever experience significant gaming-related problems. Nevertheless, for a few individuals the activity may become disordered, and at that point, gaming may harm their social, occupational, familial, educational, and psychological functioning. Thus, early diagnosis and treatment of this minority is crucial.

While there may be an extreme investment of time in gaming without it being regarded as an addiction, excessive gaming (i.e., 8–12 hours per day) is arguably the main reason for individuals with disordered gaming to be negatively impacted.

Similar to other addictions, in disordered gaming control over participation in the gaming is reduced and gamers continue this behavior despite experiencing negative consequences, typically resulting in distress and functional impairment.

Previous studies concerning disordered gaming had several types of limitations including (i) sampling problems (e.g., non-probability sampling, sampling homogeneity, low sample sizes), (ii) measurement problems (e.g., use of non-validated and modified measurements), and (iii) lack of longitudinal data (King et al. 2020; Şalvarlı and Griffiths 2019). Consistency of assessment and measurement within the officially introduced criteria in DSM-5 (APA 2013) and ICD-11 (WHO 2019) is crucial to overcome major limitations in investigating disordered gaming that are preventing researchers from studying its psychological and social detrimental effects on society more thoroughly (Kircaburun et al. 2020).

When investigating the effects of disordered gaming on behavior, it is paramount to examine certain personality characteristics, such as neuroticism, introversion, impulsivity, and aggressiveness. Using multiple measures of online addiction, examining genre preferences and underlying motives for choosing a particular game can be helpful in reaching a better understanding of the individual game player (Floros and Siomos 2012). Comorbidities of depression, anxiety, ADHD, and OCD also need to be carefully evaluated in the case of disordered gaming, as they may be important components that intervention programs need to consider and potential targets for treatments aimed at reducing disordered gaming.

Studies report similarities between certain characteristics of video games such as loot boxes and gambling. There is a debate about the efficacy of limit-setting as a policy to protect gamers with gambling problems. Griffiths and Pontes (2020) suggested that adequate cooperation between video game operators and researchers needs to be established in order to encourage the gaming industry to set up serious social responsibility policies in order to ensure player protection and minimize harm to gamers.

Since disordered gaming was included in DSM-5 (APA 2013) and ICD-11 (WHO 2019), the rise in treatment demand for this issue has made it necessary to determine the most effective treatments for different personalities and under different conditions (Stevens et al. 2019). Given that neurobiological studies are showing similarities between disordered gaming and substance use disorder, it may be relevant to apply treatment approaches similar to those proven to be effective for substance use disorder in treating disordered gaming. Although research concerning the treatment of disordered gaming is in its early stages and it is premature to reach strong conclusions at this point in time, a combination of psychological treatment (e.g., CBT) and pharmacological therapy (e.g., bupropion) together with adaptation of treatments known to be effective for other addictive behaviors, including substance use (e.g., family involvement), may show positive results.

References

Aggarwal S, Saluja S, Gambhir V et al (2020) Predicting likelihood of psychological disorders in player unknown's battlegrounds (PUBG) players from Asian countries using supervised machine learning. Addict Behav 101:106132. https://doi.org/10.1016/j.addbeh.2019.106132

American Psychiatric Association (2000) Diagnostic and statistical manual of mental disorders, 4th edn. Washington, DC

American Psychiatric Association (2013) Diagnostic and statistical manual of mental disorders, 5th edn. American Psychiatric Association, Arlington, VA

Antons S, Brand M, Potenza MN (2020) Neurobiology of cue-reactivity, craving, and inhibitory control in non-substance addictive behaviors. J Neurol Sci 415:116952. https://doi.org/10.1016/j.jns.2020.116952

Bae S, Hong JS, Kim SM et al (2018) Bupropion shows different effects on brain functional connectivity in patients with Internet-based gambling disorder and Internet gaming disorder. Front Psychiatry 9:130. https://doi.org/10.3389/fpsyt.2018.00130

Baggio S, Dupuis M, Studer J et al (2016) Reframing video gaming and internet use addiction: empirical cross-national comparison of heavy use over time and addiction scales among young users. Addiction 111:513–522. https://doi.org/10.1111/add.13192

Balhara YPS, Singh S, Saini R et al (2020) Should Internet gaming disorder be considered a subtype of generalized problematic internet use? Findings from a study among medical college students. Perspect Psychiatr Care. https://doi.org/10.1111/ppc.12558

Bányai F, Griffiths MD, Király O et al (2019) The psychology of esports: a systematic literature review. J Gambl Stud 35:351–365. https://doi.org/10.1007/s10899-018-9763-1

Bartle R (1996) Hearts, clubs, diamonds, spades: players who suit MUDs. J MUD Res 1:1–19. http://www.mud.co.uk/richard/hcds.htm. Accessed 25 Aug 2020

Bartle R (2003) Designing virtual worlds. New Riders, Indianapolis

Bertran E, Chamarro A (2016) Video gamers of league of legends: the role of passion in abusive use and in performance. Adicciones 28:28–34. https://doi.org/10.20882/adicciones.787

Billieux J, Deleuze J, Griffiths MD et al (2015) Internet gaming addiction: the case of massively multiplayer online role-playing games. In: el-Guebaly N, Carrà G, Galanter M (eds) Textbook of addiction treatment: international perspectives. Springer, Milan, pp 1515–1525

Billieux J, Van der Linden M, Achab S et al (2013) Why do you play world of Warcraft? An in-depth exploration of self-reported motivations to play online and in-game behaviours in the virtual world of Azeroth. Comput Hum Behav 29:103–109. https://doi.org/10.1016/j.chb.2012.07.021

Bioulac S, Arf L, Bouvard MP (2008) Attention defcit/hyperactivity disorder and video games: a comparative study of hyperactive and control children. Eur Psychiat 23:134–141. https://doi.org/10.1016/j.eurpsy.2007.11.002

Blasi MD, Giardina A, Giordano C et al (2019) Problematic video game use as an emotional coping strategy: evidence from a sample of MMORPG gamers. J Behav Addict 8:25–34. https://doi.org/10.1556/2006.8.2019.02

Blocker KA, Wright TJ, Boot WR (2014) Gaming preferences of aging generations. Gerontechnology 12:174–184. https://doi.org/10.4017/gt.2014.12.3.008.00

Bonnaire C, Baptista D (2019) Internet gaming disorder in male and female young adults: the role of alexithymia, depression, anxiety and gaming type. Psychiatry Res 272:521–530. https://doi.org/10.1016/j.psychres.2018.12.158

Brooks GA, Clark L (2019) Associations between loot box use, problematic gaming and gambling, and gambling-related cognitions. Addict Behav 96:26–34. https://doi.org/10.1016/j.addbeh.2019.04.009

Burleigh TL, Griffiths MD, Sumich A et al (2019) A systematic review of the co-occurrence of gaming disorder and other potentially addictive behaviors. Curr Add Rep 6:383–401. https://doi.org/10.1007/s40429-019-00279-7

Carras MC, Van Rooij AJ, Van de Mheen D et al (2017) Video gaming in a hyperconnected world: a cross-sectional study of heavy gaming, problematic gaming symptoms, and online socializing in adolescents. Comput Human Behav 68:472–479. https://doi.org/10.1016/j.chb.2016.11.060

Chang CH, Chang YC, Cheng H et al (2020) Treatment efficacy of internet gaming disorder with attention deficit hyperactivity disorder and emotional dysregulaton. Int J Neuropsychopharmacol 23:349–355. https://doi.org/10.1093/ijnp/pyaa010

Chang HY, Wong LL, Yap KZ et al (2016) Gaming preferences, motivations, and experiences of pharmacy students in Asia. Games Health J 5:40–49. https://doi.org/10.1089/g4h.2015.0028

Chia DXY, Ng CWL, Kandasami G et al (2020) Prevalence of internet addiction and gaming disorders in Southeast Asia: a meta-analysis. Int J Environ Res Public Health 17:2582. https://doi.org/10.3390/ijerph17072582

Chuang YC (2006) Massively multiplayer online role-playing game-induced seizures: a neglected health problem in Internet addiction. Cyberpsychol Behav 9:451–456

Connolly TM, Boyle EA, MacArthur E et al (2012) A systematic literature review of empirical evidence on computer games and serious games. Comput Educ 59:661–686. https://doi.org/10.1016/j.compedu.2012.03.004

Darvesh N, Radhakrishnan A, Lachance CC et al (2020) Exploring the prevalence of gaming disorder and Internet gaming disorder: a rapid scoping review. Syst Rev 9:68. https://doi.org/10.1186/s13643-020-01329-2

Demetrovics Z, Urbán R, Nagygyörgy K et al (2011) Why do you play? The development of the motives for online gaming questionnaire (MOGQ). Behav Res Methods 43:814–825. https://doi.org/10.3758/s13428-011-0091-y

Donati MA, Chiesi F, Ammannato G et al (2015) Versatility and addiction in gaming: the number of video-game genres played is associated with pathological gaming in male adolescents. Cyberpsychol Behav Soc Netw 18:129–132. https://doi.org/10.1089/cyber.2014.0342

Dong G, Wang L, Du X et al (2018) Gender-related differences in neural responses to gaming cues before and after gaming: implications for gender-specific vulnerabilities to Internet gaming disorder. Soc Cogn Affect Neurosci 13:1203–1214. https://doi.org/10.1093/scan/nsy084

Drummond A, Sauer JD, Ferguson CJ et al (2020) The relationship between problem gambling, excessive gaming, psychological distress and spending on loot boxes in Aotearoa New Zealand, Australia, and the United States-A cross-national survey. PLoS ONE 15(3):e0230378. https://doi.org/10.1371/journal.pone.0230378

Drummond A, Sauer JD, Hall LC (2019) Loot box limit-setting: a potential policy to protect video game users with gambling problems? Addiction 114:935–936. https://doi.org/10.1111/add.14583

Elliot L, Golub A, Ream G et al (2012) Video game genre as a predictor of problem use. Cyberpsychol Behav Soc Netw 15:155–161. https://doi.org/10.1089/cyber.2011.0387

Elliott L, Ream G, McGinsky E et al (2012) The contribution of game genre and other use patterns to problem video game play among adult video gamers. Int J Ment Health Addict 10:948–969. https://doi.org/10.1007/s11469-012-9391-4

Entertainment Software Association (2019) Essential facts about the computer and video game industry 2019. https://www.theesa.com/esa-research/2019-essential-facts-about-the-computer-and-video-game-industry/. Accessed 25 Aug 2020

Evren B, Evren C, Dalbudak E et al (2019a) Neuroticism and introversion mediates the relationship between probable ADHD and symptoms of Internet gaming disorder: results of an online survey. Psychiatry Clin Psychopharmacol 29:90–96. https://doi.org/10.1080/24750573.2018.1490095

Evren C, Dalbudak E, Topcu M et al (2019b) Relationship of alcohol and/or cannabis use in the last year and the severity of the Internet gaming disorder symptoms. Eur Neuropsychopharmacol 29(Supplement 1):170

Evren C, Dalbudak E, Topcu M et al (2017) The psychometric properties of the Turkish version of the Internet Gaming Disorder Scale. Dusunen Adam J Psychiatry Neurolog Sci 30:316–324. https://doi.org/10.5350/DAJPN2017300405

Evren C, Dalbudak E, Topcu M et al (2018a) Psychometric validation of the Turkish nine-item internet gaming disorder scale-short form (IGDS9-SF). Psychiatry Res 265:349–354. https://doi.org/10.1016/j.psychres.2018.05.002

Evren C, Evren B, Dalbudak E et al (2018b) Severity of Internet gaming disorder symptoms might be related with the body mass index. Psychiatry Clin Psychopharmacol 28:446–447. https://doi.org/10.1080/24750573.2018.1490097

Evren C, Evren B, Dalbudak E et al (2019c) Relationship of internet gaming disorder severity with symptoms of anxiety, depression, alexithymia, and aggression among university students. Dusunen Adam J Psychiatry Neurolog Sci 32:227–235. https://doi.org/10.14744/DAJPNS.2019.00032

Evren C, Evren B, Dalbudak E et al (2019d) Relationships of Internet addiction and Internet gaming disorder symptom severities with probable attention deficit/hyperactivity disorder, aggression and negative affect among university students. Atten Defic Hyperact Disord 11:413–421. https://doi.org/10.1007/s12402-019-00305-8

Evren C, Evren B, Dalbudak E et al (2020a) Psychometric validation of the Turkish motives for online gaming questionnaire (MOGQ) across university students and video game players. Addicta Turkish J Addict 7:81–89. https://doi.org/10.5152/ADDICTA.2020a.19093

Evren C, Evren B, Dalbudak E et al (2020b) Relationship between internet gaming disorder symptoms with attention deficit hyperactivity disorder and alexithymia symptoms among university students. Addicta Turkish J Addict 7:1–9. https://doi.org/10.5152/addicta.2020b.19076

Evren C, Evren B, Dalbudak E et al (2020c) Psychometric validation of the Turkish ten-item internet gaming disorder test (IGDT-10). Dusunen Adam J Psychiatry Neurolog Sci 33:19–28. https://doi.org/10.14744/DAJPNS.2019.00057

Evren C, Pontes HM, Dalbudak E et al (2020d) Psychometric validation of the Turkish gaming disorder test: a measure that evaluates disordered gaming according to the world health organization framework. Psychiatry Clin Psychopharmacol 30(2):144–151. https://doi.org/10.5455/PCP.20200429072430

Fam JY (2018) Prevalence of internet gaming disorder in adolescents: a meta-analysis across three decades. Scand J Psychol 59:524–531. https://doi.org/10.1111/sjop.12459

Fauth-Bühler M, Mann K (2017) Neurobiological correlates of internet gaming disorder: similarities to pathological gambling. Addict Behav 64:349–356. https://doi.org/10.1016/j.addbeh.2015.11.004

Feng W, Ramo DE, Chan SR et al (2017) Internet gaming disorder: trends in prevalence 1998–2016. Addict Behav 75:17–24. https://doi.org/10.1016/j.addbeh.2017.06.010

Ferguson CJ (ed) (2018) Video game influences on aggression, cognition, and attention. Springer, Cham

Floros G, Siomos K (2012) Patterns of choices on video game genres and Internet addiction. Cyberpsychol Behav Soc Netw 15:417–424. https://doi.org/10.1089/cyber.2012.0064

Frostling-Henningsson M (2009) First-person shooter games as a way of connecting to people: "brothers in blood." CyberPsychol Behav 12:557–562. https://doi.org/10.1089/cpb.2008.0345

Gentile DA, Choo H, Liau A et al (2011) Pathological video game use among youth: a two-year longitudinal study. Pediatrics 127:319–329. https://doi.org/10.1542/peds.2010-1353

Gioia F, Boursier V (2019) Treatment of internet addiction and internet gaming disorder in adolescence: a systematic review. In: Bozoglan B (ed) Advances in psychology, mental health, and behavioral studies (APMHBS) book series. Multifaceted approach to digital addiction and its treatment. Information Science Reference/IGI Global, pp 157–176

González-Bueso V, Santamaría JJ, Fernández D et al (2018) Association between Internet gaming disorder or pathological video-game use and comorbid psychopathology: a comprehensive review. Int J Environ Res Public Health 15:688. https://doi.org/10.3390/ijerph15040668

Granic I, Lobel A, Engels RCME (2013) The benefits of playing video games. Am Psychol 69:66–78. https://doi.org/10.1037/a0034857

Griffiths M (2017) The psychosocial impact of professional gambling, professional video gaming and eSports. Casino Gaming Int 28:59–63. https://core.ac.uk/reader/74252560 Accessed 25 Aug 2020

Griffiths MD (2019) The therapeutic and health benefits of playing video games. In: Attrill-Smith A, Fullwood C, Keep M, Kuss DJ (eds) The Oxford handbook of cyberpsychology. Oxford University Press, Oxford, pp 485–505

Griffiths MD, Pontes HM (2014) Internet addiction disorder and internet gaming disorder are not the same. J Addict Res Ther 5:e124. https://doi.org/10.4172/2155-6105.1000e124

Griffiths MD, Pontes HM (2020) The future of gaming disorder research and player protection: what role should the video gaming industry and researchers play? Int J Ment Health Addict 18:784–790. https://doi.org/10.1007/s11469-019-00110-4

Ha JH, Yoo HJ, Cho IH et al (2006) Psychiatric comorbidity assessed in Korean children and adolescents who screen positive for Internet addiction. J Clin Psychiat 67:821–826. https://doi.org/10.4088/jcp.v67n0517

Hallmann K, Giel T (2018) eSports—Competitive sports or recreational activity? Sport Manage Rev 21:14–20. https://doi.org/10.1016/j.smr.2017.07.011

Hamari J, Sjöblom M (2017) What is eSports and why do people watch it? Internet Res 27(2). https://doi.org/10.1108/IntR-04-2016-0085

Han DH, Hwang JW, Renshaw PF (2010) Bupropion sustained release treatment decreases craving for video games and cue-induced brain activity in patients with internet video game addiction. Exp Clin Psychopharmacol 18:297–304. https://doi.org/10.1037/a0020023

Han DH, Kim SM, Bae S et al (2017) Brain connectivity and psychiatric comorbidity in adolescents with Internet gaming disorder. Addict Biol 22:802–812. https://doi.org/10.1111/adb.12347

Han DH, Kim SM, Lee YS et al (2012) The effect of family therapy on the changes in the severity of on-line game play and brain activity in adolescents with on-line game addiction. Psychiatry Res Neuroimaging 202:126–131. https://doi.org/10.1016/j.pscychresns.2012.02.011

Han DH, Lee YS, Na C et al (2009) The effect of methylphenidate on internet video game play in children with attention-deficit/hyperactivity disorder. Compr Psychiatry 50:251–256. https://doi.org/10.1016/j.comppsych.2008.08.011

Han DH, Renshaw PF (2012) Bupropion in the treatment of problematic online game play in patients with major depressive disorder. J Psychopharmacol 26:689–696. https://doi.org/10.1177/0269881111400647

Han DH, Yoo M, Renshaw PF et al (2018) A cohort study of patients seeking Internet gaming disorder treatment. J Behav Addict 7:930–938. https://doi.org/10.1556/2006.7.2018.102

Han H, Jeong H, Jo SJ et al (2020a) Relationship between online game genre used and high risk of Internet gaming disorder in adolescents. Epidemiol Health 7:e2020016. https://doi.org/10.4178/epih.e2020016

Han J, Seo Y, Hwang H et al (2020b) Efficacy of cognitive behavioural therapy for internet gaming disorder. Clin Psychol Psychother 27:203–213. https://doi.org/10.1002/cpp.2419

Henchoz Y, Studer J, Deline S et al (2016) Video gaming disorder and sport and exercise in emerging adulthood: a longitudinal study. Behav Med 42:105–111. https://doi.org/10.1080/08964289.2014.965127

Huang V, Yaoung M, Fiocco AJ (2017) The association between video game play and cognitive function: does gaming platform matter? Cyberpsychol Behav Soc Netw 20:689–694. https://doi.org/10.1089/cyber.2017.0241

Hussain Z, Griffiths MD, Baguley T (2012) Online gaming addiction: classification, prediction and associated risk factors. Addict Res Theory 20:359–371. https://doi.org/10.3109/16066359.2011.640442

Interactive Games & Entertainment Association (2017) Digital Australia 2018. Eveleigh, IGEA, NSW. Retrieved from https://igea.net/2017/07/digital-australia-2018-da18/

Ioannidis K, Hook R, Goudriaan AE et al (2019) Cognitive deficits in problematic internet use: meta-analysis of 40 studies. Br J Psychiatry 215:1–8. https://doi.org/10.1192/bjp.2019.3

Jaiswal S, Asper L, Long J et al (2019) Ocular and visual discomfort associated with smartphones, tablets and computers: what we do and do not know. Clin Exp Optom 102:463–477. https://doi.org/10.1111/cxo.12851

Jung D, Shim EJ, Park H et al (2020) The association between excessive internet gaming behavior and immersive tendency, mediated by attention deficit/hyperactivity disorder symptoms, in Korean male university students. Psychiatry Investig 17:403–411. https://doi.org/10.30773/pi.2019.0173

Karhulahti VM, Koskimaa R (2019) On the prevalence of addicted or problematic gaming in Finland. Addict Behav Rep 10:100225. https://doi.org/10.1016/j.abrep.2019.100225

Kim PW, Kim SY, Shim M et al (2013) The influence of an educational course on language expression and treatment of gaming addiction for massive multiplayer online role-playing game (MMORPG) players. Comput Educ 63:208–217. https://doi.org/10.1016/j.compedu.2012.12.008

Kim SM, Han DH, Lee YS et al (2012) Combined cognitive behavioral therapy and bupropion for the treatment of problematic on-line game play in adolescents with major depressive disorder. Comput Human Behav 28:1954–1959. https://doi.org/10.1016/j.chb.2012.05.015

King DL (2018) Online gaming and gambling in children and adolescents-Normalising gambling in cyber places: a review of the literature. Victorian Responsible Gambling Foundation, Melbourne. file:///C:/Users/pc/Downloads/Online-gaming-and-gambling-in-children-and-adolescents_pdf. Accessed 25 Aug 2020

King DL, Adair C, Saunders JB et al (2018) Clinical predictors of gaming abstinence in help-seeking adult problematic gamers. Psychiatry Res 261:581–588. https://doi.org/10.1016/j.psychres.2018.01.008

King DL, Chamberlain SR, Carragher N et al (2020) Screening and assessment tools for gaming disorder: a comprehensive systematic review. Clin Psychol Rev 77:101831. https://doi.org/10.1016/j.cpr.2020.101831

King DL, Delfabbro PH (2018) The concept of "harm" in Internet gaming disorder. J Behav Addict 7:562–564. https://doi.org/10.1556/2006.7.2018.24

King D, Delfabbro P (2019a) Treatment for IGD. Internet gaming disorder, 1st edn. In: King D, Delfabbro P (eds) Theory, Assessment, Treatment, and Prevention Academic Press

King DL, Delfabbro PH (2019b) Loot box limit-setting is not sufficient on its own to prevent players from overspending: a reply to Drummond, Sauer & Hall. Addiction 114:1324–1325. https://doi.org/10.1111/add.14628

King DL, Delfabbro PH, Perales JC et al (2019) Maladaptive player-game relationships in problematic gaming and gaming disorder: a systematic review. Clin Psychol Rev 73:101777. https://doi.org/10.1016/j.cpr.2019.101777

King DL, Haagsma MC, Delfabbro PH et al (2013) Toward a consensus definition of pathological video-gaming: a systematic review of psychometric assessment tools. Clin Psych Rev 33:331–342. https://doi.org/10.1016/j.cpr.2013.01.002

King DL, Kaptsis D, Delfabbro PH et al (2017) Effectiveness of brief abstinence for modifying problematic internet gaming cognitions and behaviors. J Clin Psychol 73:1573–1585. https://doi.org/10.1002/jclp.22460

Király O, Sleczka P, Pontes HM et al (2017) Validation of the ten-item internet gaming disorder test (IGDT-10) and evaluation of the nine DSM-5 internet gaming disorder criteria. Addict Behav 64:253–260. https://doi.org/10.1016/j.addbeh.2015.11.005

Király O, Urbán R, Griffiths MD et al (2015) The mediating effect of gaming motivation between psychiatric symptoms and problematic online gaming: an online survey. J Med Internet Res 17:e88. https://doi.org/10.2196/jmir.3515

Kircaburun K, Pontes HM, Stavropoulos V et al (2020) A brief psychological overview of disordered gaming. Curr Opin Psychol 36:38–43. https://doi.org/10.1016/j.copsyc.2020.03.004

Ko CH, Yen JY, Chen CC et al (2005) Gender differences and related factors affecting online gaming addiction among Taiwanese adolescents. J Nerv Ment Dis 193:273–277. https://doi.org/10.1097/01.nmd.0000158373.85150.57

Kotyuk E, Magi A, Eisinger A et al (2020) Co-occurrences of substance use and other potentially addictive behaviors: epidemiological results from the psychological and genetic factors of the addictive behaviors (PGA) study. J Behav Addict 9:272–288. https://doi.org/10.1556/2006.2020.00033

Kristiansen S, Severin MC (2020) Loot box engagement and problem gambling among adolescent gamers: findings from a national survey. Addict Behav 103:106254. https://doi.org/10.1016/j.addbeh.2019.106254

Kühn S, Kugler DT, Schmalen K et al (2019) Does playing violent video games cause aggression? A longitudinal intervention study. Mol Psychiatry 24:1220–1234. https://doi.org/10.1038/s41380-018-0031-7

Kuss DJ, Griffiths MD (2012) Online gaming addiction in children and adolescents: a review of empirical research. J Behav Addict 1:3e22. https://doi.org/10.1556/JBA.1.2012.1.1

Kuss DJ, Lopez-Fernandez O (2016) Internet addiction and problematic Internet use: a systematic review of clinical research. World J Psychiatry 6:143–176. https://doi.org/10.5498/wjp.v6.i1.143

Kuss DJ, Louws J, Wiers RW (2012) Online gaming addiction? Motives predict addictive play behavior in massively multiplayer online role-playing games. Cyberpsychol Behav Soc Netw 15:480–485. https://doi.org/10.1089/cyber.2012.0034

Kuss DJ, Pontes HM (2019) Internet addiction, vol 41. In: Advances in psychotherapy—evidence-based practice. Hogrefe Publishing Corporation, Boston, MA

Kuss DJ, Pontes HM, Griffiths MD (2018) Neurobiological correlates in internet gaming disorder: a systematic literature review. Front Psychiatry 9:166. https://doi.org/10.3389/fpsyt.2018.00166

Lafrenière MA, Verner-Filion J, Vallerand RJ (2012) Development and validation of the gaming motivation scale (GAMS). Pers Individ Dif 53:827–831. https://doi.org/10.1016/j.paid.2012.06.013

Lam LT (2014) Internet gaming addiction, problematic use of the internet, and sleep problems: a systematic review. Curr Psychiatry Rep 16:444. https://doi.org/10.1007/s11920-014-0444-1

Larche CJ, Chini K, Lee C et al (2019) Rare loot box rewards trigger larger arousal and reward responses, and greater urge to open more loot boxes. J Gambl Stud. https://doi.org/10.1007/s10899-019-09913-5

Lee MS, Ko YH, Song HS et al (2007) Characteristics of Internet use in relation to game genre in Korean adolescents. Cyberpsychol Behav 10:278–285. https://doi.org/10.1089/cpb.2006.9958

Lee SH, Im JJ, Oh JK et al (2018a) Transcranial direct current stimulation for online gamers: a prospective single-arm feasibility study. J Behav Addict 7:1166–1170. https://doi.org/10.1556/2006.7.2018.107

Lee SY, Lee HK, Bang SY et al (2018b) Aggression and harm-avoidant trait impede recovery from internet gaming disorder. Front Psychiatry 9:263. https://doi.org/10.3389/fpsyt.2018.00263

Lemmens JS, Hendriks SJ (2016) Addictive online games: examining the relationship between game genres and internet gaming disorder. Cyberpsychol Behav Soc Netw 19:270–276. https://doi.org/10.1089/cyber.2015.0415

Lemmens JS, Valkenburg PM, Gentile DA (2015) The internet gaming disorder scale. Psychol Assess 27:567–582. https://doi.org/10.1037/pas0000062

Lemmens JS, Valkenburg PM, Peter J (2011) Psychosocial causes and consequences of pathological gaming. Comput Hum Behav 27:144–152. https://doi.org/10.1016/j.chb.2010.07.015

Li W, Mills D, Nower L (2019) The relationship of loot box purchases to problem video gaming and problem gambling. Addict Behav 97:27–34. https://doi.org/10.1016/j.addbeh.2019.05.016

López-Fernández FJ, Mezquita L, Griffiths MD et al (2020) The role of personality on disordered gaming and game genre preferences in adolescence: gender differences and person-environment transactions. Adicciones. https://doi.org/10.20882/adicciones.1370

Macey J, Hamari J (2019) The games we play: relationships between game genre, business model and loot box opening. In: GamiFIN conference 2019, Levi, Finland, Apr 8–10, pp 193–204

Mallorquí-Bagué N, Fernández-Aranda F, Lozano-Madrid M et al (2017) Internet gaming disorder and online gambling disorder: clinical and personality correlates. J Behav Addict 6:669–677. https://doi.org/10.1556/2006.6.2017.078

Maldonado-Murciano L, Guilera G, Montag C, Pontes HM (2022) Disordered gaming in esports: comparing professional and non-professional gamers. Addict Behav 132107342. https://doi.org/10.1016/j.addbeh.2022.107342

Männikkö N, Billieux J, Kääriäinen M (2015) Problematic digital gaming behavior and its relation to the psychological, social and physical health of Finnish adolescents and young adults. J Behav Addict 4:281–288. https://doi.org/10.1556/2006.4.2015.040

Männikkö N, Ruotsalainen H, Jouko Miettunen J et al (2020) Problematic gaming behaviour and health-related outcomes: a systematic review and meta-analysis. J Health Psychol 25:67–81. https://doi.org/10.1177/1359105317740414

McCaffrey M (2020) A cautious approach to public policy and loot box regulation. Addict Behav 102:106136. https://doi.org/10.1016/j.addbeh.2019.106136

McClelland DC (1985) How motives, skills, and values determine what people do. Am Psychol 40:812–825. https://doi.org/10.1037/0003-066X.40.7.812

Meng Y, Deng W, Wang H et al (2015) The prefrontal dysfunction in individuals with Internet gaming disorder: a meta-analysis of functional magnetic resonance imaging studies. Addict Biol 20:799–808. https://doi.org/10.1111/adb.12154

Mihara S, Higuchi S (2017) Cross-sectional and longitudinal epidemiological studies of Internet gaming disorder: a systematic review of the literature. Psychiatry Clin Neurosci 71:425–444. https://doi.org/10.1111/pcn.12532

Montag C, Schivinski B, Sariyska R et al (2019) Psychopathological symptoms and gaming motives in disordered gaming—a psychometric comparison between the WHO and APA diagnostic frameworks. J Clin Med 8:1691. https://doi.org/10.3390/jcm8101691

Montag C, Schivinski B, Kannen C, Pontes HM (2022) Investigating gaming disorder and individual differences in gaming motives among professional and non-professional gamers: an empirical study. Addict. Behav 107416. https://doi.org/10.1016/j.addbeh.2022.107416

Mowatt L, Gordon C, Santosh ABR et al (2018) Computer vision syndrome and ergonomic practices among undergraduate university students. Int J Clin Pract 72(1). https://doi.org/10.1111/ijcp.13035

Müller KW, Janikian M, Dreier M et al (2015) Regular gaming behavior and internet gaming disorder in European adolescents: results from a cross-national representative survey of prevalence, predictors, and psychopathological correlates. Eur Child Adol Psychiatry 24:565–574. https://doi.org/10.1007/s00787-014-0611-2

Mylona I, Deres ES, Dere GS et al (2020) The impact of internet and videogaming addiction on adolescent vision: a review of the literature. Front Public Health 8:63. https://doi.org/10.3389/fpubh.2020.00063

Myrseth H, Notelaers G, Strand LÅ et al (2017) Introduction of a new instrument to measure motivation for gaming: the electronic gaming motives questionnaire. Addiction 112:1658–1668. https://doi.org/10.1111/add.13874

Na E, Lee H, Choi I, Kim DJ (2017) Comorbidity of Internet gaming disorder and alcohol use disorder: a focus on clinical characteristics and gaming patterns. Am J Addict 26:326–334. https://doi.org/10.1111/ajad.12528

Nakayama H, Mihara S, Higuchi S (2017) Treatment and risk factors of Internet use disorders. Psychiatry Clin Neurosci 71:492–505. https://doi.org/10.1111/pcn.12493

Nam B, Bae S, Kim SM et al (2017) Comparing the effects of bupropion and escitalopram on excessive internet game play in patients with major depressive disorder. Clin Psychopharmacol Neurosci 15:361–368. https://doi.org/10.9758/cpn.2017.15.4.361

Newzoo. (2017). Global esports market report 2017. https://newzoo.com/insights/trend-reports/global-esports-market-report-2017-light/. Accessed 25 Aug 2020

Pallesen S, Lorvik IM, Bu EH et al (2015) An exploratory study investigating the effects of a treatment manual for video game addiction. Psychol Rep 117:490–495. https://doi.org/10.2466/02.PR0.117c14z9

Park JH, Lee YS, Sohn JH et al (2016) Effectiveness of atomoxetine and methylphenidate for problematic online gaming in adolescents with attention deficit hyperactivity disorder. Hum Psychopharmacol 31:427–432. https://doi.org/10.1002/hup.2559

Paulus FW, Ohmann S, von Gontard A et al (2018) Internet gaming disorder in children and adolescents: a systematic review. Dev Med Child Neurol 60:645–659. https://doi.org/10.1111/dmcn.13754

Pearcy BT, McEvoy PM, Roberts LD (2017) Internet gaming disorder explains unique variance in psychological distress and disability after controlling for comorbid depression, OCD, ADHD, and anxiety. Cyberpsychol Behav Soc Netw 20:126–132. https://doi.org/10.1089/cyber.2016.0304

Pearcy BTD, Roberts LD, McEvoy PM (2016) Psychometric testing of the personal internet gaming disorder evaluation-9: a new measure designed to assess internet gaming disorder. Cyberpsychol Behav Soc Netw 19:335–341. https://doi.org/10.1089/cyber.2015.0534

Petry NM, O'Brien CP (2013) Internet gaming disorder and the DSM-5. Addiction 108:1186–1187. https://doi.org/10.1111/add.12162

Petry NM, Rehbein F, Gentile DA et al (2014) An international consensus for assessing internet gaming disorder using the new DSM-5 approach. Addiction 109:1399–1406. https://doi.org/10.1111/add.12457

Pontes HM (2018) Making the case for video game addiction: Does it exist or not? In: Ferguson CJ (ed) Video game influences on aggression, cognition, and attention. Springer International Publishing, Cham, pp 41–57

Pontes HM, Griffiths MD (2014) Assessment of internet gaming disorder in clinical research: past and present perspectives. Clin Res Regul Aff 31:35–48. https://doi.org/10.3109/10601333.2014.962748

Pontes HM, Griffiths MD (2015) Measuring DSM-5 internet gaming disorder: development and validation of a short psychometric scale. Comput Human Behav 45:137–143. https://doi.org/10.1016/j.chb.2014.12.006

Pontes HM, Griffiths MD (2020) A new era for gaming disorder research: time to shift from consensus to consistency. Addict Behav 103:106059. https://doi.org/10.1016/j.addbeh.2019.106059

Pontes HM, Király O, Demetrovics Z et al (2014) The conceptualisation and measurement of DSM-5 internet gaming disorder: the development of the IGD- 20 Test. PLoS ONE 9:e110137. https://doi.org/10.1371/journal.pone.0110137

Pontes HM, Schivinski B, Sindermann C et al (2019) Measurement and conceptualization of gaming disorder according to the world health organization framework: the development of the gaming disorder test. Int J Ment Health Addiction. https://doi.org/10.1007/s11469-019-00088-z

Pontes HM, Schivinski B, Kannen C, Montag C (2022) The interplay between time spent gaming and disordered gaming: a large-scale world-wide study. Soc Sci Med 296114721. https://doi.org/10.1016/j.socscimed.2022.114721

Pornnoppadol C, Ratta-apha W, Chanpen S et al (2018) A comparative study of psychosocial interventions for internet gaming disorder among adolescents aged 13–17 years. Int J Ment Health Addict. https://doi.org/10.1007/s11469-018-9995-4

Przepiórka AM, Blachnio A, Miziak B et al (2014) Clinical approaches to treatment of Internet addiction. Pharmacol Rep 66:187–191. https://doi.org/10.1016/j.pharep.2013.10.001

Raneri PC, Montag C, Rozgonjuk D, Satel J, Pontes HM (2022) The role of microtransactions in internet gaming disorder and gambling disorder: a preregistered systematic review. Addictive Behaviors Reports, 15:100415. https://doi.org/10.1016/j.abrep.2022.100415

Rehbein F, Baier D (2013) Family-, media-, and school-related risk factors of video game addiction. J Media Psychol 25:118–128. https://doi.org/10.1027/1864-1105/a000093

Rehbein F, Kleimann M, Mößle T (2010) Prevalence and risk factors of video game dependency in adolescence: results of a german nationwide survey. Cyberpsychol Behav Soc Netw 13:269–277. https://doi.org/10.1089/cyber.2009.0227

Sakuma H, Mihara S, Nakayama H et al (2017) Treatment with the self-discovery camp (SDiC) improves internet gaming disorder. Addict Behav 64:357–362. https://doi.org/10.1016/j.addbeh.2016.06.013

Şalvarlı Şİ, Griffiths MD (2019) Internet gaming disorder and its associated personality traits: a systematic review using PRISMA guidelines. Int J Ment Health Add. https://doi.org/10.1007/s11469-019-00081-6

Saunders JB, Hao W, Long J et al (2017) Gaming disorder: its delineation as an important condition for diagnosis, management, and prevention. J Behav Addict 6:271–279. https://doi.org/10.1556/2006.6.2017.039

Schimmenti A, Gervasi AM, La Marca L, Costanzo A et al (2017) Personality and internet gaming disorder: a systematic review of recent literature. Curr Addict Rep 4:293–307. https://doi.org/10.1007/s40429-017-0159-6

Şengül C, Büber A (2016) Dijital oyun bağımlılığında tanı ve tedavi. (Editor Evren C), Davranışsal Bağımlılıklar. Psikiyatride Güncel 6:184–192. (Turkish)

Seong W, Hong JS, Kim S et al (2019) Personality and psychological factors of problematic internet gamers seeking hospital treatment. Front Psychiatry 10:583. https://doi.org/10.3389/fpsyt.2019.00583

Shotton M (1989) Computer addiction? A study of computer dependency. Taylor and Francis, London, UK

Sigerson L, Li AYL, Cheung MWL et al (2017) Psychometric properties of the Chinese internet gaming disorder scale. Addict Behav 74:20–26. https://doi.org/10.1016/j.addbeh.2017.05.031

Šincek D, Humer JT, Duvnjak I (2017) Correlates of problematic gaming—is there support for proneness to risky behaviour? Psychiatr Danub 29:302–312. https://doi.org/10.24869/psyd.2017.302

Smith AC, Stewart B (2010) The special features of sport: a critical revisit. Sport Manage Rev 13:1–13. https://doi.org/10.1016/j.smr.2009.07.002

Song J, Park JH, Han DH et al (2016) Comparative study of the effects of bupropion and escitalopram on internet gaming disorder. Psychiatry Clin Neurosci 70:527–535. https://doi.org/10.1111/pcn.12429

Soper WB, Miller MJ (1983) Junk time junkies: an emerging addiction among students. Sch Couns 31:40–43

Statista (2020) Leading video game genres played in the United States as of February 2019. https://www.statista.com/statistics/232332/types-of-mobile-games-played-most-often-in-the-us/. Accessed 25 Aug 2020

Stavropoulos V, Adams BLM, Beard CL et al (2019) Associations between attention deficit hyper- activity and internet gaming disorder symptoms: is there consistency across types of symptoms, gender and countries? Addict Behav Rep 9:100158. https://doi.org/10.1016/j.abrep.2018.100158

Stavropoulos V, Kuss DJ, Griffiths MD et al (2017) MMORPG gaming and hostility predict internet addiction symptoms in adolescents: an empirical multilevel longitudinal study. Addict Behav 64:294–300. https://doi.org/10.1016/j.addbeh.2015.09.001

Stetina BU, Kothgassner OD, Lehenbauer M et al (2011) Beyond the fascination of online-games: probing addictive behavior and depression in the world of online-gaming. Comput Hum Behav 27:473–479. https://doi.org/10.1016/j.chb.2010.09.015

Stevens MWR, King DL, Dorstyn D et al (2019) Cognitive-behavioral therapy for internet gaming disorder: a systematic review and meta-analysis. Clin Psychol Psychother 26:191–203. https://doi.org/10.1002/cpp.2341

Sugaya N, Shirasaka T, Takahashi K et al (2019) Bio-psychosocial factors of children and adolescents with internet gaming disorder: a systematic review. Biopsychosoc Med 13:3. https://doi.org/10.1186/s13030-019-0144-5

Thorne HT, Smith JJ, Morgan PJ et al (2014) Video game genre preference, physical activity and screen-time in adolescent boys from low-income communities. J Adolesc 37:1345–1352. https://doi.org/10.1016/j.adolescence.2014.09.012

Ünübol H, Koç AŞ, Sayar GH et al (2020) Measurement, profiles, prevalence, and psychological risk factors of problematic gaming among the Turkish community: a large-scale national study. Int J Ment Health Addict. https://doi.org/10.1007/s11469-020-00254-8

Ustinavičienė R, Škėmienė L, Lukšienė D et al (2016) Problematic computer game use as expression of Internet addiction and its association with self-rated health in the Lithuanian adolescent population. Medicina 52:199–204. https://doi.org/10.1016/j.medici.2016.04.002

van Hilvoorde I (2016) Sport and play in a digital world. Sport Ethic Philos 10:1–4. https://doi.org/10.1080/17511321.2016.1171252

van Hilvoorde I, Pot N (2016) Embodiment and fundamental motor skills in eSports. Sport Ethic Philos 10:14–27. https://doi.org/10.1080/17511321.2016.1159246

Van Rooij AJ, Schoenmakers TM, van de Mheen D (2017) Clinical validation of the C-VAT 2.0 assessment tool for gaming disorder: a sensitivity analysis of the proposed DSM-5 criteria and the clinical characteristics of young patients with 'video game addiction.' Add Beh 64:269–274. https://doi.org/10.1016/j.addbeh.2015.10.018

Van Rooij AJ, Schoenmakers TM, Vermulst AA et al (2011) Online video game addiction: identification of addicted adolescent gamers. Addiction 106:205e21. https://doi.org/10.1111/j.1360-0443.2010.03104.x

Walther B, Morgenstern M, Hanewinkel R (2012) Co-occurrence of addictive behaviours: personality factors related to substance use, gambling and computer gaming. Eur Addict Res 18:167–174. https://doi.org/10.1159/000335662

Wang HY, Cheng C (2022) The associations between gaming motivation and internet gaming disorder: systematic review and meta-analysis. JMIR Ment Health 9(2): e23700. https://doi.org/10.2196/23700

Wartberg L, Kriston L, Zieglmeier M et al (2019) A longitudinal study on psychosocial causes and consequences of Internet gaming disorder in adolescence. Psychol Med 49:287–294. https://doi.org/10.1017/S003329171800082X

Weaver JB 3rd, Mays D, Sargent Weaver S et al (2009) Health-risk correlates of video-game playing among adults. Am J Prev Med 37:299–305. https://doi.org/10.1016/j.amepre.2009.06.014

Weinstein A, Lejoyeux M (2015) New developments on the neurobiological and pharmaco-genetic mechanisms underlying internet and videogame addiction. Am J Addict 24:117–125. https://doi.org/10.1111/ajad.12110

Winkler A, Dorsing B, Rief W et al (2013) Treatment of internet addiction: a meta-analysis. Clin Psychol Rev 33:317–329. https://doi.org/10.1016/j.cpr.2012.12.005

World Health Organization (2018) International classification of diseases, 11th Rev (ICD-11). In: ICD-11 beta draft: gaming disorder. www.who.int/classifications/icd/. Accessed 25 Aug 2020

World Health Organization (2019). 6C51 gaming disorder 2019. Available from https://www.who.int/features/qa/gaming-disorder/en/. Accessed 25 Aug 2020

Wu AM, Lai MH, Yu S et al (2017) Motives for online gaming questionnaire: its psychometric properties and correlation with internet gaming disorder symptoms among Chinese people. J Behav Addict 6:11–20. https://doi.org/10.1556/2006.6.2017.007

Yao YW, Liu L, Ma SS et al (2017) Functional and structural neural alterations in internet gaming disorder: a systematic review and meta-analysis. Neurosci Biobehav Rev 83:313–324. https://doi.org/10.1016/j.neubiorev.2017.10.029

Yee N (2006a) The demographics, motivations and derived experiences of users of massively-multiuser online graphical environments. PRESENCE Virt Augment Real 15:309–329. https://doi.org/10.1162/pres.15.3.309

Yee N (2006b) Motivations for play in online games. CyberPsychol Behav 9:772–775. https://doi.org/10.1089/cpb.2006.9.772

Yen JY, Liu TL, Wang PW et al (2017) Association between Internet gaming disorder and adult attention defcit and hyperactivity disorder and their correlates: impulsivity and hostility. Addict Behav 64:308–313. https://doi.org/10.1016/j.addbeh.2016.04.024

Yu F, Sariyska R, Lachmann B et al (2020) Convergent cross-sectional and longitudinal evidence for gaming-cue specific posterior parietal dysregulations in early stages of internet gaming disorder. Addict Biol e12933. https://doi.org/10.1111/adb.12933

Zajac K, Ginley MK, Chang R (2020) Treatments of internet gaming disorder: a systematic review of the evidence. Expert Rev Neurother 20:85–93. https://doi.org/10.1080/14737175.2020.1671824

Zajac K, Ginley MK, Chang R et al (2017) Treatments for Internet gaming disorder and Internet addiction: a systematic review. Psychol Addict Behav 31:979–994. https://doi.org/10.1037/adb0000315

Zanetta Dauriat F, Zermatten A, Billieux J et al (2011) Motivations to play specifically predict excessive involvement in massively multiplayer online role-playing games: evidence from an online survey. Eur Addict Res 17:185–189. https://doi.org/10.1159/000326070

Zendle D (2019) Problem gamblers spend less money when loot boxes are removed from a game: a before and after study of *Heroes of the Storm*. PeerJ 7:e7700. https://doi.org/10.7717/peerj.7700

Zendle D, Cairns P (2019) Loot boxes are again linked to problem gambling: Results of a replication study. PLoS ONE 14:e0213194. https://doi.org/10.1371/journal.pone.0213194

Zendle D, Meyer R, Over H (2019) Adolescents and loot boxes: links with problem gambling and motivations for purchase. R Soc Open Sci 6:190049. https://doi.org/10.1098/rsos.190049

Zheng H, Hu Y, Wang Z et al (2019) Meta-analyses of the functional neural alterations in subjects with Internet gaming disorder: similarities and differences across different paradigms. Prog Neuropsychopharmacol Biol Psychiatry 94:109656. https://doi.org/10.1016/j.pnpbp.2019.109656

Zhou F, Montag C, Sariyska R et al (2019a) Orbitofrontal gray matter deficits as marker of Internet gaming disorder: converging evidence from a cross-sectional and prospective longitudinal design. Addict Biol 24:100–109. https://doi.org/10.1111/adb.12570

Zhou X, Wu R, Liu C et al. (2019b) Dimensional neuroimaging: (internet) gaming disorder symptoms according to the WHO and APA frameworks associate with lower striatal volume. bioRxiv:852855. https://doi.org/10.1101/852855

Chapter 3
Social Media Addiction

Tayana Panova and Xavier Carbonell

Abstract The immense popularity of social networks such as Facebook has led to concerns about their potentially addictive nature and the ways in which they may be negatively affecting users, especially adolescents. However, despite the fact that "Facebook addiction" and "social media addiction" have become common terms in the media and social dialogue, the empirical evidence at this time does not support the existence of such a psychological affliction for several reasons: (1) The majority of studies on social media addiction are correlational and use self-report questionnaires which are not suitable for diagnosis; (2) Most studies employ non-standardized measures, cut-off scores, and criteria, and (3) There is an absence of case studies, experimental studies, longitudinal studies, and clinical studies in the field. Social interaction is a fundamental human need which social networks facilitate. Therefore, their widespread appeal is understandable. However, although an *addiction* to social media might not exist, there are still various problems that have been associated with social media use, including lower self-esteem, Fear of Missing Out (FOMO), bullying, anxiety, and depression, among others. In this chapter, we review the research on social media addiction, analyze how it fulfills the psychological criteria that define a true addiction, discuss the various problems associated with social media use outside of the addiction framework, and explore how these problems develop as well as look at potential treatments and prevention strategies for them.

Keywords Social media sites · Facebook · Addiction · FOMO

3.1 What is Social Media?

Social media sites are defined as "websites and computer programs that allow people to communicate and share information on the internet using a computer or mobile phone" (Cambridge Dictionary 2020). The most well-known and popular social media websites in recent years have been Facebook, Instagram, Twitter, and

T. Panova · X. Carbonell (✉)
FPCEE Blanquerna, Universitat Ramon Llull, Císter 34, 08022 Barcelona, Spain
e-mail: xaviercs@blanquerna.url.edu

H. M. Pontes (ed.), *Behavioral Addictions*, Studies in Neuroscience, Psychology and Behavioral Economics, https://doi.org/10.1007/978-3-031-04772-5_3

LinkedIn. Others were popular in the past but are no longer as relevant today (e.g., Myspace) and new ones will most likely emerge in the future and become more popular than the current favorites. Regardless of the specific platforms and the differences between them, the two defining features of social media are the profiles they allow users to create and the networks of contacts they allow users to develop.

A social media user profile almost always includes a picture which is meant to represent the user, some semi-permanent information about them such as their name, a brief biography, interests, and relationship status. Importantly, social media websites allow the user to publish dynamic content which is seen in the network's "feeds" for a day or two, such as text, pictures or videos. A user can be active on a social media website without ever uploading content, putting a profile picture or sharing information about themselves as these actions are not mandatory. However, even with minimal input from the user, a social media website can still be used to create a social network and monitor the content produced by its members. Therefore, developing, monitoring, and interacting with a network of contacts can be considered the fundamental purpose of social networking sites (SNS).

User "feeds" are at the heart of social media websites. This is the part of the site which displays the dynamic content from members of the user's network such as pictures or status updates. The user can interact with this content by using 'likes' (certain social networking sites also offer the possibility to leave dislikes or to display a range of emotions such as anger or surprise), and/or commenting on it. This gives each user feedback about how their content is perceived by the people in their network. Users can also sometimes see content uploaded by people not in their network which introduces them to potential new contacts. On Facebook and LinkedIn this occurs when the user sees posts that their friends have liked or shared. On Instagram this is done via the "Explore page" and on Twitter this is achieved via "retweets" and "hashtags" that people can follow.

These two primary facets of social networking sites—the external facet (i.e., the content from others that the user sees and interacts with) and the internal facet (i.e., the user's profile and the content that the user posts) are at the root of the various benefits and drawbacks of social media use. Arguably, SNSs are so popular and feared to be addictive because of how reinforcing they are. Humans are fundamentally social creatures who seek to have regular social experiences and SNSs offer them just that through instant, hassle-free, and regular interactions with a network of social contacts. Additionally, people crave a sense of control over their lives and interactions which SNSs provide for them by allowing them to craft their profile and consequently to exert control over how their network perceives them. Regarding the negative consequences of these two facets, the ability to see what one's peers are doing at all times can lead to detrimental social comparisons, can encourage a pursuit of positive feedback and validation, and can lead to lower life-satisfaction if one does not feel that they "measure up" to their peers. The ability to control one's online identity can lead to discrepancies between the real self and the idealized self with consequent lower life satisfaction.

These social media-related problems, among others, along with the high levels of social media use that many people engage in despite some negative consequences, have triggered concerns about social media addiction. Whether social media addiction exists as a disorder or not is still under discussion among researchers. Some argue that the behaviors and mental states of SNS users in regards to social media are analogous to those of substance addicts and SNSs should therefore be considered addictive. Others argue that the low severity of SNS use consequences do not pass the threshold necessary to merit the diagnosis of addiction. Regardless of what consensus the field eventually comes to on this matter, the fact is that certain kinds of social media use are associated with various negative consequences at varying degrees of severity. The focus of concern should therefore be on why and how these negative consequences arise from social media use and what can be done to minimize them.

3.2 Why Has Social Media Been Perceived as Addictive?

The advent of Internet addiction in the late 1990s (Young 1998) opened the door to the possibility of specific online addictions such as online gaming, social media, and smartphone addiction. This occurred because after the Internet was determined to be potentially addictive, the next question naturally turned to what aspect of it was the addictive one (Widyanto and Griffiths 2006). After all, the Internet is simply a medium which holds countless activities (e.g., such as navigation, banking, emailing, and information seeking) and the engagement with them is overall unproblematic. Although generalized Internet addiction is believed to exist (the idea that one can be addicted to being online in a general sense) (Brand et al. 2014; Kraut et al. 1998; Young 1998), the recent literature on the subject has tended to lean towards the existence of specific Internet addictions (i.e. addictions to specific online activities such as video gaming or social media use) (Montag et al. 2014; Lowe-Calverley and Pontes 2020). These activities are suspected to be addictive because of how reinforcing they are, leading to high levels of use.

In order to understand why people do particular activities excessively, it is necessary to identify why they are reinforcing in the first place. Regarding social media use, the simple answer is that humans are social creatures. For millennia, we have lived in groups and human interaction has been vital to our survival, wellbeing, and prosperity. When we interact with people that we like we generally feel happier, safer, and healthier. When we do not interact with other people for prolonged periods of time, we tend to feel lonely, which is a negative state that has been correlated with health risks such as heart disease and stroke (Valtorta et al. 2006), depression (Erzen and Çikrikci 2018), suicidal ideation (Stravynski and Boyer 2001), among others. It is therefore natural for us to seek out regular social interactions with other people. Social media offers us such interactions in a quick, free, easy, and fun way, which is why people around the world spend so much time on them. In short, SNSs fulfill a basic human need.

The concerns about technological addictions such as smartphone and social media addiction arose from the observation that people use these technologies for a substantial amount of time on a regular basis and continue to do so even when negative consequences related to their use emerge. Andreassen and Pallessen explain that online addictions occur when people are "overly concerned about online activities, driven by an uncontrollable motivation to perform the behavior, and devoting so much time and effort to it that it impairs other important life areas" (Andreassen and Pallesen 2014, p. 4054). Since a day only has twenty-four hours, time devoted to any one activity naturally leads to less time devoted to another, thereby impairing it to some degree. Therefore, if a person spends a significant amount of time on social media, certain activities must be sacrificed to allow for this. Some of the most often cited being quality sleep (Vernon et al. 2015; Wolniczak et al. 2013; Xanidis and Brignell 2016), studying (Al-Menayes 2014; Lau 2017; Maqableh et al. 2015), and job performance (Brooks and Califf 2017; Moqbel and Kock 2017; Zivnuska et al. 2019). However, the severity of these negative consequences is not especially high, nor are they always understood by the user as being a direct consequence of their social media use, thereby not triggering the levels of concern within the user which is necessary to motivate a decrease in use (Turel et al. 2014). If a person does not believe that sufficiently serious consequences are directly related to their behavior, they cannot be expected to want or to try to minimize that behavior. Thus, the criterion of "continues to engage in the behavior despite negative consequences" cannot reliably diagnose a social media "addiction" because it implies an understanding that one's behavior is responsible for serious negative consequences in one's life.

Additionally, there is a tendency to pathologize the act of carrying out pleasurable behaviors over healthy or productive behaviors, even though to do so is human nature. In a typical cost–benefit analysis of whether one should log on to Facebook or Instagram or not, the benefits include seeing interesting images, satisfying one's curiosity about what other users are doing, and engaging in pleasant social interactions. The costs include wasting a bit of time that could otherwise be devoted to sleeping, studying, working or other productive but not necessarily enjoyable pursuits. Given the choice between doing work or having fun, choosing not to do work is not indicative of a disorder, it is a preference for a short-term pleasure over a less pleasurable investment in long-term benefit that people make on a daily basis and have done since the beginning of time (Carbonell and Panova 2017). That which is beneficial to us is often not what we choose to do because we are prone to choose pleasure over sensibility. Before social media existed, this tendency manifested in whatever other distractors were available at the time. The issue is more notable now because technological distractors like social media are more readily available and easily accessible than any distractors before them, thereby making it easier to ignore healthier or productive alternatives in their presence.

If we want to use the argument of an individual choosing to do a behavior despite negative consequences in order to illustrate problematic use, then we must examine situations where the potential negative consequence is sufficiently severe. A good example of this in the context of social media use is when people choose to risk their lives by using social media while driving. There are laws in place to forbid such dangerous behavior and accidents often occur due to distracted phone use, yet many people continue to use SNSs while driving. However, this does not necessarily indicate social media addiction. This behavior can be attributed to inaccurate judgment of risk probability and an inability to activate the required levels of impulse control, often due to symptoms of attention deficit (Turel and Bechara 2016). People tend to use their phones while driving when they believe they are in sufficient control of the car (perhaps because they are on a familiar route, there is a traffic jam, it is a straight drive or they are at a traffic light) and feel that they are able to spare the attentional resources. What they often fail to do is calculate for external factors that can suddenly appear and require their immediate attention that is otherwise engaged by the phone, which is what often causes accidents. However, these errors in judgment and lapses in control can and do exist outside the addiction framework.

The pathologization of new technologies goes back hundreds of years with notable examples including the radio, television, and even the printing press. When new technologies become widely popular, they are met with suspicion because of how they reshape existing social and cultural structures in unfamiliar ways (Surrat 1999). While it is possible that certain technologies can trigger serious negative consequences or a problematic attachment that constitutes an addiction, as video gaming has been closer to demonstrating, this is not the case for all technologies that are popular and frequently used. In order to determine whether a certain behavior is a significant problem or an addiction, we must leave behind premature judgments about how good or bad the behavior is from our subjective point-of-view and explore its effects on the users' behaviors and wellbeing in context.

3.3 Prevalence Rates of Social Media Addiction

Correlational studies on social media addiction (Andreassen et al. 2012, 2013; Floros and Siomos 2013; Hormes et al. 2014; Koc and Gulyagci 2013; Masur et al. 2014; Moreau et al. 2015; Wang et al. 2015; Wegmann et al. 2015; Wolniczak et al. 2013) use a variety of questionnaires adapted from Internet addiction tests, DSM-IV alcohol dependence, pathological gambling or other addiction criteria to measure the prevalence of social media addiction. The studies in the field usually start off by accepting the existence of social media addiction a priori (Billieux et al. 2015), based on observation and perceived similarities, then continue by measuring its prevalence in a particular population, usually university students, via self-report measures. The problem is that self-report is based on participants' perceptions of their own behavior, perceptions which are significantly influenced by social dialogue, social pressure, popular culture, and the media. Considering that there is a tendency for society and the media to pathologize the use of new technologies, the term "addiction" is often used liberally, suggesting a bigger problem than research supports. Surrounded by such concerns, many people can develop a belief that they are addicted to SNSs, primarily because of the amount of time they spend on them.

However, time spent conducting a behavior is not a sufficient criterion of addiction (Charlton and Danforth 2007; Wood et al. 2007). People spend a lot of time doing activities for various reasons, and sometimes the activity is required for success in their job (in the case of emailing), other times it is an indicator of positive life changes such as making more friends (in the case of messaging), and sometimes it is necessary to do the behavior often in order to improve certain skills (in the case of esports or taking online courses). The important aspect is not how much time is spent doing the activity, but rather the manner, context, and consequences of its use. If the context gives adequate justification for the behavior and if there are no significantly negative consequences related to its use, then even if the behavior is engaged frequently and on a regular basis, it would not be considered addictive.

Prevalence of social media addiction is tricky to report on because the majority of studies use non-representative samples of participants and employ a range of measures to identify "addicts" rather than using standardized screening methods, thus making it difficult to compare findings. There are also differences in prevalence rates depending on ease of internet access, smartphone penetration, culture of the sample, and socioeconomic context. Despite these issues, prevalence rates of social media addiction has been found to be as low as 1.6% (Alabi 2013) in some studies and as high as 36.9% in others (Ramesh Masthi et al. 2018). Assessment tools used include The Bergen Facebook Addiction Scale (BFAS) (Andreassen et al. 2012), Social Networking Website Addiction Scale (SNWAS) (Turel and Senko 2012), Facebook Dependence Questionnaire (FDQ) (Wolniczak et al. 2013), the Addictive Tendencies Scale (ATS) (Wilson et al. 2010), Facebook Intrusion Questionnaire (Elphinston and Noller 2011), and the Facebook Use Disorder scale (FUD-S) (Sindermann et al. 2020). After the levels of "addiction" are gauged, these studies usually correlate said addiction with comorbid disorders such as depression and anxiety, or personality traits such as self-esteem, and narcissism (see Kuss and Griffiths 2011; Ryan et al. 2014 for reviews). Critiques of the social media addiction literature refer to the lack of longitudinal studies and clinical samples, the large probability of false positives, and the lack of consistency in methodology, terminology, cut-off scores and diagnostic criteria (Andreassen 2015; Carbonell and Panova 2017; Haagsma et al. 2013). These limitations illustrate a lack of construct validity for social media addiction (Ryan et al. 2014) that do not support its nature as a true disorder at this time.

3.4 Arguments Against the Existence of Social Media Addiction

The concern about the existence of social media addiction persists because intensive social media use often presents in an analogous way to substance use disorder. Griffiths developed one of the most widely accepted definitions of addiction based on the following components: (1) Mood modification (i.e., the substance/behavior leads to positive emotional states), (2) Tolerance (i.e., more and more of the substance/behavior is required over time to achieve the same rewarding effect), (3) Salience (i.e., the substance/behavior dominates thinking even when it is not being engaged with), (4) Withdrawal (i.e., strong negative reaction when engagement is not possible), (5) Conflict (i.e., the substance/behavior leads to interpersonal or intrapsychic conflict), and (6) Relapse (i.e., reverting back to substance/behavior after an abstinence period) (Griffiths 1995, 2005). Using this framework, it is easy to see how intense social media use fulfils these criteria. Social media makes users feel good (i.e., mood modification), some users start to use it more and more over time (i.e., tolerance), even when a user is not online, they often think about checking their social media accounts (i.e., salience), when a user is not on social media for some time, they can get anxious (i.e., withdrawal), using social media often can cause arguments with family members or friends (i.e., conflict), and finally, even if a user abstains from logging on their social media for some time, they generally go back to it eventually (i.e., relapse).

However, this addiction framework on its own does not take into account the severity of consequences required in order for a problematic behavior to be considered a disorder. For example, a young football enthusiast may feel happy while playing football (i.e., mood modification), want to play more and more as time passes (i.e., tolerance), always think about playing even when off the field (i.e., salience), feel anxious when unable to play for an extended period (i.e., withdrawal), get into arguments with parents who are tired of taking him to the park to play (i.e., conflict), and go back to playing after a period of giving it up (i.e., relapse), but this young man does not suffer from a football addiction. Similarly, although there are similarities between intense social media use and addiction, the consequences directly associated with excessive social media use so far reported have been mild whereas those of addiction are severe. This is one of the reasons why social media addiction has not been included in the DSM up to now, whereas other technology-related behaviors that exhibit more severe consequences, such as disordered gaming, have. Petry and O'Brien (2013), who reviewed the DSM process, argued that including conditions that do not cause significant distress and impairment (e.g., dance addiction or chocolate addiction) lowers the credibility of mental disorders. The problems related to social media use regarding sleep, school, and/or work for instance, do not constitute clinical-level impairment, whereas "addiction" is a clinical-level disorder. Furthermore, it is challenging to successfully argue that a tool which satisfies a basic human need (i.e., socialization), is normalized in society and has only mild negative consequences creates addictions.

The novelty effect is another factor that makes it difficult to determine whether social media addiction is a true disorder. The novelty effect is when a new device or technology is highly used primarily because it is new and interesting, much like the way a new toy receives a child's full attention for a short period after being gifted (Jeno et al. 2019; Silvia 2005). When the object loses its novelty and its ability to captivate attention, its use becomes normalized and integrated into a user's regular functioning. This has occurred throughout history with other technologies that have been put under the microscope as being potentially addictive before their use became routine. However, since social media sites have not been around for very long, as well as the fact that new social media sites emerge every few years and old ones are regularly updated with new features, it can be argued that the novelty effect of social media has not yet worn off. What's more, there is no longitudinal research to support the persistence of social media "addiction" over time. It would appear from cross-sectional and demographic studies that social media use peaks during adolescence when teenagers get online for the first time to develop their social networks and identities, activities that are characteristic of their developmental stage, then use becomes less intense and less problematic with age (Andreassen 2015; Carbonell et al. 2018; Van Deursen et al. 2015).

One of the main reasons why social media addiction has not been recognized as a true construct at this time is because it has not yet been disproven to be a secondary effect of a separate primary disorder. In order for a problematic behavior to be considered an addiction, said addiction must be able to exist without the presence of another psychological disorder, otherwise it would be better considered a symptom or secondary effect of that disorder. For example, the DSM-5 states that if gambling disorder is better explained by a manic episode rather than the gambling activity, then it should not be diagnosed as a gambling addiction. Problematic social media is likely a manifestation of a primary disorder, much like other problematic communication behaviors conducted by a person suffering from a psychological disorder. Billieux et al. (2014) presented a pedagogical case of a "mobile phone addict" and determined that the subject's behavior would be better explained by her psychological profile, characterized by irrational beliefs, dependent relationship-maintenance style, insecure attachment, low impulse control, repetitive negative thoughts and reassurance behaviors. They concluded that the subject's mobile phone "addiction" should not be directly addressed by a treatment but rather would improve as the primary psychological condition was treated. Such is likely the case with problematic social media use as well.

3.5 If It is not an *Addiction*, then What is It?

It is clear that certain people use social media in a problematic way and consequently experience negative consequences from it. However, labeling this situation as an addiction (whether it is or is not) does not give us much insight into the specific mechanisms by which SNSs can lead to these negative consequences (Panova et al. 2019). Using the framework of addiction likens social media to a homogeneous substance which is used in much the same way and has more-or-less the same effects on everybody. However, it is not and this framework does not account for the countless different ways in which social media can be used depending on a user's needs, interests, personality, comorbid disorders, sociocultural context, and social environment. From our perspective, it would benefit the research to move away from the addiction framework in the pursuit of deeper understanding on this phenomenon because the addiction framework occludes the specific pathways between problematic social media use and negative consequences. Alternatively, a focus on motivations, needs, and gratifications would offer more insight about the specific ways in which social media use influences well-being. At this time, there is research support for three primary motivations at the root of problematic social media use: Fear of Missing Out (FOMO), validation seeking, and avoidance coping.

FOMO is defined as the feeling that one is missing interesting or important information or that other people are doing something fun or interesting without them, leading to frustration and anxiety associated with the feeling of being left out (Przybylski et al. 2013). Social media often creates this experience for two main reasons, as there is a constant stream of information being produced on them at all times and because social media users mainly post about the highlights of their lives. The awareness that there is always something happening online, some interesting information being shared that others know about but you do not, creates anxiety and an urge to continuously "stay in the loop" in order to not be left out. This causes frequent and often compulsive checking so that one can keep updated on what is going on. Once online, users see all the interesting activities that their peers are doing and automatically compare these experiences with their own experiences (consciously or subconsciously) which can lead to dissatisfaction with the self. This dissatisfaction occurs because people online post primarily their most beautiful pictures and most interesting experiences. When a viewer sees only these carefully-selected and edited positive excerpts, it is easy to extrapolate from them and acquire a mistaken belief that these highlights are representative of their peers' lives. The viewer then automatically compares their own life (e.g., tedium, struggles and all) to this falsely positive impression of others' lives which leads to a belief that their own is somehow worse. This can precipitate feelings of life dissatisfaction (Przybylski et al. 2013), anxiety, and depression (Baker et al. 2016; Oberst et al. 2017). For many people, it also creates a competitive urge to "catch up" and reach the perceived level of their peers.

Validation-seeking online is the act of posting flattering pictures, updates on accomplishments or other attention-capturing content in the pursuit of positive feedback from one's network. Validation-seeking is problematic because it bases a person's value of themselves and their experiences on what others think (i.e., self-value stops being intrinsic and coming from within the self), and starts being extrinsic (i.e., dependent on others' perceptions). When self-value is extrinsic, it is unstable and transient and can lead to a constant pursuit of more and more validation in order to remain in a positive affective state. Validation-seeking online has been shown to be exhibited in narcissism (Barry et al. 2017; Casale and Fiorvanti 2018; Miller and Campbell 2008; Pincus and Roche 2011) and low in self-esteem (Marshall et al. 2015), those high in neuroticism (Marshall et al. 2015), and those high in perfectionism (Flett et al. 2014). Neurotic individuals are anxious, self-conscious, and sensitive to threat. They tend to use social media to acquire the attention, social support, and validation they do not receive enough of in their offline lives (Marshall et al. 2015; Ross et al. 2009). Finally, trait perfectionism and perfectionistic self-presentation have been found to be strongly associated with validation-seeking. Perfectionists are highly concerned with presenting an image of flawlessness and have an uncertain sense of self which motivate the pursuit of frequent validation

(Flett et al. 2014). SNSs allow users to craft whatever messages about themselves they desire. Therefore, perfectionists can create their ideal self-presentation in the pursuit of validation and the evasion of criticism and negative feedback.

The third motivation found to be associated with problematic social media use is that of employing SNSs for avoidance coping. Avoidance coping is the act of distracting oneself from an unpleasant internal state or external situation instead of addressing it and dealing with it directly. Avoidance coping via technology can be mild, such as when people turn to their smartphones to avoid the aversive state of boredom or it can be more intense, such as when people use technology to distract themselves from feelings of anxiety, stress or depression. It has been suggested that with the constant availability of endless entertainment that technology permits, people have grown accustomed to always being cognitively stimulated and do not allow themselves to experience states of "cognitive downtime" which are necessary for self-reflection, emotional processing, and mental rest (Immordino-Yang et al. 2012). Since social media is one of the most used features on personal electronic devices (Marketing Charts 2017), this means that when people feel bored, they usually open Instagram, Facebook, Twitter or the like to cope with the boredom by entertaining themselves. This compulsive social media checking to escape boredom can be detrimental as it leads to both higher social media exposure as well as training the brain to be intolerant of cognitive downtime, creating what some have called a "distraction addiction" (Oraison et al. 2020; Pang 2013). The brain of the intensive digital technology (e.g., internet/SNS/smartphone) user may lose its ability to stimulate *itself* successfully because of the constant availability of external stimulation which makes internal stimulation largely unnecessary. Further research needs to be conducted to ascertain how the absence of sufficient cognitive downtime may affect psychological and emotional wellbeing in the long term.

People also tend to use digital technology to escape aversive negative states such as stress. In this day and age, when a person finds themselves in an uncomfortable situation, it is typical for them to look down at their smartphones and scroll through something distracting such as their Instagram or Twitter feed to try to disengage from aversive experiences. While doing so is a normal reflex, ignoring a problem closes the possibility of learning from it and developing active coping skills that can be employed to deal with similar stressors in the future. The small stressors of life are training grounds for resilience and problem-solving skills that help us effectively handle more disturbing stressors, a process sometimes called "hardship inoculation" (Alter 2017). Regularly avoiding such small discomforts or inconveniences in favor of pleasant distraction can lead to underdeveloped resilience and coping abilities which may leave people more vulnerable and sensitive to stress down the line. It is also common to engage in avoidance coping when the discomfort is internal such as stress, anxiety or depression. Instead of processing the psychological discomfort and making efforts to actively deal with and overcome it, many people prefer to distract themselves with digital technology. This, again, is an understandable urge and often provides a welcome relief from psychological suffering. However, avoidance coping via digital technology has shown to be a potentially problematic behavior (Kircaburun and Griffiths 2018; Kuss and Griffiths 2011) and has been associated with anxiety and depression (Panova and Lleras 2016) due to the fact that it is a maladaptive coping mechanism (Blalock and Joiner 2000; Holahan et al. 2005). Not addressing a problem does not resolve or eliminate the issue, it simply covers it up in the short term. Active coping has been found to be much more effective at resolving stress and improving mental health in the long run (Frydenberg and Lewis 2009; LeDoux and Gorman 2002). However, active coping requires reflection and the kind of cognitive effort which is often hindered by the constant availability of distractors. Therefore, people must be conscious of why they are pulling out their digital devices (i.e., for functionality or for escape) and be aware that the escapism may be leading to the development of a maladaptive coping mechanism.

All in all, FOMO, validation-seeking, and avoidance coping are three promising pathways in which social media use can become problematic and lead to negative consequences. Social media use rooted in functional motives (e.g., finding or exchanging information) has not been shown to create significant problems. The negative effects of social media use often arise when SNSs are used as compensatory tools in response to psychological needs, often due to the existence of a comorbid disorder or certain personality profiles.

3.6 Comorbid Disorders

Problematic social media use is often an indicator or side-effect of a separate primary disorder. SNSs provide a platform on which the primary disorder is expressed and may exacerbate the disorder or alternatively, help users cope with it. Rosen et al. (2013) conducted an in-depth study exploring how different types of social media behaviors function as predictors of mood and personality disorders. They found that clinical symptoms of bipolar-mania, narcissism, and histrionic personality disorder were predicted by more general Facebook use, more Facebook impression management, and more Facebook friends. More Facebook general use and impression management also predicted more signs of antisocial and compulsive disorders whereas more Facebook general use predicted paranoid disorder and fewer Facebook friends predicted schizoid disorder. Compulsive disorder and histrionic disorder were predicted by the belief that emotional support was available online. Narcissism, antisocial, and compulsive disorders were predicted by anxiety about not checking Facebook. Andreassen et al. (2016) conducted another study to explore psychopathological comorbidities with problematic social media use and found that obsessive compulsive disorder (OCD), attention deficit hyperactivity disorder (ADHD), and anxiety were all associated with addictive use of social media. They hypothesized that OCD and ADHD are linked to problematic social media use for similar reasons, mainly because both disorders involve problems with compulsive behavior and difficulty inhibiting urges. On the other hand, they argued that anxiety is associated with problematic social media use because anxious people are more likely to have difficulties with social interactions in the real world and may compensate for it online.

Numerous studies have found social media use to be associated with depression (Appel et al. 2016; Brailovskaia et al. 2019; Lin et al. 2016) and anxiety (Lee-Won et al. 2015; Panova et al. 2019; Shaw et al. 2015), with some suggesting that social media use causes or exacerbates these disorders while others suggest that social media can be employed as a tool to alleviate the symptoms of these pre-existing conditions. Social media is simply a channel for accessing entertainment, information, and one's social network, therefore its effects can be positive or negative depending on how it is utilized. On the one hand, social media can trigger social comparisons which can deepen anxiety, depression, lower self- and life satisfaction, create a fertile ground for bullying and negative peer pressure, and allow one to escape the important processing of offline problems. On the other hand, social media provides an easy access point to one's support system in times of need, mental health resources, groups for people experiencing similar problems who can help, and it creates opportunities for users to anonymously share their pain and process it in a way that may not be possible offline. Dodemaide et al. (2019) found that for young people with suicidal ideation, social media had the potential to be significantly beneficial as it offered five therapeutic affordances: (1) Connection with peers or professionals; (2) The ability to explore and gather information; (3) The ability to tell one's story; (4) The ability to interact and collaborate with others; and (5) Introspection and perspective. In their meta-analysis, they found a reduction in suicidal ideation for Facebook users in the majority of studies. Like any other tool, social media is not inherently bad or good as its effects are dependent on how it is employed.

One of the primary arguments against social media is that although it facilitates social interaction, this kind of interaction is fundamentally different and less valuable than in-person interaction. It may be better to simply think of it as a different type of social interaction, one that can complement real life interactions in a healthy way. If a person meets regularly with friends and family and has a satisfactory real-world social life, then social media can simply provide an efficient way to maintain those connections when people are physically apart (Weiqin et al. 2016). Furthermore, social media can provide a valuable social channel for people who have limited opportunities for real-life social interactions. Many people around the world live in isolated communities that make it difficult for them to engage in regular social interactions with other people. However, this does not lower their needs for socialization. For example, in the United States, suburbs are highly dispersed communities which necessitate the use of a vehicle to get around. Since children and adolescents cannot drive, their real-life social opportunities are limited by their physical circumstances. Moreover, there are rarely cafés, plazas, town centers or any kind of real social nexus within walking distance of suburban residential districts. Therefore, for young people in these communities who have high socializing needs but few socializing opportunities, social media functions as a much-needed social space (Boyd 2014) while also providing a virtual place for them to interact, joke, flirt, make new friends, show off, organize events, and all things that they would be doing in person if they could, but they often cannot.

It is also accepted that SNSs can impede healthy socializing. A common scenario observed in the age of digital technology is a group of friends going out together when one or more starts scrolling through their social media, ignoring others around them. This often leads the other members of the group to feel snubbed, unappreciated or disrespected. This behavior has been described in the literature as "phubbing" (i.e., phone snubbing) and research shows that it negatively affects relationships between romantic partners, friends and even colleagues (Roberts and David 2016, 2017). Additionally, phubbing is detrimental not only for those on the receiving end but also for those doing it. If you are in the midst of a real-world social interaction but you sacrifice it to pursue a virtual one on your phone, you are substituting a rich and multi-dimensional social experience for a one-dimensional social experience. In this kind of social media use, we can argue that something of value is truly being lost. However, not all social media use is done at the cost of real-life social interactions and sometimes it can even enhance them.

3.7 Personalities Susceptible to Problematic Social Media Use

It has been suggested that attachment styles play an important role in the development of psychological disorders and addictions. Attachment styles describe the relationship between a young child and parent, a relationship that sets the tone for the individual's sense of self, personality, needs, fears, and vulnerabilities later in life (Ainsworth and Bowlby 1991). D'Arienzo et al. (2019) conducted a meta-analysis of 32 attachment and social media addiction studies. They found a significant positive association between people with insecure attachment, especially those with anxious attachment, and problematic social media use. Insecure attachment during childhood develops when a parent alternates between being nurturing/attentive and insensitive/emotionally unavailable. This kind of inconsistent behavior on the part of the caretaker confuses the child, leading them to feel insecure and distrustful of the parent while simultaneously desperate for their affection which they try to achieve. Children with this attachment style grow into adults who frequently seek reassurance and approval and who feel insecure about themselves and their relationships. D'Arienzo et al. argued that people with insecure attachment may use social media sites as a replacement and compensation for the affection that they crave from family and peers. They can also use social media as a tool to fulfill their attention-seeking needs and feed off of the feedback they receive on social media (Hart et al. 2015).

Various personality types have been found to be associated with problematic social media use. Kicaburun et al. (2018) found that extraverts use social media more to maintain existing relationships, pass their time, and manage tasks. Less conscientious people use it to present themselves as being more popular while neurotic people use it more often than others to pass their time. People open to experience use it more to maintain relationships and for informational and educational reasons. Regarding problematic social media use, they found that introversion, agreeableness, neuroticism, and lower conscientiousness were all associated personality traits. It is important to note that findings often differ depending on the specific SNS studied as well as the culture of the participants. Facebook use has been positively associated with extraversion and negatively associated with openness and conscientiousness in Norwegian users (Andreassen et al. 2013) and similarly with Polish users (Błachnio and Przepiorka 2016). However, a Taiwanese study found Facebook addiction to be negatively correlated with agreeableness and neuroticism as well as conscientiousness (Tang et al. 2015). Regarding Twitter addiction, agreeableness, conscientiousness, and extraversion were significantly and negatively associated (Kırcaburun 2016) and regarding Instagram addiction, agreeableness and conscientiousness were negatively associated (Kircaburun and Griffiths 2018). In the case of narcissism, it has been suggested that vulnerable narcissists may prefer platforms that are less reciprocal such as Twitter because they fear receiving negative evaluations, whereas grandiose narcissists may prefer more reciprocal platforms such as Facebook because they do not fear negative feedback and Facebook is better suited for self-promotion (Casale and Fiorvanti 2018).

It has been found that both introversion (Amichai-Hamburger et al. 2002) and extraversion (Correa et al. 2010; Wilson et al. 2010) are associated with higher use of the internet for social purposes. This seemingly conflicting finding has led to the "rich get richer" and the "poor get richer" theory, which is based on the fact that introverts have difficulty socializing in the real world, however they still have social needs that must be met. Social media provides them with an option to create and maintain a network of peers, interact and socialize without forcing them out of their comfort zone. The "rich get richer" theory argues that extraverts, with their advanced social skills and high socializing needs utilize social media to further develop relationships. They enjoy frequent interactions with people which are facilitated by SNSs and they use SNSs to maintain relationships with people they meet offline, thus strengthening those connections. Weiqin et al. (2016) found that adding and contacting friends on Facebook fully mediates the link between extraversion and bonding social capital (i.e., deeper relationships which offer emotional, social, and physical support) as well as bridging social capital (i.e., weaker social relationships that can offer information and new perspectives but not emotional support) (Granovetter 1983; Putnam 2000).

When people with certain vulnerable personality profiles get online, they may be more likely to develop problematic social media use because of the unique rewards social media can provide in response to their wants and needs. The consequent high levels of use can then exacerbate pre-existing issues such as low self-esteem, need for attention, insecurities, and under-developed real-world social skills. The worsening of these issues can serve to stimulate even more social media use as an attempt to alleviate or cater to these issues, creating a downward cycle for mental health. However, the main problem that needs to be addressed in this situation is the primary comorbid disorder and maladaptive cognitions, because social media is simply the channel through which they are expressed, not the origin of the problem.

3.8 Biological Bases

Although not many neurobiological studies have been conducted on how the brain behaves during social media use, there has been buzz around the idea that receiving "likes" can activate the reward system in a similar way that substances do. However, it is normal for pleasurable activities to activate the reward system and this occurrence does not need to be pathologized. Receiving likes on social media is rewarding because it is an indication of social acceptance and social connection, something highly valuable to our social brains. It has been found that acquiring social rewards activates similar brain regions as receiving money (Izuma et al. 2008) because both money and social status are important for an adaptive and prosperous human experience. It has been found that giving "likes" activates the reward circuitry as well because doing so contributes to forming new social relationships, strengthening social bonds, and eventually eliciting reciprocal prosocial behavior (Sherman et al. 2018).

When comparing technology addictions to traditional addictions (e.g., substance use disorders), important differences in neurobiology have been found. Unlike with substance use disorders, in which the impulsive system is hyperactive and the inhibition system is hypoactive, high scores on Facebook "addiction" tests have been associated only with the activation of the amygdala-striatal (impulsive) brain system but not with the prefrontal cortex (inhibition) brain system (He et al. 2017; Turel et al. 2014). This is suggested to occur because the adverse consequences for technology-related "addictions", such as negative effects on schoolwork or sleep, are generally much less severe than the consequences of substance use disorders and gambling addictions which include major health risks, financial problems, and legal issues. Another study found that symptom severity of problematic social media use was primarily associated with attentional impulsivity (Wegmann et al. 2020). They found that general executive functions and specific inhibitory control had no direct effect on symptom severity, however, increased symptom severity was associated with higher attentional impulsivity especially if there were reductions in executive functions or specific inhibitory control. With this in mind, the authors suggest treatments for problematic social media use that aim to strengthen general executive functioning for individuals with higher attentional impulsivity as well as to improve specific inhibitory control when confronted with auditory cues like message pings and SNS notifications.

More research on the neurobiology of social media use should be conducted as it uncovers interesting pathways between how social media content can influence thought and behavior. For example, Sherman et al. (2016) found that people look at pictures with many likes in a qualitatively different way than pictures with few likes. Certain regions of the brain are activated more for highly liked images such as the visual cortex, which suggests that people look with greater care at pictures they believe many others like, further demonstrating how online popularity influences interest, attention, and consequently memory. Other activated regions include those related to social cognition and social memories, including the medial prefrontal cortex, precuneus and hippocampus, along with the inferior frontal gyrus, which is implicated in imitation. They also found that looking at pictures of risky behaviors such as drug use and drinking decreased activation of the cognitive control network in adolescents, possibly illustrating how peer behaviors that are observed online disinhibit cognitive control in high-risk situations and increase the likelihood of taking similar risks.

3.9 Problems Associated with Social Media Use

Addictive or not, certain behaviors on social media have been associated with negative consequences. In their study on Facebook use, Fox and Moreland (2015) identified five themes related to negative Facebook experiences: (1) Managing inappropriate or annoying content; (2) Being tethered; (3) Lack of privacy and control; (4) Social comparison and jealousy, and (5) Relationship tension and conflict. Cyberbullying has also emerged as a serious problem on social media, especially among adolescents. These issues often mediate the relationship between social media use and negative psychological consequences such as anxiety, stress, low self-esteem, and depression.

One of the most significant negative experiences online is that of feeling tethered which emerges because of the constant availability the social network makes possible. Everybody knows that if they send a message, comment or like a post on social media, the recipient will see it soon after because portable digital devices such as smartphones make SNSs available at all times. Not only that, but people usually receive notifications when somebody interacts with them on their social network, making it unlikely for them to not know it occurred. There is no longer the option of giving the excuse "*I'm sorry, I wasn't home, so I missed your call*" like people did in the past. Nowadays, the message sender knows that the recipient has seen their message and if they do not respond to it in a timely manner, it can cause the sender to feel ignored or snubbed which can lead to interpersonal tension and even conflicts. Knowing that this can occur creates a pressure within the message recipient to respond quickly, even if they do not wish to or are not in the emotional state to do so (Thomee et al. 2010). This creates a sense of obligation that can increase stress and anxiety levels, especially if a person is juggling various conversations online, they can begin to feel overwhelmed by the expectation to be responsive and can even begin to feel like they do not have time enough for themselves and their own thoughts. Since mental rest is very important for psychological health, this burden of feeling tethered to one's online contacts can become highly problematic for wellbeing.

The second negative aspect of social media use is the lack of privacy and control which occurs because of the public nature of social media. Everybody in one's network can see what they post, have an opinion on it, share it, give feedback on it, and even save it to their own devices. Although people are mostly in control of what they put online, this is not completely the case. On certain SNSs, other people can "tag" users in photos and posts, meaning that the content will be made visible to the user's network as well as theirs even though they did not create or approve the content. If the content is something that users do not like, either because it is unflattering, it is a secret or it goes against the way they want to be perceived, feelings of anxiety can be triggered. On SNSs, people want to present their best selves, and so the control that SNSs give users in regards to crafting their self-presentation is one of the aspects that makes using social media so appealing, especially to adolescents who are developing their identities. However, crafting the ideal self implies keeping certain aspects of one's identity private, which can lead to anxiety when those aspects become public.

One of the most significant problems related to social media use is that of social comparison. Using SNSs means being exposed to other people's lives on a regular basis. When users perceive those lives as better than theirs in some way, feelings of envy, inadequacy, low self-esteem and mental distress can be triggered. Verduyn et al. (2015) found that passive Facebook use (i.e., scrolling rather than commenting and posting) predicted envy, and that envy predicted decline in affective wellbeing. Envy occurs when there is a discrepancy between users' real selves and their desired selves, a state which predicts lower life satisfaction and depression (Higgins et al. 1985). Regularly seeing pictures or posts about what users perceive to be "better" lives can also serve to lower their self-esteem (Vogel et al 2014). Low self-esteem has been shown to mediate the relationship between problematic social media use and decreased mental health (Hou et al. 2019). Studies show that the use of Instagram, an SNS where people are known to post the most carefully selected and edited pictures to show an idealized version of themselves, is correlated with depressive symptoms, self-esteem, anxiety about physical appearance, and body dissatisfaction (Sherlock and Wagstaff 2019) with the relationship between Instagram use and each other variable mediated by social comparison. Being inundated with pictures of people who users may consider to be more attractive than they are or have more appealing lives on a regular basis leads to feelings of inadequacy and dissatisfaction with the self (Brown and Tiggemann 2016; Fardouly and Vartanian 2016).

Social comparison is so detrimental because it externalizes the judgement of value. When the value users put on themselves and their life is intrinsic (i.e., originating within them and their personal perceptions, experiences and opinions), it is stable and within users' control. However, when users externalize their value by making it dependent on what others think about them, it becomes susceptible to external forces and therefore conditional and unstable, no longer within their own control. There will always be somebody who is or who users perceive to be more attractive, more successful, more intelligent or more talented than them, both online and offline. This does not invalidate or devalue their gifts or achievements. Personal value is not a zero-sum game where only the objectively "best" (as measured by likes and follows) are valued and valuable. This mindset is exacerbated by overly competitive societies which make it appear that the "good life" is reserved for those who have money, beauty and/or success, which is not the case. Social comparison has been associated with lower life satisfaction, lower self-esteem, and depression (Appel et al. 2016; Vogel et al. 2014). Social comparison on social media is especially detrimental for young women (Kleemans et al. 2016; Åberga et al. 2020) who are usually burdened with more pressure to live up to societal expectations of physical beauty.

The final negative experience Fox and Moreland (2015) identified is that of relationship tension and conflict among friends, family members, and romantic couples.

Since social media has become a prominent platform for social communication, problematic interactions or conflicts that occur on the platform can and often do carry over into real life. For example, the failure to give positive feedback on a post by a friend or partner can lead to arguments if it is seen as a snub or indicator of some deeper relational unrest. A problematic comment online that somebody in one's network does not agree with can trigger discussions offline that uncover unfavorable opinions and lead to relationship tensions. The online world is not separate from the real world even though it may sometimes seem that way, especially because of the layers of anonymity it affords and because feedback to social media interactions is not instant or particularly salient. Online communication, being one-dimensional and lacking other communication indicators besides words (e.g. facial expressions, body language, and vocal tone) may also facilitate misunderstandings which can create conflicts offline as well.

The age of technology also brings unique challenges for romantic couples. As SNSs provide users with quick and easy access to thousands of other people, the probability of encountering somebody that one is attracted to is greater than ever before, as is the temptation to reach out to them, which has never been easier. This situation is stressful for many couples who must fight such temptations in order to maintain a healthy relationship. Even the thought alone that a virtual infidelity may occur can hamper trust and relationship stability. Social media use has been associated with romantic disengagement, relationship dissatisfaction, infidelity, and a higher risk of divorce (Abbasi and Alghamdi 2017a, b; Abbasi 2018; Valenzuela et al. 2014). Prioritizing spending time online over spending time with a partner takes a toll by lowering the partner's trust, increasing suspicion, jealousy and perceived respect received. However, it is necessary to understand why the preference for social media over one's partner occurs in the first place in order to uncover issues at the heart of the unsuccessful relationship which may be leading to the problematic social media use.

3.10 Treatment Approaches

Arguably, problematic social media use may be conceptualized as a side effect of a comorbid disorder such as anxiety, ADHD, and/or OCD or the dysfunctional cognitions and behavioral patterns characteristic of a particular kind of personality profile such as narcissistic or neurotic. Therefore, in order to treat problematic social media use, the primary disorder must be identified and addressed first. When the primary disorder is successfully treated, symptoms of problematic social media use will decrease along with it. For example, working through a user's need for attention, approval or reassurance due to their insecure attachment style will decrease validation-seeking behaviors offline as well as online. Treating a user's anxiety will automatically influence the SNS behaviors they engage in to self-medicate the anxiety. Teaching healthier cognitive processes to replace maladaptive ones inherent to a personality disorder will be reflected in a healthier relationship to social media as well as a healthier overall mental state. However, treating problematic social media use as the primary issue will likely be ineffective or have only short-term success if the root problem is not identified and treated first (Billieux et al. 2014). With this in mind, certain techniques have been suggested to help the general population use social media in a healthier way and to avoid the negative consequences associated with it.

3.10.1 Self-help Interventions

In response to people who desire to lower their social media use because of the anxiety or loss of productivity it causes, self-help applications for portable digital devices (e.g., smartphones) have been developed with this goal in mind. These applications (e.g., Flipd, Offtime, Cold Turkey, and Stay Focused) allow users to block certain SNSs or to set time limits on their use. These applications often require the use of a secret code to change the settings so that the user will not be tempted to simply turn them off the moment they want to go online. This option serves as an external control mechanism for users who find it difficult to exert internal control over their social media use (and are mainly targeting social media use within smartphones as opposed to web browser usage through a computer).

For those who do not want to use an external control tool but prefer to develop internal control, mindfulness has been suggested as an important skill for healthier social media use (Charoensukmongkol 2016). Mindfulness involves being in the moment and sitting with one's thoughts and perceptions rather than seeking out cognitive distraction such as via SNSs. Mindfulness is also valuable in handling emotional discomfort so that the SNS does not become a tool for avoidance coping (Shonin et al. 2014).

3.10.2 Therapeutic Interventions

Cognitive-Behavioral Therapy techniques have been developed for Internet addiction (Young 2007) and may also be helpful regarding problematic social media use. These techniques involve exploring and reconstructing dysfunctional cognitions and behaviors regarding social media use. Alternative ways of thinking and strategies are taught for coping with emotional discomfort and needs. Behavior management for online and offline behavior can be employed involving behavioral rehearsal, recovery, modeling, self-instruction, and the development of adaptive skills (Andreassen 2015).

Motivational interviewing has also shown to be an effective treatment for behavioral addictions and has been proposed as a treatment for problematic social media use (Andreassen 2015). The objective of motivational interviewing is to help the user understand the negative effects of their behavior and to stimulate their motivation for change by creating a discrepancy between the user's current and desired state.

3.10.3 Psychoeducation

One of the main reasons why negative consequences arise from social media use is that people, especially children and adolescents who are among the most vulnerable users, are simply not aware of how social media use can be harmful. Therefore, one of the most effective ways to mitigate problematic social media use is to educate people about the risks inherent to using SNSs, such as how SNS use interacts with pre-existing disorders to create a downward cycle of exacerbated symptoms, how it can be a potent tool for cyberbullying or breach of privacy, how social comparisons can trigger negative affect and deteriorate mental health, and how using it as a coping mechanism is ineffective for truly coping with distress, especially in the long run (Carbonell and Panova 2017). Educating users about the risks and problems related to social media use and encouraging dialogue about these issues will increase the likelihood that they will approach SNS use with care, caution, and conscientiousness.

3.11 Conclusion

SNSs are first and foremost tools to fulfill the basic human need of social interaction. Since social interaction is a fundamental human need, the widespread popularity of SNSs is understandable, especially among young people who are in a developmental stage characterized by higher social and identity development needs. Considering that the negative consequences of social media use are not sufficiently severe, the measures to gauge "addiction" are mostly self-report, neither standardized nor suitable for diagnosis, there are no clinical or longitudinal studies to support the existence of social media addiction, and the necessary DSM criteria for a disorder are not fulfilled, there is insufficient evidence to support that social media addiction is a true disorder at this time. It would therefore benefit research to move away from the addiction framework because using such a framework to address social media use may be counterproductive for deepening the current understanding on this subject, as it imposes a set of criteria that may not be the best fit for the construct. Finding out what percent of a given population is "addicted" to social media based on fairly arbitrary cut-off scores on non-diagnostic measures that do not account for a user's context, needs or desires, offers little in terms of understanding why people engage in problematic social media use in the first place and how to help them. Social media use is best understood within the context of a user's psychological profile, their motivations for use, and the gratifications received from that use, their culture, social needs and environment, socioeconomic status and the consequences they experience from their social media activity in order to see how problematic use comes about and how to prevent it.

Social media use often becomes problematic when its use is underpinned by emotionally centered motives (as opposed to functional) such as for alleviating

FOMO, validation-seeking or avoidance coping. It can lead to negative consequences such as low self-esteem and lower life-satisfaction as well as deepening anxiety, stress, and depression levels because of how it facilitates social comparison, interpersonal conflict, and cyberbullying. Individuals most likely to engage in problematic social media use are those with comorbid disorders including anxiety, ADHD, OCD, bipolar-mania, paranoia, narcissism, antisocial, and histrionic personality disorders or those with extraverted, introverted, agreeable, unconscientious and neurotic personality types. However, susceptible personality profiles may vary depending on the user's culture and context. Preventing problematic social media use is a question of finding an adaptive fit for social media in the user's life, helping them develop control over compulsive SNS checking, educating them about the negative consequences of social media use, and focusing treatment on the comorbid primary disorders of which problematic social media use is often a secondary effect.

References

Abbasi IS, Alghamdi NG (2017a) The pursuit of romantic alternatives online: social media friends as potential alternatives. J Sex Marital Ther 44(1):16–28. https://doi.org/10.1080/0092623X.2017.1308450

Abbasi IS, Alghamdi N (2017b) When flirting turns into infidelity: the Facebook dilemma. Am J Family Therapy 45(1):1–14. https://doi.org/10.1080/01926187.2016.1277804

Abbasi IS (2018) Social media and committed relationships: what factors make our romantic relationship vulnerable? Soc Sci Comput Rev 1–10.https://doi.org/10.1177/0894439318770609

Åberg E, Koivula A, Kukkonen I (2020) A feminine burden of perfection? Appearance-related pressures on social networking sites. Telematics Informatics 46.https://doi.org/10.1016/j.tele.2019.101319

Ainsworth MDS, Bowlby J (1991) An ethological approach to personality development. Am Psychol 46(4):331–341

Alabi F (2013) A survey of Facebook addiction level among selected Nigerian university undergraduates. New Media Mass Commun 10:70–80

Al-Menayes JJ (2014) The relationship between mobile social media use and academic performance in university students. New Media Mass Commun 25:23–29. https://doi.org/10.7176/NMMC.vol2523-29

Alter A (2017) Irresistible: the rise of addictive technology and the business of keeping us hooked. Penguin, NY

Andreassen CS (2015) Online social network site addiction: a comprehensive review. Curr Addict Rep 2:175–184. https://doi.org/10.1007/s40429-015-0056-9

Andreassen CS, Pallesen S (2014) Social network site addiction—an overview. Curr Pharm Des 20(25):4053–4061. https://doi.org/10.2174/13816128113199990616

Andreassen CS, Billieux J, Griffiths MD, Kuss DJ, Demetrovics Z, Mazzoni E, Pallesen S (2016) The relationship between addictive use of social media and video games and symptoms of psychiatric disorders: a large-scale cross-sectional study. Psychol Addict Behav 30(2):252–262. https://doi.org/10.1037/adb0000160

Andreassen CS, Griffiths MD, Gjertsen SR, Krossbakken E, Kvam S, Pallesen S (2013) The relationships between behavioral addictions and the five-factor model of personality. J Behav Addict 2(2):90–99. https://doi.org/10.1556/JBA.2.2013.003

Andreassen CS, Torsheim T, Brunborg GS, Pallesen S (2012) Development of a Facebook addiction scale. Psychol Rep 110(2):501–517. https://doi.org/10.2466/02.09.18.PR0.110.2.501-517

Amichai-Hamburger Y, Wainapel G, Fox S (2002) "On the Internet no one knows I'm an introvert": extroversion, neuroticism, and Internet interaction. Cyberpsychol Behav 5(2):125–128. https://doi.org/10.1089/109493102753770507

Appel H, Gerlach AL, Crusius J (2016) The interplay between Facebook use, social comparison, envy, and depression. Curr Opin Psychol 9:44–49. https://doi.org/10.1016/j.copsyc.2015.10.006

Baker ZG, Krieger H, LeRoy AS (2016) Fear of missing out: relationships with depression, mindfulness, and physical symptoms. Trans Issues Psychol Sci 2(3):275–282. https://doi.org/10.1037/tps0000075

Barry CT, Doucette H, Loflin DC, Rivera-Hudson N, Herrington LL (2017) "Let me take a selfie": associations between self-photography, narcissism, and self-esteem. Psychol Pop Media Cult 6(1):48–60. https://doi.org/10.1037/ppm0000089

Billieux J, Philippot P, Schmid C, Maurage P, De Mol J, Van der Linden M (2014) Is dysfunctional use of the mobile phone a behavioural addiction? Confronting symptom-based versus process-based approaches. Clin Psychol Psychother 22(5):460–468. https://doi.org/10.1002/cpp.1910

Billieux J, Schimmenti A, Khazaal Y, Maurage P, Heeren A (2015) Are we overpathologizing everyday life? A tenable blueprint for behavioral addiction research. J Behav Addict 4(3):119–123. https://doi.org/10.1556/2006.4.2015.009

Błachnio A, Przepiorka A (2016) Personality and positive orientation in internet and Facebook addiction. an empirical report from Poland. Comput Hum Behav 59:230–236. https://doi.org/10.1016/j.chb.2016.02.018

Blalock JA, Joiner TE Jr (2000) Interaction of cognitive avoidance coping and stress in predicting depression/anxiety. Cogn Ther Res 24(1):47–65. https://doi.org/10.1023/A:1005450908245

Boyd DM (2014) It's complicated: the social lives of networked teens. Yale University Press, New Haven

Brailovskaia J, Velten J, Margaf J (2019) Relationship between daily stress, depression symptoms, and Facebook addiction disorder in Germany and in the United States. Cyberpsychol Behav Soc Netw 22(9):610–614. https://doi.org/10.1089/cyber.2019.0165

Brand M, Laier C, Young K (2014) Internet addiction: coping styles, expectancies, and treatment implications. Front Psychol 11. https://doi.org/10.3389/fpsyg.2014.01256

Brooks S, Califf C (2017) Social media-induced technostress: its impact on the job performance of it professionals and the moderating role of job characteristics. Comput Netw 114:143–153. https://doi.org/10.1016/j.comnet.2016.08.020

Brown Z, Tiggemann M (2016) Attractive celebrity and peer images on Instagram: effect on women's mood and body image. Body Image 19:37–43. https://doi.org/10.1016/j.bodyim.2016.08.007

Carbonell X, Chamarro A, Oberst U, Rodrigo B, Prades M (2018) Problematic use of the internet and smartphones in university students: 2006–2017. Int J Environ Res Public Health 15(3):1–13

Carbonell X, Panova T (2017) A critical consideration of social networking sites' addiction potential. Addict Res Theory 25(1):48–57. https://doi.org/10.1080/16066359.2016.1197915

Casale S, Fioravanti G (2018) Why narcissists are at risk for developing Facebook addiction: the need to be admired and the need to belong. Addict Behav 76:312–318. https://doi.org/10.1016/j.addbeh.2017.08.038

Charlton JP, Danforth IDW (2007) Distinguishing addiction and high engagement in the context of online game playing. Comput Hum Behav 23:1531–1548. https://doi.org/10.1016/j.chb.2005.07.002

Charoensukmongkol P (2016) Mindful Facebooking: the moderating role of mindfulness on the relationship between social media use intensity at work and burnout. J Health Psychol 21(9):1966–1980. https://doi.org/10.1177/1359105315569096

Correa T, Hinsley AW, de Zúñiga HG (2010) Who interacts on the web? the intersection of users' personality and social media use. Comput Hum Behav 26(2):247–253. https://doi.org/10.1016/j.chb.2009.09.003

D'Arienzo MC, Boursier V, Griffiths M (2019) Addiction to social media and attachment styles: a systematic literature review. Int J Ment Heal Addict 17:1094–1118. https://doi.org/10.1007/s11469-019-00082-5

Dodemaide P, Joubert L, Merolli M, Hill N (2019) Exploring the therapeutic and nontherapeutic affordances of social media use by young adults with lived experience of self-harm or suicidal ideation: a scoping review. Cyberpsychol Behav Soc Netw 22(10):622–633. https://doi.org/10.1089/cyber.2018.0678

Elphinston RA, Noller P (2011) Time to face it! Facebook intrusion and the implications for romantic jealousy and relationship satisfaction. Cyberpsychol Behav Soc Netw 14(11):631–635. https://doi.org/10.1089/cyber.2010.0318

Erzen E, Çikrikci Ö (2018) The effect of loneliness on depression: a meta-analysis. Int J Soc Psychiatry 64(5):427–435. https://doi.org/10.1177/0020764018776349

Fardouly J, Vartanian LR (2016) Social media and body image concerns: current research and future directions. Curr Opin Psychol 9:1–5. https://doi.org/10.1016/j.copsyc.2015.09.005

Flett GL, Besser A, Hewitt PL (2014) Perfectionism and interpersonal orientations in depression: an analysis of validation seeking and rejection sensitivity in a community sample of young adults. Psychiatry 77(1):67–85. https://doi.org/10.1521/psyc.2014.77.1.67

Floros G, Siomos K (2013) The relationship between optimal parenting, internet addiction and motives for social networking in adolescence. Psychiatry Res 209(3):1–6. https://doi.org/10.1016/j.psychres.2013.01.010

Fox J, Moreland JJ (2015) The dark side of social networking sites: an exploration of the relational and psychological stressors associated with Facebook use and affordances. Comput Hum Behav 45:168–176. https://doi.org/10.1016/j.chb.2014.11.083

Frydenberg E, Lewis R (2009) Relations among well-being, avoidant coping, and active coping in a large sample of Australian adolescents. Psychol Rep 104(3):745–758. https://doi.org/10.2466/PR0.104.3.745-758

Griffiths MD (1995) Technological addictions. Clin Psychol Forum 76:14–19

Granovetter M (1983) The strength of weak ties: a network theory revisited. Sociol Theory 1(1):201–233

Griffiths MD (2005) A "components" model of addiction within a biopsychosocial framework. J Substance Use 10(4):191–197. https://doi.org/10.1080/14659890500114359

Haagsma MC, Caplan SE, Peters O, Pieterse ME (2013) A cognitive-behavioral model of problematic online gaming in adolescents aged 12–22 years. Comput Hum Behav 29(1):202–209. https://doi.org/10.1016/j.chb.2012.08.006

Hart J, Nailling E, Bizer G, Collins C (2015) Attachment theory as a framework for explaining engagement with Facebook. Personality Individ Differ 77:33–40. https://doi.org/10.1016/j.paid.2014.12.016

Holahan CJ, Moos RH, Holahan CK, Brennan PL, Schutte KK (2005) Stress generation, avoidance coping, and depressive symptoms: a 10-year model. J Consult Clin Psychol 73(4):658–666. https://doi.org/10.1037/0022-006X.73.4.658

Hormes JM, Kearns B, Timko CA (2014) Craving Facebook? Behavioral addiction to online social networking and its association with emotion regulation deficits. Addiction 109(12):2079–2088. https://doi.org/10.1111/add.12713

He Q, Turel O, Brevers D, Bechara A (2017) Excess social media use in normal populations is associated with amygdala-striatal but not with prefrontal morphology. Psychiatry Res Neuroimaging 269:31–35. https://doi.org/10.1016/j.pscychresns.2017.09.003

Higgins ET, Klein R, Strauman T (1985) Self-concept discrepancy theory: a psychological model for distinguishing among different aspects of depression and anxiety. Soc Cogn 3(1):51–76. https://doi.org/10.1521/soco.1985.3.1.51

Hou Y, Xiong D, Jiang T, Song L, Wang Q (2019) Social media addiction: its impact, mediation, and intervention. Cyberpsychol: J Psychosoc Res Cyberspace 13(1). https://doi.org/10.5817/CP2019-1-4

How are consumers spending their time online? Marketing Charts (2017, July 5). https://www.marketingcharts.com/featured-66742

Immordino-Yang MH, Christodoulou JA, Singh V (2012) Rest is not idleness: implications of the brain's default mode for human development and education. Perspect Psychol Sci 7(4):352–364. https://doi.org/10.1177/1745691612447308

Izuma K, Saito DN, Sadato N (2008) Processing of social and monetary rewards in the human striatum. Neuron 58(2):284–294. https://doi.org/10.1016/j.neuron.2008.03.020

Jeno LM, Vandvik V, Eliassen S, Grytnes JA (2019) Testing the novelty effect of an m-learning tool on internalization and achievement: a self-determination theory approach. Comput Educ 128:398–413. https://doi.org/10.1016/j.compedu.2018.10.008

Kircaburun K (2016) Effects of gender and personality differences on Twitter addiction among Turkish undergraduates. J Educ Pract 7(24):33–42

Kırcaburun K, Alhabash S, Tosuntaş SB, Griffiths MD (2018) Uses and gratifications of problematic social media use among university students: a simultaneous examination of the Big Five of personality traits, social media platforms, and social media use motives. Int J Mental Health Addict 18:525–547. https://doi.org/10.1007/s11469-018-9940-6

Kircaburun K, Griffiths MD (2018) Instagram addiction and the big five of personality: the mediating role of self-liking. J Behav Addict 7(1):158–170. https://doi.org/10.1556/2006.7.2018.15

Kleemans M, Dalmaans S, Carbaat I, Anschütz D (2016) Picture perfect: the direct effect of manipulated Instagram photos on body image in adolescent girls. Media Psychol 21(1):93–110. https://doi.org/10.1080/15213269.2016.1257392

Koc M, Gulyagci S (2013) Facebook addiction among Turkish college students: the role of psychological health, demographic, and usage characteristics. Cyberpsychol Behav Soc Netw 16(4):279–284. https://doi.org/10.1089/cyber.2012.0249

Kraut R, Patterson M, Lundmark V, Kiesler S, Mukopadhyay T, Scherlis W (1998) Internet paradox. A social technology that reduces social involvement and psychological well-being? Am Psychologist 53(9):1017–1031. https://doi.org/10.1037//0003-066x.53.9.1017

Kuss DJ, Griffiths MD (2011) Online social networking and addiction—a review of the psychological literature. Int J Environ Res Public Health 8(9):3528–3552. https://doi.org/10.3390/ijerph8093528

Lau W (2017) Effects of social media usage and social media multitasking on the academic performance of university students. Comput Hum Behav 68:286–291. https://doi.org/10.1016/j.chb.2016.11.043

Ledoux J, Gorman J (2002) A call to action: overcoming anxiety through active coping. Am J Psychiatry 158(12):1953–1955. https://doi.org/10.1176/appi.ajp.158.12.1953

Lee-Won R, Herzog L, Park S (2015) Hooked on Facebook: the role of social anxiety and need for social assurance in problematic use of Facebook. Cyberpsychol Behav Soc Netw 18(10):567–574. https://doi.org/10.1089/cyber.2015.0002

Lin LY, Sidani JE, Shensa A, Radovic A, Miller E, Colditz JB, Hoffman BL, Giles LM, Primack BA (2016) Association between social media use and depression among U.S. young adults. Depression Anxiety 33(4):323–331. https://doi.org/10.1002/da.22466

Lowe-Calverley E, Pontes H (2020) Challenging the concept of smartphone addiction: an empirical pilot study of smartphone usage patterns and psychological well- being. Cyberpsychol Behav Social Networking 23(8). https://doi.org/10.1089/cyber.2019.0719.

Maqableh M, Rajab L, Quteshat W, Masa'deh RMT, Khatib T, Karajeh H (2015) The impact of social media networks websites usage on students' academic performance. Commun Network 7:159–171.https://doi.org/10.4236/cn.2015.74015

Marshall TC, Lefringhausen K, Ferenczi N (2015) The big five, self-esteem, and narcissism as predictors of the topics people write about in Facebook status updates. Personal Individ Differ 85:35–40. https://doi.org/10.1016/j.paid.2015.04.039

Masur PK, Reinecke L, Ziegele M, Quiring O (2014) The interplay of intrinsic need satisfaction and Facebook specific motives in explaining addictive behavior on Facebook. Comput Hum Behav 39:376–386. https://doi.org/10.1016/j.chb.2014.05.047

Miller JD, Campbell WK (2008) Comparing clinical and social-personality conceptualizations of narcissism. J Pers 76(3):449–476. https://doi.org/10.1111/j.1467-6494.2008.00492.x

Montag C, Bey K, Sha P, Li M, Chen YF, Liu WY, Zhu YK, Li CB, Markett S, Keiper J, Reuter M (2014) Is it meaningful to distinguish between generalized and specific Internet addiction? Evidence from a cross-cultural study from Germany, Sweden, Taiwan and China. Asia-Pacific Psychiatry 7(1):20–26.https://doi.org/10.1111/appy.12122

Moqbel M, Kock N (2017) Unveiling the dark side of social networking sites: personal and work-related consequences of social networking site addiction. Inform Manage 55(1):109–119. https://doi.org/10.1016/j.im.2017.05.001

Moreau A, Laconi S, Delfour M, Chabrol H (2015) Psychopathological profiles of adolescent and young adult problematic Facebook users. Comput Hum Behav 44:64–69. https://doi.org/10.1016/j.chb.2014.11.045

Oberst U, Wegmann E, Stodt B, Brand M, Chamarro A (2017) Negative consequences from heavy social networking in adolescents: the mediating role of fear of missing out. J Adolesc 55:51–60. https://doi.org/10.1016/j.adolescence.2016.12.008

Oraison H, Nash-Dolby O, Wilson B, Malhotra R (2020) Smartphone distraction-addiction: examining the relationship between psychosocial variables and patters of use. Australian J Psychol 72(2). https://doi.org/10.1111/ajpy.12281

Pang ASK (2013) The distraction addiction. Little Brown and Company, New York

Panova T, Carbonell X, Chamarro A, Puerta-Cortés DX (2019) Specific smartphone uses and how they relate to anxiety and depression in university students: a cross-cultural perspective. Behav Inform Technol 39(9):944–956. https://doi.org/10.1080/0144929x.2019.1633405

Panova T, Lleras A (2016) Avoidance or boredom: negative mental health outcomes associated with use of information and communication technologies depend on users' motivations. Comput Hum Behav 58:249–258. https://doi.org/10.1016/j.chb.2015.12.062

Petry NM, O'Brien CP (2013) Internet gaming disorder and the DSM-5. Addiction 108(7):1186–1187. https://doi.org/10.1111/add.12162

Pincus AL, Roche MJ (2011) Narcissistic grandiosity and narcissistic vulnerability. In: Campbell WK, Miller JD (eds) The handbook of narcissism and narcissistic personality disorder: theoretical approaches, empirical findings, and treatments. Wiley, pp 31–40

Przybylski AK, Murayama K, DeHaan CR, Gladwell V (2013) Motivational, emotional, and behavioral correlates of fear of missing out. Comput Hum Behav 29:1841–1848. https://doi.org/10.1016/j.chb.2013.02.014

Putnam RD (2000) Bowling alone: the collapse and revival of American community. Simon and Schuster, New York, NY

Ramesh Masthi NR, Pruthvi S, Phaneendra MS (2018) A comparative study on social media usage and health status among students studying in pre-university colleges of urban Bengaluru. Indian J Commun Med 43(3):180–184. https://doi.org/10.4103/ijcm.IJCM_285_17

Roberts JA, David ME (2016) My life has become a major distraction from my cell phone: partner phubbing and relationship satisfaction among romantic partners. Comput Hum Behav 54:134–141. https://doi.org/10.1016/j.chb.2015.07.058

Roberts JA, David ME (2017) Put down your phone and listen to me: how boss phubbing undermines the psychological conditions necessary for employee engagement. Comput Hum Behav 75:206–217. https://doi.org/10.1016/j.chb.2017.05.021

Rosen LD, Whaling K, Rab S, Carrier LM, Cheever NA (2013) Is Facebook creating '"iDisorders"'? The link between clinical symptoms of psychiatric disorders and technology use, attitudes and anxiety. Comput Hum Behav 29(3):1243–1254. https://doi.org/10.1016/j.chb.2012.11.012

Ross C, Orr ES, Sisic M, Arseneault JM, Simmering MG, Orr R (2009) Personality and motivations associated with Facebook use. Comput Hum Behav 25(2):578–586. https://doi.org/10.1016/j.chb.2008.12.024

Ryan T, Chester A, Reece J, Xenos S (2014) The uses and abuses of Facebook: a review of Facebook addiction. J Behav Addict 3(3):133–148. https://doi.org/10.1556/JBA.3.2014.016

Shaw AM, Timpano KR, Tran TB, Joormann J (2015) Correlates of Facebook usage patterns: the relationship between passive Facebook use, social anxiety symptoms, and brooding. Comput Hum Behav 48:575–580. https://doi.org/10.1016/j.chb.2015.02.003

Sherlock M, Wagstaff DL (2019) Exploring the relationship between frequency of Instagram use, exposure to idealized images, and psychological well-being in women. Psychol Pop Media Cult 8(4):482–490. https://doi.org/10.1037/ppm0000182

Sherman L, Hernandez L, Greenfield P, Dapretto M (2018) What the brain "likes:" neural correlates of providing feedback on social media. Soc Cogn Affect Neurosci 13(7):699–707. https://doi.org/10.1093/scan/nsy051

Sherman LE, Payton AA, Hernandez LM, Greenfield PM, Dapretto M (2016) The power of the like in adolescence: effects of peer influence on neural and behavioral responses to social media. Psychol Sci 27(7):1027–1035. https://doi.org/10.1177/0956797616645673

Shonin E, Van Gordon W, Griffiths MD (2014) Practical tips for teaching mindfulness to children and adolescents in school-based setting. Educ Health 32(2):69–72

Silvia PJ (2005) What is interesting? Exploring appraisal structure of interest. Emotion 5(1):89–102. https://doi.org/10.1037/1528-3542.5.1.89

Sindermann C, Duke É, Montag C (2020) Personality associations with Facebook use and tendencies towards Facebook use disorder. Addict Behav Rep 11:100264. https://doi.org/10.1016/j.abrep.2020.100264

SOCIAL MEDIA: meaning in the Cambridge english dictionary (2020) Retrieved July 18, 2020, from https://dictionary.cambridge.org/dictionary/english/social-media

Stravynski A, Boyer R (2001) Loneliness in relation to suicide ideation and parasuicide: a population-wide study. Suicide Life-Threatening Behav 31(1):32–40. https://doi.org/10.1521/suli.31.1.32.21312

Surrat CG (1999) Netaholics? The creation of a pathology. Nova Science, New York

Tang J-H, Chen M-C, Yang H-Y, Chung T-Y, Lee Y-A (2015) Personality traits, interpersonal relationships, online social support, and Facebook addiction. Telematics Inform 33(1):102–108. https://doi.org/10.1016/j.tele.2015.06.003

Thomée S, Dellve L, Harenstam A, Hagberg M (2010) Perceived connections between information and communication technology use and mental symptoms among young adults—a qualitative study. BMC Public Health 10(1):66. https://doi.org/10.1186/1471-2458-10-66

Turel O, Bechara A (2016) Social networking site use while driving: ADHD and the mediating roles of stress, self-esteem and craving. Front Psychol 7:1–10. https://doi.org/10.3389/fpsyg.2016.00455

Turel O, He Q, Xue G, Xiao L, Bechara A (2014) Examination of neural systems sub-serving Facebook "addiction." Psychol Rep 115(3):675–695. https://doi.org/10.2466/18.PR0.115c31z8

Turel O, Serenko A (2012) The benefits and dangers of enjoyment with social networking websites. Eur J Inf Syst 21(5):512–528. https://doi.org/10.1057/ejis.2012.1

Valenzuela S, Halpern D, Katz JE (2014) Social network sites, marriage well-being and divorce: survey and state-level evidence from the United States. Comput Hum Behav 36:94–101. https://doi.org/10.1016/j.chb.2014.03.034

Valtorta NK, Kanaan M, Gilbody S, Ronzi S, Hanratty B (2006) Loneliness and social isolation as risk factors for coronary heart disease and stroke: systematic review and meta-analysis of longitudinal observational studies. Heart 102:1009–1016. https://doi.org/10.1136/heartjnl-2015-308790

Van Deursen AJAM, Bolle CL, Hegner SM, Kommers PAM (2015) Modeling habitual and addictive smartphone behavior. The role of smartphone usage types, emotional intelligence, social stress, self-regulation, age, and gender. Comput Hum Behav 45:411–420. https://doi.org/10.1016/j.chb.2014.12.039

Verduyn P, Lee DS, Park J, Shablack H, Orvell A, Bayer J, Ybarra O, Jonides J, Kross E (2015) Passive Facebook usage undermines affective well-being: experimental and longitudinal evidence. J Exp Psychol Gen 144(2):480–488. https://doi.org/10.1037/xge0000057

Vernon L, Barber BL, Modecki KL (2015) Adolescent problematic social networking and school experiences: the mediating effects of sleep disruptions and sleep quality. Cyberpsychol Behav Soc Netw 18(7):386–392. https://doi.org/10.1089/cyber.2015.0107

Vogel E, Rose J, Roberts L, Eckles K (2014) Social comparison, social media, and self-esteem. Psychol Pop Media Cult 3(4):206–222. https://doi.org/10.1037/ppm0000047

Wang C-W, Ho RTH, Chan CLW, Tse S (2015) Exploring personality characteristics of Chinese adolescents with internet-related addictive behaviors: trait differences for gaming addiction and social networking addiction. Addict Behav 42:32–35. https://doi.org/10.1016/j.addbeh.2014.10.039

Wegmann E, Mueller S, Turel O, Brand M (2020) Interactions of impulsivity, general executive functions, and specific inhibitory control explain symptoms of social-networks-use disorder: an experimental study. Sci Rep 10.https://doi.org/10.1038/s41598-020-60819-4

Wegmann E, Stodt B, Brand M (2015) Addictive use of social networking sites can be explained by the interaction of Internet use expectancies, internet literacy, and psychopathological symptoms. J Behav Addict 4(3):155–162. https://doi.org/10.1556/2006.4.2015.021

Weiqin EL, Campbell M, Kimpton M, Wozencroft K, Orel A (2016) Social capital on Facebook: the impact of personality and online communication behaviors. J Educ Comput Res 54(6):747–786. https://doi.org/10.1177/0735633116631886

Wilson K, Fornasier S, White KM (2010) Psychological predictors of young adults' use of social networking sites. Cyberpsychol Behav Soc Netw 13(2):173–177. https://doi.org/10.1089/cyber.2009.0094

Wolniczak I, Cáceres-DelAguila JA, Palma-Ardiles G, Arroyo KJ, Solís-Visscher R, Paredes-Yauri S, Mego-Aquije K, Bernabe-Ortiz A (2013) Association between Facebook dependence and poor sleep quality: a study in a sample of undergraduate students in Peru. PLoS ONE 8(3):e59087. https://doi.org/10.1371/journal.pone.0059087

Wood RTA, Griffiths MD, Parke A (2007) Experiences of time loss among videogame players: an empirical study. Cyberpsychol Behav 10(1):38–44. https://doi.org/10.1089/cpb.2006.9994

Widyanto L, Griffiths MD (2006) Internet addiction: a critical review. Int J Ment Heal Addict 4(1):31–51. https://doi.org/10.1007/s11469-006-9009-9

Xanidis N, Brignell C (2016) The association between the use of social network sites, sleep quality and cognitive function during the day. Comput Hum Behav 55:121–126. https://doi.org/10.1016/j.chb.2015.09.004

Young KS (1998) Internet addiction: the emergence of a new clinical disorder. Cyberpsychol Behav 1(3):237–244. https://doi.org/10.1089/cpb.1998.1.237

Young KS (2007) Cognitive behavior therapy with internet addicts: treatment outcomes and implications. Cyberpsychol Behav 10(5):671–679. https://doi.org/10.1089/cpb.2007.9971

Zivnuska S, Carlson J, Carlson D, Harris R, Harris K (2019) Social media addiction and social media reactions: the implications for job performance. J Soc Psychol 159(6):746–760. https://doi.org/10.1080/00224545.2019.1578725

Chapter 4
Smartphone Addiction

Dmitri Rozgonjuk, Christian Montag, and Jon D. Elhai

Abstract Since the proliferation of smartphones, the general public as well as researchers have shown a growing interest in their potentially addictive effects. Over the past years, smartphone use research has grown exponentially. While scholars have moved away from "smartphone addiction" terminology in contemporary research and opted for alternative terminology (e.g., "problematic smartphone use" and "smartphone use disorder") and frameworks, excessive smartphone use has been found to be correlated with depression and anxiety, as well as loss of productivity at work and school. This chapter provides a broader view of this research domain by exploring the following issues: is "smartphone addiction" a real phenomenon? How does it develop? Are some people more prone to experiencing it than others? How can it be assessed, and if one experiences this issue, how could it be treated? We aim to answer these questions in the current work. In addition, we outline some concerns with the construct of "smartphone addiction" while also providing insights into the latest research on this topic.

Keywords Smartphone addiction · Problematic smartphone use · Smartphone use disorder · Internet addiction · Internet use disorder · Social media addiction

D. Rozgonjuk (✉) · C. Montag
Department of Molecular Psychology, Institute of Psychology and Education, Ulm University, Helmholtzstraße 8/1, 89081 Ulm, Germany
e-mail: dmitri.rozgonjuk@uni-ulm.de

C. Montag
e-mail: mail@christianmontag.de

D. Rozgonjuk
Institute of Mathematics and Statistics, University of Tartu, Tartu, Estonia

J. D. Elhai
Department of Psychology, University of Toledo, Toledo, OH, USA
e-mail: contact@jon-elhai.com

Department of Psychiatry, University of Toledo, Toledo, OH, USA

H. M. Pontes (ed.), *Behavioral Addictions*, Studies in Neuroscience, Psychology and Behavioral Economics, https://doi.org/10.1007/978-3-031-04772-5_4

4.1 Introduction

As with the Internet in general, smartphones have many features that can enhance productivity as well as other aspects of everyday life activities. Yet there are also findings that link excessive smartphone use to various detrimental intra- and interpersonal outcomes (Duke and Montag 2017a; Kates et al. 2018). As with the Internet earlier (Young 1998), researchers as well as the general public started discussing potentially addictive effects of smartphones (Choi et al. 2012), giving birth to a field of "smartphone addiction" studies. It should be noted that before "smartphone addiction", research first focused on "mobile phone addiction" (Park 2005). Therefore, one may view smartphone research also somewhat (but not completely) synonymous with mobile phone use research. Duke and Montag (2017b) proposed that mobile phone use could include "smartphone use", with the latter reflecting the latest technological developments in which digital mobile devices became more sophisticated due to the wider range of utilities and diverse applications.

In this chapter, we aim to provide a broader overview of the "smartphone addiction" field. We discuss what is meant by this phenomenon, how it is conceptualized theoretically, and some challenging aspects regarding this line of research. In addition, we outline key findings regarding the biological bases of this phenomenon, comorbidities, and individual differences implicated in this condition. We then discuss prevalence rates, as well as assessment and potential interventions. Finally, some concluding remarks are made.

4.1.1 What is "Smartphone Addiction" or Problematic Smartphone Use?

"Smartphone addiction" as a term has generated a lot of controversies over recent years. But what is it exactly? This phenomenon has been described through several terms (see below) but, in essence, it could be treated as an umbrella term for functional adversities associated with excessive smartphone use (Billieux et al. 2015a; De-Sola Gutierrez et al. 2016). However, this notion needs some background, as this terminology is controversial in the scientific literature.

As it happened with the Internet (Young 1996) approximately a decade before smartphones were adopted in the wider population, notions on potential "addictive" effects of smartphone use were raised (Choi et al. 2012). Therefore, many researchers started using the term "smartphone addiction" (Kwon et al. 2013a; Cho et al. 2017). However, since this was (and still is not) an officially recognized mental health diagnosis, researchers have also adopted alternative terms, such as "smartphone overuse" (Hwang et al. 2012) and "proneness to smartphone addiction" (Randler et al. 2016; Rozgonjuk et al. 2016).

As mentioned earlier, essentially, this phenomenon (despite slightly different terms used) involves two key components: (a) excessive frequency of smartphone use and (b) the experience of daily-life adversities as a consequence of excessive smartphone use. Although the term "smartphone addiction" may be popular in the media, researchers have argued whether using addiction terminology is appropriate in this context. Among a wide range of criticisms, researchers showed concerns about how easily everyday life behavior can become overpathologized (Billieux et al. 2015b; Kardefelt-Winther et al. 2017), as well as the notion that there are no standardized criteria for this "addiction" (Panova and Carbonell 2018).

Therefore, researchers have proposed using the term "problematic smartphone use" (PSU) instead (Panova and Carbonell 2018). Furthermore, it has been argued, that the construct of PSU in itself is confusing as it does not specify whether the person experiencing this condition is on the process of "moving" from a healthy to pathological state, or whether the person is already at the pathological end of this condition (Rozgonjuk et al. 2019a; Montag et al. 2019b). Alternatively, recent works have implemented the "smartphone use disorder" terminology (Lachmann et al. 2018; Montag et al. 2019b). This latter term is at least partly inspired by the inclusion of Gaming Disorder to ICD-11 as a diagnosable condition (World Health Organization 2018; Montag et al. 2021d), indicating that it is not unreasonable to foresee other forms of technology overuse being diagnosable in the future (Elhai et al. 2019b). Moreover, the term smartphone use *disorder* strives at unification of terminology in the field by its orientation with the WHO terminology, not necessarily being meant as a final judgement about the actual nature of "smartphone addiction". Furthermore, smartphone use disorder/PSU as a construct belongs to the larger research field studying Internet Use Disorders, whereas one could describe excessive smartphone use to fall in the realm of a mobile form of Internet Use Disorder (Montag et al. 2019b).

The current debate on the appropriate nomenclature shows that there is plenty of controversy in this line of research. For consistency with the proposed terminology in Panova and Carbonell (2018), we will use the term PSU throughout the chapter when discussing the literature on the so-called phenomenon of "smartphone addiction".

Research on PSU can be considered as relatively young in comparison to other disciplines within psychology and psychiatry. Obviously, one of the reasons lies in the fact that these digital technologies are relatively new in the historical context. If we tentatively consider the diffusion of the first smartphones in the population as a rough start of PSU research, then we can conclude that field of research has been around for only little more than a decade. Therefore, it is also understandable that there are several questions that are still waiting to be solved in the field.

4.1.2 Problematic Smartphone Use Theoretical Frameworks

Throughout the recent decades, both excessive Internet and smartphone use research have had several theoretical frameworks aiming to explain why some people develop problematic digital technology usage patterns. A more detailed overview about some of these models and theories can be found in Rozgonjuk (2019). In this subsection, we will also provide a brief overview of some of these approaches, as follows.

One of the most popular approaches has historically been the components model of addiction (CMA) developed by Griffiths (2005). The main factors within the CMA are, in essence, similar to the ones measured by the Smartphone Addiction Scale (SAS; Kwon et al. 2013b, see below for more information on this scale). Another popular theory in digital technology (and media) use literature is the *uses and gratifications theory* (UGT; Blumler 1979) that outlines the (psychological) need satisfactions sought for by digital media users. There are different types of gratifications (e.g., content, process, and social) that may drive people to use some specific medium and/or media (Stafford et al. 2004; Song et al. 2004).

Moreover, according to the *compensatory Internet use theory* (CIUT; Kardefelt-Winther et al. 2017), people tend to use digital technology for mood regulation and stress management purposes. One could then conceive smartphone use as part of a self-medication coping strategy to assist in overcoming negative affect. In some cases, individuals may end up using digital technology in problematic ways. Another highly influential theoretical framework in Internet addictions research is the *cognitive-behavioral model (CBM) of pathological Internet use* by Davis (2001). This is a process model, encompassing several causes for problematic (or as the author calls it, pathological) Internet use, as well as distinguishing between generalized and specific problematic Internet use (see also Montag et al. 2015a). The former is reflected in a more general and multidimensional overuse of Internet, while the latter is the dependency on some specific function of the Internet (e.g., gambling, cybersex, etc.). Importantly, Davis (2001) also acknowledged that excessive Internet use may not necessarily develop into problematic behavior.

A further expansion of this approach was provided by the model for Internet Use Disorders and the *interaction of person-affect-cognition-execution model of excessive internet communications use model* (Brand et al. 2014, 2019, 2016). In general, the I-PACE emphasizes the role of predisposing factors (e.g., genetics, personality traits, pre-existing psychopathology, etc.) in how a person copes with situations and emotions, as well as perceives and reacts to certain stimuli. Depending on how situations are perceived, one of the reactions could be choosing a specific application or device (e.g., social media) that may lead to satisfying one's needs (e.g., social connectedness). Over time, this kind of coping with negative affect inducing situations may develop into a habitual response to similar situations and stimuli, potentially leading to excessive use of that application or device, and in some cases leading to problematic behavior. The I-PACE model appears to present with elements of other abovementioned frameworks (Rozgonjuk 2019). It has also been used in several studies to conceptualize PSU in relation to various factors (Elhai et al. 2020; Wolniewicz et al. 2019). Although the I-PACE model does not explicitly mention smartphones, recent studies have shown that there may be significant overlap between smartphone and social media use (Rozgonjuk et al. 2020c, 2021b; Sha et al. 2019), potentially suggesting that PSU could represent a mobile version of Internet use disorder (Montag et al. 2019b; see also above).

4.1.3 Confusion with the Construct of Problematic Smartphone Use

As mentioned earlier, one important element of PSU is that adversities in one's everyday life are caused by excessive smartphone use. However, from a conceptual point of view, it is not clear what is meant by "excessive" smartphone use. One may even ask if there could be a cutoff point for "excessive use" in terms of screen time minutes and/or phone-checking behavior. This is a challenging question to be answered because the world seems to be moving towards higher levels of digitalization and virtualization. It can be conceived that time spent on one's smartphone five years ago differs from time spent on one's device now. For example, using one's smartphone for two hours a day may have been widely judged as excessive in 2008, but this would not be the case today. In this realm, we mention that associations between key variables such as productivity and smartphone use might not be necessarily linear, but could be also curvilinear. Montag and Walla (2016) mentioned that the right amount/type of smartphone use might enhance productivity, but there comes a point where interruptions strongly impair productivity. Such an inverted curvilinear function was also discussed in Montag and Diefenbach (2018). Beyond that, it makes a difference if a person uses the phone for business or personal purposes and perhaps the number of interruptions is of more importance to determine PSU than the mere number of minutes to hours spent on the phone.

Another challenging question that has been discussed decades ago within the context of problematic Internet use is related to the following issue: is the problematic behavior related *to* the use of the digital device or the features used *on* the technology (Griffiths 2000; Spada 2014)? It has been argued that the device (i.e., smartphone) as a physical entity is not as relevant as the activity conducted on one's smartphone. It has been recently demonstrated that PSU has significant overlap with WhatsApp use (Rozgonjuk et al., 2020c; Sha et al. 2019); see also the study by Montag et al. (2015b) relating real-world WhatsApp and Facebook consumption to the duration of smartphone use. Therefore, it could be argued that smartphones enable problematic social media use. If that is the case, should we still be referring to PSU or should we be referring to problematic social media use instead? Panova and Carbonell (2018) mentioned the analogy that an alcoholic is also not addicted to the bottle (here the smartphone), but rather the alcohol (here social media or Freemium game applications; Montag et al. (2019b).

Additionally, the chicken-or-the-egg question is important to be considered. It is quite difficult to establish a causal link in associations between PSU and other factors of interest. Currently, there is relatively little research that can show robust longitudinal findings demonstrating that either PSU causes X or that X causes PSU. In fact, perhaps there is no unidirectional association, and causality is in the form of feedback loops? This notion is at least partially hypothesized in the I-PACE model (Brand et al. 2016, 2019). Considering that the vast majority of research relies on linear modelling, including non-linearity and feedback loops into modeling relations between PSU and other variables could be an interesting avenue that can provide further answers to this question.

Some of the key methodological issues in the field tend to involve cross-sectional study design and specific sample-related issues. The former means that while several studies have analyzed theoretically causal models, in practice, this is done on cross-sectional data. However, this approach is still useful, as these models provide some empirical evidence for theoretical hypotheses that may be worth pursuing with more resource-demanding approaches (e.g., experimental and longitudinal studies). Of note, a recent study by Chen et al. (2020) implemented a longitudinal design, demonstrating a link between PSU severity and psychological distress. Despite this, more studies of this nature are needed in order to solve some of the pressing issues in this line of research. Furthermore, sample limitations typically include adopting convenience sampling of university students where gender distribution tends to be in favor of more female participants. This kind of sampling may pose some limitations to generalizing the findings of studies.

4.2 Problematic Smartphone Use and Biological Bases

The study of PSU is currently dominated by self-report research and unfortunately only a small number of studies until now tried to shed light on the biological underpinnings of PSU. As with the study of Internet Use Disorders, the most often applied neuroscientific technique to understand the actual nature of smartphone addiction represents magnetic resonance imaging (MRI; Becker and Montag 2019). MRI research can be divided into a structural and functional parts. Structural MRI provides insights into individual differences in brain volume or integrity of white matter tracts which also can be linked to individual differences in PSU severity.

Lee et al. (2019) contrasted in their work 39 people with high levels of PSU with 49 control persons and observed that the PSU (with a focus on social media) was associated with lower gray matter volume in the right orbitofrontal cortex, a brain region of relevance in controlling emotions. This finding is highly interesting because recent work investigating individual differences in tendencies towards Gaming Disorder showed in a longitudinal MRI design that loss of gray matter volume in this brain area might be a result of six of weeks of gaming (Zhou et al. 2019). Whether such an observation can be transferred to PSU is not clear because Lee et al. (2019) investigated brain-PSU correlates at one time point. A new study by Horvath et al. (2020) contrasted 22 persons who reported high levels of PSU and 26 controls. Perhaps of highest interest among several findings from this study is again a negative association between gray matter volume of the orbitofrontal cortex (OFC) (this time the left hemisphere) and PSU. Moreover, the authors observed an inverse association between gray matter volume of the anterior cingulate cortex and PSU severity, an association which was also visible in the work by Montag et al. (2018), although the latter investigated associations with problematic WeChat use. As the smartphone often is a vehicle to access social media applications (Montag et al. 2019a, b), the findings of Horvath et al. (2020) might be explained by excessive use of social media as supported with the findings by Montag et al. (2018). The anterior cingulate cortex (ACC) itself plays an important role in processing of conflicting information and top-down control (Bush et al. 2000). A study by Hu et al. (2017) investigated the white matter tract architecture by applying diffusion tensor imaging (DTI) and observed lower integrity of white matter in several areas of the brain including the superior longitudinal fasciculus. Of high importance the white matter tract integrity was negatively correlated with severity of PSU in the stria terminalis and the internal capsule.

In the domain of functional MRI (fMRI) different approaches to study the human brain of people with high levels of PSU have been used. A study by Chun et al. (2018) applied resting state fMRI to understand PSU while contrasting 38 young persons characterized by high PSU and 42 control persons. Among others, they observed "a lower functional connectivity between the right OFC and nucleus accumbens (NAcc)" (p. 1) in the high-PSU group when compared to healthy controls. Of high interest, the functional connectivity patterns between those two brain areas were inversely associated with both withdrawal symptoms and cortisol levels. Please note, that the already mentioned Horvath et al. (2020) study also presented findings from resting state fMRI, where the right ACC was associated with lower intrinsic activity. For reasons of completeness we also mention a resting state fMRI study, which was presented at a conference and revealed that persons with high levels of PSU showed lower functional connectivity in the anterior insula and primary motor cortex (Kim et al. 2016).

Beyond resting state fMRI, also task-based fMRI studies have been published showing how the brain responds to smartphone related cues. In a study by Schmitgen et al. (2020) participants underwent two different picture contrast conditions in the brain scanner. In the contrast "smartphone vs. neutral pictures" differences between high-PSU group and controls were observed in "medial prefrontal (MPFC), occipital, temporal, and anterior cingulate (ACC) cortices, in temporoparietal regions, and cerebellum" (p. 1). When participants processed photos of "active versus inactive smartphones", differences between groups occurred in the "frontal operculum/anterior insula and precentral gyrus" (p. 1). Note that some of the brain results were also associated with different aspects of PSU as reported in self-report questionnaires. An earlier study by Chun et al. (2017) is also noteworthy, as they observed lower blood-oxygen-level-dependent (BOLD) activity in the dorsolateral prefrontal cortex and the dorsal ACC in high-PSU group vs. control persons when they were processing angry faces or emotional transition.

Finally, we report that few other studies applied electroencephalography (EEG) technology (Chen et al. 2016) and transcranial magnetic stimulation (TMS) (Hadar et al. 2017), and found evidence for either deficits in inhibitory control or reduced excitability of the prefrontal cortex in people with high levels of PSU. As one can see from the mentioned studies, the molecular basis of PSU is completely understudied with the exception of the work by Chun et al. (2018) taking into account cortisol levels (stress hormone) and new work by Seo et al. (2020) who observed that "Brain parenchymal and gray matter volume–adjusted g -aminobutyric acid-to-creatine ratios were higher in subjects with Internet and smartphone addiction" (p. 1293), whereas these brain markers decreased after cognitive behavioral therapy. Although the different approaches from brain imaging (and molecular psychology) point to differences in top-down control (e.g., inhibition) and bottom-up processes (cue reactivity), when contrasting people with high levels of PSU to controls in diverse setups, a systemic understanding of PSU is still missing. For example, in many cases it is not clear what differences in the structure of brain anatomy mean for the functionality of the brain. Of highest importance, future studies will need longitudinal approaches to assess how the brain changes due to smartphone use (see an exemplarily work in the realm of Gaming Disorder by Yu et al. (2020)), something which is mostly lacking. And again, also in this area it is of importance to take more into account the actual usage of certain smartphone features to ultimately understand its potential impact on the human brain (Montag et al. 2017; Montag 2019; see also new research in the field of digital biomakers to be of possible interest here: Montag et al. 2021a, b).

4.3 Comorbidities and Individual Differences in Problematic Smartphone Use

As mentioned earlier, research on excessive smartphone use related adversities has grown substantially over the past years. One may even go as far as noting that the vast majority of literature involving smartphone use regards its associations with (decreased) psychological well-being.

One of the more robust findings regarding comorbidities is the relationship between depression and PSU symptoms. In addition, anxiety has also been found to correlate with PSU severity. It should be noted, however, that these associations yield small-to-medium effect sizes. Studies have also demonstrated PSU's associations with other anxiety-related disorder symptoms, such as social anxiety (Peterka-Bonetta et al. 2019; Elhai et al. 2018a, b, c, d; Enez Darcin et al. 2016), generalized anxiety disorder (GAD) (Rozgonjuk et al. 2020a), and post-traumatic stress disorder (PTSD; Contractor et al. 2017) symptoms.

As mentioned earlier, the I-PACE model (Brand et al. 2016, 2019) outlines several factors of importance in the development of PSU. Indeed, research has shown that many of these mediating and moderating characteristics are associated with higher levels of PSU, and among these factors are emotion dysregulation (Rozgonjuk and Elhai 2019; Pancani et al. 2019), boredom proneness (Elhai et al. 2017; Wolniewicz et al. 2019), worry and anger (Elhai et al. 2019a, b), intolerance of uncertainty (Rozgonjuk et al. 2019b), lower distress tolerance (Elhai et al. et al. 2018a, b, c, d), negative affectivity (Montag et al. 2016; Elhai et al. 2018a), procrastination

(Rozgonjuk et al. 2018a), and fear of missing out (FoMO) on rewarding experiences (Elhai et al. 2016, 2020). FoMO could be of particular interest since it tends to show high correlations with PSU severity.

Importantly, PSU also correlates with one's daily-life functioning as studies have demonstrated that people reporting higher levels of PSU also report lower levels of productivity at school or within the workplace (Duke and Montag 2017a), poorer academic outcomes (Lepp et al. 2015; Kates et al. 2018), and more surface approach to learning (Rozgonjuk et al. 2018c). One may hypothesize that loss of productivity or declines in performance at work or school may further fuel PSU as well as additional comorbidities. Also note that recent studies provided evidence that the problematic use of social media applications, abundantly used on smartphones, result in lower productivity (Rozgonjuk et al. 2020b).

Last but not least, the Big Five personality traits have also been researched in relation to PSU. Of more consistent findings is—perhaps not surprisingly—the relationship between higher trait neuroticism (tendency to experience negative emotions, such as anger, depression, and anxiety; Costa and McCrae 1985) and PSU (Marengo et al. 2020; Hussain et al. 2017; Gao et al. 2020). Other findings are more mixed and covariates (e.g., gender, age) could also play a pivotal role.

In addition to the positive correlation between neuroticism and PSU, negative associations between PSU and conscientiousness (tendency to be self-disciplined and organized) and agreeableness (tendency for pro-social behavior) have also been demonstrated (Arpaci and Kocadag Unver 2020). In another study, a negative correlation between openness to experience and PSU was found (Peterka-Bonetta et al. 2019). Yet another study showed that conscientiousness, openness to experience, and emotional stability (inverse Neuroticism) were negatively correlated with PSU (Marengo et al. 2020; Hussain et al. 2017; Gao et al. 2020).

4.4 Prevalence Rates of Problematic Smartphone Use—And Why They May Be Troublesome

How big of a problem is PSU in a society? This is a difficult question since PSU is (i) not a formal diagnosis and that it (ii) presents with several conceptual and assessment-related problems. A recent study demonstrated that there are 78 validated self-report questionnaires aiming to measure mobile phone/smartphone use, with many of them focusing on assessing PSU-like phenomena (Harris et al. 2020). This may be an indication on the general lack of consensus on what exactly is PSU, and, therefore, aiming to diagnose individuals as "smartphone addicts" may be highly misleading.

However, there is some work that aimed to provide prevalence estimates for this phenomenon. Based on the discussion above, we therefore suggest the reader to take these findings with a grain of salt. It should also be noted that some studies have used the "mobile phone addiction" terminology, and we do acknowledge that mobile phones may differ from smartphones in terms of functionalities. This said, prevalence rates for PSU across some studies are presented in Table 4.1.

Table 4.1 shows that prevalence rates of PSU across different studies vary from 10 to 36.5%. Importantly, although some of the sample sizes are relatively large, it should be noted that most of the studies are clearly not representative of the population investigated. Typically, these studies include either school or university students. In addition, one may observe that out of these ten studies, eight studies used different scales for assessing PSU. This, as mentioned earlier, further raises questions regarding comparability as well as the validity and reliability of these studies reporting the prevalence rates observed in Table 4.1.

In conclusion, prevalence rates in the field of PSU are highly unreliable as they are largely derived from the lack of consensus in the concept of this phenomenon, as well as the lack of a standardized way of measuring PSU. Therefore, we do not

Table 4.1 Studies outlining the prevalence of problematic smartphone use (ordered chronologically)

Study	Prevalence (%)	Sample origin	Sample	Sample size	Measure
Lopez-Fernandez et al. (2014)	10	United Kingdom	11–18 years old students	1026	MPPUS
Smetaniuk (2014)[a]	10–25	USA	Undergraduate students and non-students	301 and 362	AMPUH & ACPAT
Haug et al. (2015)	16.9	Switzerland	Vocational school students	1519	SAS-SV
Nikhita et al. (2015)	31.33	India	8–10th grade students	415	MPD
Long et al. (2016)	21.3	Mainland China	Undergraduate students	1121	PCPUQ
de-Sola et al. (2017)	20.5	Spain	Adults (aged 16–65 years)	1126	MPPUS
Lee and Lee (2017)	35.6	South Korea	Adolescents	3000	SAPS
Alhazmi et al. (2018)	36.5	Saudi Arabia	Medical students	181	SAS
Cha and Seo (2018)	30.9	South Korea	Middle school students	1824	SAPS
Chang et al. (2019)	15.2	Taiwan	Fifth-grade students	2621	SPAI-SF

[a] Work by Smetaniuk (2014) included two studies; information for (a) undergraduate students and (b) non-students is presented

MPPUS mobile phone problem use scale, *PCPUQ* problematic cellular phone use questionnaire, *SAS-SV* smartphone addiction scale for adolescents, *AMPUH* adapted mobile phone use habits, *ACPAT* adapted cell phone addiction test, *MPD* mobile phone dependence questionnaire, *SAPS* smartphone addiction proneness scale, *SAS* smartphone addiction scale, *SPAI-SF* smartphone addiction inventory short form recommend relying on these prevalence rates in one's research nor in policymaking. Nevertheless, we believe that these issues do not disqualify this line of work from further investigation. On the contrary, specifying how many people are in peril and experience problems in everyday life due to PSU is an important issue be investigated but firstly, outstanding conceptual issues need to be solved.

4.5 Assessment Approaches for Problematic Smartphone Use

We would like to emphasize once again that "smartphone addiction" or PSU is not a formally recognized diagnosis. Despite that, there is epidemiological research investigating its prevalence rates as well as labeling study participants as "smartphone addicts" primarily based on measures with questionable clinical validity. In this subsection, we provide a brief overview on how questionnaires are used in this line of research on the example of the Smartphone Addiction Scale (SAS; Kwon et al. 2013b). This is followed by discussing how objectively measured smartphone use variables correlate with self-reports (e.g., PSU score) as well as other variables of interest (e.g., self-reported depression).

Arguably the most popular self-report measure of PSU is the 33-item Smartphone Addiction Scale (SAS; Kwon et al. 2013b), however, its shorter 10-item version (Kwon et al. 2013a) is also be found frequently employed in this line of research. The SAS is a multidimensional scale initially aimed to measure the frequency of PSU symptoms that one may experience. In addition to providing the possibility to form a single PSU score, it is also possible to treat the long version of the scale multidimensionally with its subscales, as it is aimed to measure different facets of PSU. The concept has been inspired from the components model of addiction (CMA; Griffiths 2005).

The SAS includes the following dimensions: daily-life disturbances, positive anticipation, withdrawal, cyberspace-oriented relationships, overuse, and tolerance (Kwon et al. 2013b). *Daily-life disturbances* seem to be more self-explanatory and this domain means that one's everyday life is disrupted due to smartphone use. This may involve experiencing conflicts with others (e.g., problems with relationships), oneself (e.g., feeling stressed out due to perceiving oneself as being "weak" in self-control, or feelings of losing self-control), or one's tasks (e.g., missing planned activities due to smartphone use). *Positive anticipation* relates to *salience* as defined within the CMA (Griffiths 2005), and one may interpret this domain as craving for activities related to smartphone use. *Withdrawal* means that the person may experience negative psychological and even physiological symptoms after the decrease or discontinuation of smartphone use. In other words, the person may experience distress, irritability, anger, as well as sadness when not able to use their smartphone with the preferred frequency and/or duration. *Cyberspace-oriented relationships* in this context mean that social connections formed online may be prioritized over one's social network in the offline world. *Overuse* is also self-explanatory and involves using one's smartphone excessively or too frequently. It may reflect in others complaining to the user about "being always on the smartphone" or the user him or herself may feel that they spend too much time on their device. On a related note, what constitutes "too much" smartphone use is not crystal clear to this day. Finally, *tolerance*, which may be experienced when a person needs to increase the use their smartphone over time in order to achieve the same gratification. As can be seen from this information, the SAS is a multi-faceted measure that aims to capture several significant aspects in smartphone use that could be described as addictive.

While there are many "smartphone addiction" or PSU scales, the SAS (whether the 33- or 10-item scale) seems to be among the most popular ones used across different cultures. The original SAS was developed in Korea, although published in English (Kwon et al. 2013b). The SAS, its shorter version or other adapted spin-off scales have been used in English-speaking cultures including the United States (Wolniewicz et al. 2019) as well as Australia (Winskel et al. 2019) and in several European countries (Duke and Montag 2017a; Montag 2018; Haug et al. 2015; Lopez-Fernandez 2017; Rozgonjuk et al. 2018a, b, c), as well as other parts of Asia (Ching et al. 2015; Samaha and Hawi 2016; Luk et al. 2018). As evidenced above, it is clear that the SAS has had major impact on the field of PSU research.

Because of its length, the shorter 10-item SAS-SV may be preferred in investigation of PSU. Importantly, the SAS-SV also has suggested cutoff scores for diagnosing "smartphone addiction" in adolescents. The theoretical range of scores is from 10 to 60, and the authors suggested the cutoff value of 31 for boys and 33 for girls when diagnosing "smartphone addiction" (Kwon et al. 2013a). Again, we emphasize two important issues. First, the study was published in 2013, which means that smartphone use patterns may have changed considerably since that time. Second, "smartphone addiction" is not yet a formal diagnosis, further implying that the proposed cutoff scores may be highly misleading, especially if a clinical practitioner decides to use this scale to screen clients for "smartphone addiction".

An important problem in this line of research—and assessment of PSU—is that self-reported PSU correlates relatively poorly with objectively-measured smartphone use (Rozgonjuk et al. 2018b, 2020a; Ellis et al. 2019; Loid et al. 2020). As mentioned earlier in this chapter, it is unclear what "excessive" smartphone use really means. Is it the time spent on a smartphone (smartphone use duration, or screen time), the frequency of smartphone use (phone-checking, screen activation, etc.), the combination of these behaviors, or something else? For a recent study trying to answer some of these questions, see Marengo et al. (2021).

There is still currently relatively little research that has implemented objectively measured smartphone use in combination with self-reports. A recent study by Ellis et al. (2019) demonstrated that the relationships between self-reported PSU measures and objectively measured smartphone use may also depend on a specific scale used to assess PSU. Overall, both objectively measured smartphone use duration and frequency were typically positively correlated with self-reports (Ellis et al. 2019). However, the effect sizes were small to medium, and in some cases PSU was not associated with smartphone pick-ups/use frequency. Interestingly, some studies have found that more objectively measured screen time was positively associated with PSU, while the correlation between PSU and objectively measured smartphone frequency was not statistically significant (Elhai et al. 2018a, b, c, d; Loid et al. 2020; Rozgonjuk et al. 2018a, b, c, 2021a). Finally, there are also studies that demonstrate the lack of significant relationship between self-reported PSU and actual smartphone use behavior (Andrews et al. 2015; Wilcockson et al. 2018). As can be seen from these findings, making robust interpretations is quite difficult due to mixed results. Therefore, assessment of PSU should also be put in the context of these findings as only gauging someone's daily smartphone use may not be entirely sufficient to determine whether the person experiences daily-life adversities due to smartphone use or even PSU.

4.6 Treatment Approaches for Problematic Smartphone Use

If a person is experiencing daily-life adversities due to smartphone use, what and how exactly should be treated?

Firstly, the target of such intervention could be the excessive smartphone use itself and be focused on reducing actual smartphone use in order to decrease PSU scores and symptoms. Recent studies have looked into some means of intervention with the aim of reducing excessive smartphone use. For instance, Loid et al. (2020) conducted a two-month experimental study aiming to find out if prompting notifications about excessive smartphone use would reduce PSU scores, as well as objectively measured smartphone use frequency (phone-checking) and duration (screen time). The results of this study showed that these notifications do not really reduce smartphone use nor self-reported PSU. Holte and Ferraro (2020) conducted an experiment that entailed making smartphones less appealing by changing the phones to grayscale for 8–10 days. They found that total screen time was reduced by an average of almost 40 min per day, suggesting that by targeting the appearance (or, essentially, the functionality) of smartphones may reduce the usage of these devices. Lowe-Calverley and Pontes (2020) found that smartphone users reported potentially checking their smartphones less if their top-three applications—mainly communication-based apps—were inaccessible to them. On the one hand, these results suggest that by limiting the functionality of smartphones may reduce actual use of smartphones. However, this is yet to be tested with objectively measured behavioral data. On the other hand, these results are in line with findings suggesting that the smartphone is a medium that facilitates the use of specific content (e.g., social networking services applications) and that people may be "hooked" to those applications and not smartphone devices themselves (Rozgonjuk et al., 2020c; Sha et al. 2019). This needs to be seen also against the background of the data business model behind many smartphone-apps, which led to highly immersive applications

(Montag and Hegelich 2020) and where corporate social responsibility is of rising importance (Montag et al. 2021c).

Secondly, underlying psychopathological conditions as well as dysfunctional coping mechanisms that may fuel PSU could be targeted in treatment approaches for PSU. In this case, it may be useful to consider the literature reporting on the successful treatment approaches employed for other problematic behaviors such as Internet Gaming Disorder (IGD) given the potential commonalities between these phenomena. Recently, some authors have proposed the PIPATIC (*Individualized psychotherapy program for addiction to information and communication technologies*) Program (Torres-Rodríguez et al. 2018a), which is an intervention program for adolescents with IGD. The goal of this program is to reduce IGD symptoms and to improve the well-being of people experiencing adversities due to IGD. However, one may also notice that the pillars upon which this approach is founded (e.g., cognitive-behavioral therapy) might as well be applicable to adults who are struggling with excessive Internet and smartphone use. The PIPATIC program is a synthesis of many domains of intervention and is structured within six therapeutic work modules: (1) psychoeducation and motivations; (2) addictions treatment as usual adapted to IGD; (3) intrapersonal skills; (4) interpersonal skills; (5) family intervention; and (6) development of a new lifestyle. The PIPATIC Program has shown promising results in being an effective treatment approach against IGD as well as comorbidity symptoms and intra- and inter-personal functioning (Torres-Rodríguez et al. 2018b). While PSU does not constitute an officially recognized addictive disorder, we believe that these principles may be also helpful in reducing daily-life adversities due to (excessive) smartphone use. Yet it should be noted that the guidelines for IGD treatment are also currently under debate (King et al. 2020).

4.7 Final Conclusions

To summarize, PSU research is growing and interest in this phenomenon has increased both within and outside the scientific domain due to its public relevance. Despite the numerous media reports describing the many potential harmful effects stemming from smartphones, investigating such issue or domain is far more complex as the existing findings are mixed when it comes to actual observed smartphone use behavior. Although PSU research presents with important limitations, the new emerging field of digital phenotyping has a lot of promise with regards to human behavior and psychology research (Baumeister and Montag 2019; Montag and Rumpf 2021).

References

Alhazmi AA, Alzahrani SH, Baig M et al (2018) Prevalence and factors associated with smartphone addiction among medical students at King Abdulaziz University, Jeddah. Pak J Med Sci 34. https://doi.org/10.12669/pjms.344.15294

Alosaimi F, Alyahya H, Alshahwan H et al (2016) Smartphone addiction among university students in Riyadh, Saudi Arabia. Saudi Med J 37:675–683. https://doi.org/10.15537/smj.2016.6.14430

Andrews S, Ellis DA, Shaw H, Piwek L (2015) Beyond self-report: tools to compare estimated and real-world smartphone use. PLoS ONE 10:e0139004. https://doi.org/10.1371/journal.pone.0139004

Arpaci I, Kocadag Unver T (2020) Moderating role of gender in the relationship between big five personality traits and smartphone addiction. Psychiatr Q 91:577–585. https://doi.org/10.1007/s11126-020-09718-5

Baumeister H, Montag C (2019) Digital phenotyping and mobile sensing. Springer International Publishing

Becker B, Montag C (2019) Psychological and neuroscientific advances to understand internet use disorder. Neuroforum 25:99–107. https://doi.org/10.1515/nf-2018-0026

Billieux J, Maurage P, Lopez-Fernandez O et al (2015a) Can disordered mobile phone use be considered a behavioral addiction? An update on current evidence and a comprehensive model for future research. Curr Addict Rep 2:156–162

Billieux J, Schimmenti A, Khazaal Y et al (2015b) Are we overpathologizing everyday life? A tenable blueprint for behavioral addiction research. J Behav Addict 4:119–123. https://doi.org/10.1556/2006.4.2015.009

Blumler JG (1979) The role of theory in uses and gratifications studies. Commun Res 6:9–36. https://doi.org/10.1177/009365027900600102

Brand M, Rumpf H-Jü, Demetrovics Z et al (2020) Which conditions should be considered as disorders in the International Classification of Diseases (ICD-11) designation of "other specified disorders due to addictive behaviors"? J Behav Addict.https://doi.org/10.1556/2006.2020.00035

Brand M, Wegmann E, Stark R et al (2019) The Interaction of Person-Affect-Cognition-Execution (I-PACE) model for addictive behaviors: update, generalization to addictive behaviors beyond internet-use disorders, and specification of the process character of addictive behaviors. Neurosci Biobehav Rev 104:1–10. https://doi.org/10.1016/j.neubiorev.2019.06.032

Brand M, Young KS, Laier C (2014) Prefrontal control and internet addiction: a theoretical model and review of neuropsychological and neuroimaging findings. Front Hum Neurosci 8:375. https://doi.org/10.3389/fnhum.2014.00375

Brand M, Young KS, Laier C et al (2016) Integrating psychological and neurobiological considerations regarding the development and maintenance of specific Internet-use disorders: an Interaction of Person-Affect-Cognition-Execution (I-PACE) model. Neurosci Biobehav Rev 71:252–266. https://doi.org/10.1016/j.neubiorev.2016.08.033

Bush G, Luu P, Posner MI (2000) Cognitive and emotional influences in anterior cingulate cortex. Trends Cogn Sci 4:215–222. https://doi.org/10.1016/S1364-6613(00)01483-2

Cha S-S, Seo B-K (2018) Smartphone use and smartphone addiction in middle school students in Korea: prevalence, social networking service, and game use. Health Psychol Open 5:205510291875504. https://doi.org/10.1177/2055102918755046

Chang F-C, Chiu C-H, Chen P-H et al (2019) Children's use of mobile devices, smartphone addiction and parental mediation in Taiwan. Comput Hum Behav 93:25–32. https://doi.org/10.1016/j.chb.2018.11.048

Chen I-H, Pakpour AH, Leung H et al (2020) Comparing generalized and specific problematic smartphone/internet use: longitudinal relationships between smartphone application-based addiction and social media addiction and psychological distress. J Behav Addict 9:410–419. https://doi.org/10.1556/2006.2020.00023

Chen J, Liang Y, Mai C et al (2016) General deficit in inhibitory control of excessive smartphone users: evidence from an event-related potential study. Front Psychol 7:511. https://doi.org/10.3389/fpsyg.2016.00511

Ching SM, Yee A, Ramachandran V et al (2015) Validation of a Malay version of the smartphone addiction scale among medical students in Malaysia. PLoS ONE 10:e0139337. https://doi.org/10.1371/journal.pone.0139337

Cho HY, Kim DJ, Park JW (2017) Stress and adult smartphone addiction: mediation by self-control, neuroticism, and extraversion. Stress Health 33:624–630. https://doi.org/10.1002/smi.2749

Choi H-S, Lee H-K, Ha J-C (2012) The influence of smartphone addiction on mental health, campus life and personal relations - Focusing on K university students. J Korean Data Inf Sci Soc 23:1005–1015. https://doi.org/10.7465/jkdi.2012.23.5.1005

Chun J-W, Choi J, Cho H et al (2018) Role of frontostriatal connectivity in adolescents with excessive smartphone use. Front Psychiatry 9:437. https://doi.org/10.3389/fpsyt.2018.00437

Chun J-W, Choi J, Kim J-Y et al (2017) Altered brain activity and the effect of personality traits in excessive smartphone use during facial emotion processing. Sci Rep 7:12156. https://doi.org/10.1038/s41598-017-08824-y

Contractor AA, Weiss NH, Tull MT, Elhai JD (2017) PTSD's relation with problematic smartphone use: mediating role of impulsivity. Comput Hum Behav 75:177–183. https://doi.org/10.1016/j.chb.2017.05.018

Costa PT, McCrae RR (1985) Hypochondriasis, neuroticism, and aging: when are somatic complaints unfounded? Am Psychol 40:19–28. https://doi.org/10.1037/0003-066X.40.1.19

Davis R (2001) A cognitive-behavioral model of pathological internet use. Comput Hum Behav 17:187–195

de-Sola J, Talledo H, Rodríguez de Fonseca F, Rubio G (2017) Prevalence of problematic cell phone use in an adult population in Spain as assessed by the Mobile Phone Problem Use Scale (MPPUS). PLOS ONE 12:e0181184. https://doi.org/10.1371/journal.pone.0181184

De-Sola Gutierrez J, Rodriguez de Fonseca F, Rubio G (2016) Cell-phone addiction: a review. Front Psychiatry 7:175. https://doi.org/10.3389/fpsyt.2016.00175

Duke E, Montag C (2017a) Smartphone addiction, daily interruptions and self-reported productivity. Addict Behav Rep 6:90–95. https://doi.org/10.1016/j.abrep.2017.07.002

Duke E, Montag C (2017b) Smartphone addiction and beyond: initial insights on an emerging research topic and its relationship to Internet addiction. In: Internet addiction. Springer, Berlin, pp 359–372

Elhai JD, Levine JC, Alghraibeh AM et al (2018a) Fear of missing out: testing relationships with negative affectivity, online social engagement, and problematic smartphone use. Comput Hum Behav 89:289–298. https://doi.org/10.1016/j.chb.2018.08.020

Elhai JD, Levine JC, O'Brien KD, Armour C (2018b) Distress tolerance and mindfulness mediate relations between depression and anxiety sensitivity with problematic smartphone use. Comput Hum Behav 84:477–484. https://doi.org/10.1016/j.chb.2018.03.026

Elhai JD, Tiamiyu M, Weeks J (2018c) Depression and social anxiety in relation to problematic smartphone use: the prominent role of rumination. Internet Res 28:315–332. https://doi.org/10.1108/IntR-01-2017-0019

Elhai JD, Tiamiyu MF, Weeks JW et al (2018d) Depression and emotion regulation predict objective smartphone use measured over one week. Personal Individ Differ 133:21–28. https://doi.org/10.1016/j.paid.2017.04.051

Elhai JD, Levine JC, Dvorak RD, Hall BJ (2016) Fear of missing out, need for touch, anxiety and depression are related to problematic smartphone use. Comput Hum Behav 63:509–516. https://doi.org/10.1016/j.chb.2016.05.079

Elhai JD, Rozgonjuk D, Yildirim C et al (2019a) Worry and anger are associated with latent classes of problematic smartphone use severity among college students. J Affect Disord 246:209–216. https://doi.org/10.1016/j.jad.2018.12.047

Elhai JD, Yang H, Montag C (2019b) Cognitive- and emotion-related dysfunctional coping processes: transdiagnostic mechanisms explaining depression and anxiety's relations with problematic smartphone use. Curr Addict Rep. https://doi.org/10.1007/s40429-019-00260-4

Elhai JD, Vasquez JK, Lustgarten SD et al (2017) Proneness to boredom mediates relationships between problematic smartphone use with depression and anxiety severity. Soc Sci Comput Rev. https://doi.org/10.1177/0894439317741087

Elhai JD, Yang H, Rozgonjuk D, Montag C (2020) Using machine learning to model problematic smartphone use severity: the significant role of fear of missing out. Addict Behav 103:106261. https://doi.org/10.1016/j.addbeh.2019.106261

Ellis DA, Davidson BI, Shaw H, Geyer K (2019) Do smartphone usage scales predict behavior? Int J Hum-Comput Stud 130:86–92. https://doi.org/10.1016/j.ijhcs.2019.05.004

Enez Darcin A, Kose S, Noyan CO et al (2016) Smartphone addiction and its relationship with social anxiety and loneliness. Behav Inf Technol 35:520–525. https://doi.org/10.1080/0144929x.2016.1158319

Gao L, Zhai S, Xie H et al (2020) Big five personality traits and problematic mobile phone use: a meta-analytic review. Curr Psychol. https://doi.org/10.1007/s12144-020-00817-x

Griffiths MD (2005) A 'components' model of addiction within a biopsychosocial framework. J Subst Use 10:191–197. https://doi.org/10.1080/14659890500114359

Griffiths MD (2000) Internet addiction-time to be taken seriously? Addict Res 8:413–418

Hadar A, Hadas I, Lazarovits A et al (2017) Answering the missed call: initial exploration of cognitive and electrophysiological changes associated with smartphone use and abuse. PLoS ONE 12:e0180094. https://doi.org/10.1371/journal.pone.0180094

Harris B, Regan T, Schueler J, Fields SA (2020) Problematic mobile phone and smartphone use scales: a systematic review. Front Psychol 11:672. https://doi.org/10.3389/fpsyg.2020.00672

Haug S, Castro RP, Kwon M et al (2015) Smartphone use and smartphone addiction among young people in Switzerland. J Behav Addict 4:299–307. https://doi.org/10.1556/2006.4.2015.037

Holte AJ, Ferraro FR (2020) True colors: grayscale setting reduces screen time in college students. Soc Sci J 1–17.https://doi.org/10.1080/03623319.2020.1737461

Horvath J, Mundinger C, Schmitgen MM et al (2020) Structural and functional correlates of smartphone addiction. Addict Behav 105:106334. https://doi.org/10.1016/j.addbeh.2020.106334

Hu Y, Long X, Lyu H et al (2017) Alterations in white matter integrity in young adults with smartphone dependence. Front Hum Neurosci 11:532. https://doi.org/10.3389/fnhum.2017.00532

Hussain Z, Griffiths MD, Sheffield D (2017) An investigation into problematic smartphone use: the role of narcissism, anxiety, and personality factors. J Behav Addict 6:378–386. https://doi.org/10.1556/2006.6.2017.052

Hwang K-H, Yoo Y-S, Cho O-H (2012) Smartphone overuse and upper extremity pain, anxiety, depression, and interpersonal relationships among college students. J Korea Contents Assoc 12:365–375. https://doi.org/10.5392/jkca.2012.12.10.365

Kardefelt-Winther D, Heeren A, Schimmenti A et al (2017) How can we conceptualize behavioural addiction without pathologizing common behaviours? addiction 112:1709–1715. https://doi.org/10.1111/add.13763

Kates AW, Wu H, Coryn CLS (2018) The effects of mobile phone use on academic performance: a meta-analysis. Comput Educ 127:107–112. https://doi.org/10.1016/j.compedu.2018.08.012

Kim DJ, Kim JY, Pyeon A (2016) PM306. Altered functional connectivity related smartphone overuse in adolescent. Int J Neuropsychopharmacol 19:9–9. https://doi.org/10.1093/ijnp/pyw041.306

King DL, Wölfling K, Potenza MN (2020) Taking gaming disorder treatment to the next level. JAMA Psychiat. https://doi.org/10.1001/jamapsychiatry.2020.1270

Kwon M, Kim DJ, Cho H, Yang S (2013a) The smartphone addiction scale: development and validation of a short version for adolescents. PLoS ONE 8:e83558. https://doi.org/10.1371/journal.pone.0083558

Kwon M, Lee JY, Won WY et al (2013b) Development and validation of a smartphone addiction scale (SAS). PLoS ONE 8:e56936. https://doi.org/10.1371/journal.pone.0056936

Lachmann B, Sindermann C, Sariyska RY et al (2018) The role of empathy and life satisfaction in internet and smartphone use disorder. Front Psychol 9:398. https://doi.org/10.3389/fpsyg.2018.00398

Lee C, Lee S-J (2017) Prevalence and predictors of smartphone addiction proneness among Korean adolescents. Child Youth Serv Rev 77:10–17. https://doi.org/10.1016/j.childyouth.2017.04.002

Lee D, Namkoong K, Lee J et al (2019) Lateral orbitofrontal gray matter abnormalities in subjects with problematic smartphone use. J Behav Addict 8:404–411. https://doi.org/10.1556/2006.8.2019.50

Lepp A, Barkley JE, Karpinski AC (2015) The relationship between cell phone use and academic performance in a sample of U.S. college students. Sage Open 5. https://doi.org/10.1177/2158244015573169

Loid K, Täht K, Rozgonjuk D (2020) Do pop-up notifications regarding smartphone use decrease screen time, phone checking behavior, and self-reported problematic smartphone use? Evidence from a two-month experimental study. Comput Hum Behav 102:22–30. https://doi.org/10.1016/j.chb.2019.08.007

Long J, Liu T-Q, Liao Y-H et al (2016) Prevalence and correlates of problematic smartphone use in a large random sample of Chinese undergraduates. BMC Psychiatry 16:408. https://doi.org/10.1186/s12888-016-1083-3

Lopez-Fernandez O (2017) Short version of the smartphone addiction scale adapted to Spanish and French: towards a cross-cultural research in problematic mobile phone use. Addict Behav 64:275–280. https://doi.org/10.1016/j.addbeh.2015.11.013

Lopez-Fernandez O, Honrubia-Serrano L, Freixa-Blanxart M, Gibson W (2014) Prevalence of problematic mobile phone use in British adolescents. Cyberpsychology Behav Soc Netw 17:91–98. https://doi.org/10.1089/cyber.2012.0260

Lowe-Calverley E, Pontes HM (2020) Challenging the concept of smartphone addiction: an empirical pilot study of smartphone usage patterns and psychological well-being. Cyberpsychology Behav Soc Netw Cyber 23(8):550-556. https://doi.org/10.1089/cyber.2019.0719

Luk TT, Wang MP, Shen C et al (2018) Short version of the smartphone addiction scale in Chinese adults: psychometric properties, sociodemographic, and health behavioral correlates. J Behav Addict 7:1157–1165. https://doi.org/10.1556/2006.7.2018.105

Marengo D, Sindermann C, Häckel D, et al (2020) Associations between the big five personality traits and smartphone use disorder: a meta-analysis. J Behav Addict (3):534–550. https://doi.org/10.1556/2006.2020.00069

Marengo D, Sariyska R, Schmitt HS, Messner EM, Baumeister H, Brand M, Kannen C, Montag C (2021) Exploring the associations between self-reported tendencies toward smartphone use disorder and objective recordings of smartphone, instant messaging, and social networking app usage: correlational study. J Med Internet Res 23(9):e27093. https://doi.org/10.2196/27093

Montag C (2018) Homo Digitalis. Springer Fachmedien Wiesbaden

Montag C (2019) The neuroscience of smartphone/social media usage and the growing need to include methods from 'psychoinformatics'. In: Davis F, Riedl R, vom Brocke J, Léger PM, Randolph A (eds) Information systems and neuroscience. Lecture notes in information systems and organisation. Springer, Berlin

Montag C, Diefenbach S (2018) Towards homo digitalis: important research issues for psychology and the neurosciences at the dawn of the internet of things and the digital society. Sustainability 10:415. https://doi.org/10.3390/su10020415

Montag C, Hegelich S (2020) Understanding detrimental aspects of social media use: will the real culprits please standup? Front Soc 94

Montag C, Rumpf HJ (2021) The potential of digital phenotyping and mobile sensing for psycho-diagnostics of internet use disorders. Curr Addic Rep 8(3):422–430

Montag C, Walla P (2016) Carpe diem instead of losing your social mind: beyond digital addiction and why we all suffer from digital overuse. Cogent Psychol 3.https://doi.org/10.1080/23311908.2016.1157281

Montag C, Bey K, Sha P, Li M, Chen YF, Liu WY, Zhu YK, Li CB, Markett S, Keiper J, Reuter M (2015a) Is it meaningful to distinguish between generalized and specific Internet addiction? Evidence from a cross–cultural study from Germany, Sweden, Taiwan and China. Asia–Pacific Psychiatry 7(1):20–26

Montag C, Blaszkiewicz K, Sariyska R et al (2015b) Smartphone usage in the 21st century: who is active on WhatsApp? BMC Res Notes 8:331. https://doi.org/10.1186/s13104-015-1280-z

Montag C, Markowetz A, Blaszkiewicz K et al (2017) Facebook usage on smartphones and gray matter volume of the nucleus accumbens. Behav Brain Res 329:221–228. https://doi.org/10.1016/j.bbr.2017.04.035

Montag C, Lachmann B, Herrlich M, Zweig K (2019a) Addictive features of social media/messenger platforms and freemium games against the background of psychological and economic theories. Int J Environ Res Public Health 16:2612. https://doi.org/10.3390/ijerph16142612

Montag C, Wegmann E, Sariyska R, et al (2019b) How to overcome taxonomical problems in the study of Internet use disorders and what to do with "smartphone addiction"? J Behav Addict 1–7. https://doi.org/10.1556/2006.8.2019b.59

Montag C, Elhai JD, Dagum P (2021a) Show me your smartphone… and then I will show you your brain structure and brain function. Hum Behav Emerg Technol 3(5):891–897. https://doi.org/10.1002/hbe2.272

Montag C, Elhai JD, Dagum P (2021b) On blurry boundaries when defining digital biomarkers: how much biology needs to be in a digital biomarker? Front Psychiatry 1690. https://doi.org/10.3389/fpsyt.2021.740292

Montag C, Hegelich S, Sindermann C, Rozgonjuk D, Marengo D, Elhai JD (2021c) On corporate responsibility when studying social media use and well-being. Trends Cogn Sci 25(4):268–270. https://doi.org/10.1016/j.tics.2021.01.002

Montag C, Schivinski B, Pontes HM (2021d) Is the proposed distinction of gaming disorder into a predominantly online vs. offline form meaningful? Empirical evidence from a large German speaking gamer sample. Addict Behav Rep 14:100391

Montag C, Sindermann C, Becker B, Panksepp J (2016) An affective neuroscience framework for the molecular study of internet addiction. Front Psychol 7:1906. https://doi.org/10.3389/fpsyg.2016.01906

Montag C, Zhao Z, Sindermann C et al (2018) Internet communication disorder and the structure of the human brain: initial insights on WeChat addiction. Sci Rep 8:2155. https://doi.org/10.1038/s41598-018-19904-y

Nikhita CS, Jadhav PR, Ajinkya (2015) Prevalence of mobile phone dependence in secondary school adolescents. J Clin Diagn Res.https://doi.org/10.7860/JCDR/2015/14396.6803

Pancani L, Preti E, Riva P (2019) The psychology of smartphone: the development of the smartphone impact scale (SIS). Assessment.https://doi.org/10.1177/1073191119831788

Panova T, Carbonell X (2018) Is smartphone addiction really an addiction? J Behav Addict 1–8.https://doi.org/10.1556/2006.7.2018.49

Park WK (2005) Mobile phone addiction. In: Mobile communications. Computer supported cooperative work. Springer, London

Peterka-Bonetta J, Sindermann C, Elhai JD, Montag C (2019) Personality associations with smartphone and internet use disorder: a comparison study including links to impulsivity and social anxiety. Front Public Health 7:127. https://doi.org/10.3389/fpubh.2019.00127

Randler C, Wolfgang L, Matt K et al (2016) Smartphone addiction proneness in relation to sleep and morningness–eveningness in German adolescents. J Behav Addict 5:465–473. https://doi.org/10.1556/2006.5.2016.056

Rozgonjuk D (2019) Problematic smartphone use: behavioral, psychopathological, dispositional, and educational correlates. University of Tartu Press, Tartu, Estonia

Rozgonjuk D, Elhai JD (2019) Emotion regulation in relation to smartphone use: process smartphone use mediates the association between expressive suppression and problematic smartphone use. Curr Psychol 1–10.https://doi.org/10.1007/s12144-019-00271-4

Rozgonjuk D, Rosenvald V, Janno S, Täht K (2016) Developing a shorter version of the Estonian Smartphone Addiction Proneness Scale (E-SAPS18). Cyberpsychology J Psychosoc Res Cyberspace 10.https://doi.org/10.5817/cp2016-4-4

Rozgonjuk D, Kattago M, Täht K (2018a) Social media use in lectures mediates the relationship between procrastination and problematic smartphone use. Comput Hum Behav 89:191–198. https://doi.org/10.1016/j.chb.2018.08.003

Rozgonjuk D, Levine JC, Hall BJ, Elhai JD (2018b) The association between problematic smartphone use, depression and anxiety symptom severity, and objectively measured smartphone use over one week. Comput Hum Behav 87:10–17. https://doi.org/10.1016/j.chb.2018.05.019

Rozgonjuk D, Saal K, Täht K (2018c) Problematic smartphone use, deep and surface approaches to learning, and social media use in lectures. Int J Environ Res Public Health 15.https://doi.org/10.3390/ijerph15010092

Rozgonjuk D, Elhai JD, Hall BJ (2019a) Studying psychopathology in relation to smartphone use. In: Baumeister H, Montag C (eds) Digital phenotyping and mobile sensing. Springer International Publishing, Cham, pp 185–199

Rozgonjuk D, Elhai JD, Täht K et al (2019b) Non-social smartphone use mediates the relationship between intolerance of uncertainty and problematic smartphone use: evidence from a repeated-measures study. Comput Hum Behav. https://doi.org/10.1016/j.chb.2019.02.013

Rozgonjuk D, Pruunsild P, Jürimäe K et al (2020a) Instagram use frequency is associated with problematic smartphone use, but not with depression and anxiety symptom severity. Mob Media Commun. https://doi.org/10.1177/2050157920910190

Rozgonjuk D, Sindermann C, Elhai JD, Montag C (2020b) Fear of missing out (FoMO) and social media's impact on daily-life and productivity at work: do WhatsApp, Facebook, Instagram and Snapchat use disorders mediate that association? Addict Behav 110:106487. https://doi.org/10.1016/j.addbeh.2020.106487

Rozgonjuk D, Sindermann C, Elhai JD et al (2020c) Associations between symptoms of problematic smartphone, Facebook, WhatsApp, and Instagram use: an item-level exploratory graph analysis perspective. J Behav Addict 9(3):686–697. https://doi.org/10.1556/2006.2020.00036

Rozgonjuk D, Elhai JD, Sapci O, Montag C (2021a) Discrepancies between self-reports and behavior: fear of missing out (fomo), self-reported problematic smartphone use severity, and objectively measured smartphone use. Digit Psychol 2(2):3–10. https://doi.org/10.24989/dp.v2i2.2002

Rozgonjuk D, Sindermann C, Elhai JD, Montag C (2021b) Comparing smartphone, WhatsApp, Facebook, Instagram, and Snapchat: which platform elicits the greatest use disorder symptoms? Cyberpsychol Behav Soc Network 24:129–134. https://doi.org/10.1089/cyber.2020.0156

Samaha M, Hawi NS (2016) Relationships among smartphone addiction, stress, academic performance, and satisfaction with life. Comput Hum Behav 57:321–325

Schmitgen MM, Horvath J, Mundinger C et al (2020) Neural correlates of cue reactivity in individuals with smartphone addiction. Addict Behav 108:106422. https://doi.org/10.1016/j.addbeh.2020.106422

Seo HS, Jeong E-K, Choi S et al (2020) Changes of neurotransmitters in youth with internet and smartphone addiction: a comparison with healthy controls and changes after cognitive behavioral therapy. Am J Neuroradiol 41:1293–1301. https://doi.org/10.3174/ajnr.A6632

Sha P, Sariyska R, Riedl R et al (2019) Linking internet communication and smartphone use disorder by taking a closer look at the Facebook and WhatsApp applications. Addict Behav Rep 9:100148. https://doi.org/10.1016/j.abrep.2018.100148

Smetaniuk P (2014) A preliminary investigation into the prevalence and prediction of problematic cell phone use. J Behav Addict 3:41–53. https://doi.org/10.1556/JBA.3.2014.004

Song I, LaRose R, Eastin MS, Lin CA (2004) Internet gratifications and internet addiction: on the uses and abuses of new media. Cyberpsychol Behav 7:384–394. https://doi.org/10.1089/cpb.2004.7.384

Spada MM (2014) An overview of problematic internet use. Addict Behav 39:3–6. https://doi.org/10.1016/j.addbeh.2013.09.007

Stafford TF, Stafford MR, Schkade LL (2004) Determining uses and gratifications for the internet. Decis Sci 35:259–288. https://doi.org/10.1111/j.00117315.2004.02524.x

Torres-Rodríguez A, Griffiths MD, Carbonell X (2018a) The treatment of internet gaming disorder: a brief overview of the PIPATIC program. Int J Ment Health Addict 16:1000–1015. https://doi.org/10.1007/s11469-017-9825-0

Torres-Rodríguez A, Griffiths MD, Carbonell X, Oberst U (2018b) Treatment efficacy of a specialized psychotherapy program for internet gaming disorder. J Behav Addict 7:939–952. https://doi.org/10.1556/2006.7.2018.111

Wilcockson TDW, Ellis DA, Shaw H (2018) Determining typical smartphone usage: what data do we need? Cyberpsychology Behav Soc Netw 21:395–398. https://doi.org/10.1089/cyber.2017.0652

Winskel H, Kim T-H, Kardash L, Belic I (2019) Smartphone use and study behavior: a Korean and Australian comparison. Heliyon 5:e02158. https://doi.org/10.1016/j.heliyon.2019.e02158

Wolniewicz CA, Rozgonjuk D, Elhai JD (2019) Boredom proneness and fear of missing out mediate relations between depression and anxiety with problematic smartphone use. Hum Behav Emerg Technol. https://doi.org/10.1002/hbe2.159

World Health Organization (2018) International classification of diseases and related health problems, 11th edn. World Health Organization, Geneva, Switzerland

Young KS (1998) Internet addiction: the emergence of a new clinical disorder. Cyberpsychol Behav 1:237–244. https://doi.org/10.1089/cpb.1998.1.237

Young KS (1996) Psychology of computer use: XL. Addictive use of the internet: a case that breaks the stereotype. Psychol Rep 79:899–902. https://doi.org/10.2466/pr0.1996.79.3.899

Yu F, Sariyska R, Lachmann B et al (2020) Convergent cross-sectional and longitudinal evidence for gaming-cue specific posterior parietal dysregulations in early stages of internet gaming disorder. Addict Biol. https://doi.org/10.1111/adb.12933

Zhou F, Montag C, Sariyska R et al (2019) Orbitofrontal gray matter deficits as marker of Internet gaming disorder: converging evidence from a cross-sectional and prospective longitudinal design: Orbitofrontal gray matter deficits as marker of Internet gaming disorder: prospective evidence. Addict Biol 24:100–109. https://doi.org/10.1111/adb.12570

Chapter 5
Internet Addiction

Halley M. Pontes, Jason Satel, and Almuth McDowall

Abstract This chapter reviews the current literature on internet addiction (IA) and provides a comprehensive summary regarding: (i) potential positive and negative effects of internet and technology use, (ii) main conceptual frameworks, (iii) biological bases, (iv) comorbidity factors, (v) prevalence rates, (vi) assessment methodologies, and (vii) treatment approaches. Although the current evidence suggests a relatively low prevalence rate of IA worldwide, and that several scholars remain doubtful about the validity and utility of IA as a clinical phenomenon, the existing evidence indicates that further research is required in order to facilitate greater understanding of this intricate issue and to tackle a range of challenges identified in the literature. Furthermore, the current scientific trend points toward the adoption of more specific terms that underscore the role of specific online activities in eliciting addictive usage, as opposed to the adoption of the broad and unspecific umbrella term IA.

Keywords Internet addiction · Problematic internet use · Compulsive internet use · Behavioral addiction · Treatment

5.1 Introduction

5.1.1 Background

In January 2021, nearly 60% of the global population (i.e., 4.66 billion individuals) were active internet users (Statista 2021). Furthermore, about 4.28 billion were unique mobile internet users while 4.14 billion were active social media users (Statista 2021). These figures indicate that the internet is a critical tool in today's society permeating all aspects of people's lives.

H. M. Pontes (✉) · A. McDowall
Department of Organizational Psychology, Birkbeck, University of London, London, UK
e-mail: contactme@halleypontes.com

J. Satel
School of Psychological Sciences, University of Tasmania, Launceston, TAS, Australia

H. M. Pontes (ed.), *Behavioral Addictions*, Studies in Neuroscience, Psychology and Behavioral Economics, https://doi.org/10.1007/978-3-031-04772-5_5

Through the use of the internet, people can easily communicate with each other regardless of their physical locations, search for information online, check their email, engage in synchronous or asynchronous communications, use social networking sites (SNSs), carry out work and/or study, play video games, manage their finances through online banking, order goods online, find romantic partners, watch movies, sports, and/or shows, etc. The internet provides many tools that can enhance our lives with beneficial effects at several levels including but not confined to greater reach and connectivity. However, due to its pervasiveness, scholars have become interested in investigating the potential positive and negative effects stemming from internet use at psychological, social, and biological level (Griffiths et al. 2016a; Kuss and Pontes 2019; Marin et al. 2021; Pontes et al. 2015).

5.1.2 *Internet Use: Positive Effects*

When it comes to the benefits people experience in the context of internet use, several studies have reported a many of the advantages brought about by internet use. The internet presents the potential to promote healthy lifestyles and to support coping and management strategies of comorbidities. In a recent nation-wide study conducted among older age groups in Germany, Quittschalle et al. (2020) found that internet use among older people was associated with higher quality of life and lower levels of depressive symptoms. Moreover, additional benefits for older users reported in the literature include improvements in social connectivity (Choi and DiNitto 2013), prevention of social isolation (Chen and Schulz 2016), and greater access to information about leisure activities (Näsi et al. 2012). For younger users, the current evidence suggests that the use of SNSs can lead to lower levels of depression (Wang et al. 2019) as well as positive educational, social, and psychological outcomes for students (Rayan et al. 2017). Finally, in terms of positive psychological well-being effects, internet adoption has been shown to boost life satisfaction (Lissitsa and Chachashvili-Bolotin 2016).

5.1.3 *Internet Use: Negative Effects*

Notwithstanding the many positive effects that internet use can produce, a large body of evidence has reported deleterious effects (further discussed in Sect. 5.3). One of the most commonly investigated negative effects stemming from internet use includes aggression, especially in the context of violent video games (Coyne and Stockdale 2020; Verheijen et al. 2021). The well-known issue at the heart of this chapter is related to addiction in the context of excessive or problematic internet use (PIU), also termed 'internet addiction' (IA) (Griffiths 1996; Young 1996), which refers to detrimental use of the internet with negative consequences to the users (see Sect. 5.1.4: *Conceptual frameworks* for a review of the current definitions). Throughout this chapter, the term IA will be adopted for the sake of consistency.

Although research on IA is rapidly evolving, it is not a new field of research. The first reports on IA were published over two decades ago. For instance, Griffiths (1996) argued that researchers should focus on the particular addictive nature of the internet by identifying the primary behaviors associated with such addiction (e.g., gaming, emailing, information seeking, pornography, socializing, etc.). In his seminal work, Griffiths (1996) suggested that the internet should be conceptualized as a medium in which the addictive process takes place. In that same year, Young (1996) published a case report of a 43 year old woman who exhibited symptoms of IA and proposed defining and assessing IA by adapting the criteria for substance dependence from the Diagnostic and Statistical Manual of Mental Disorders (4th ed.) DSM-IV (American Psychiatric Association 1994).

Although this chapter treats IA as an addictive disorder, we advocate a critical perspective as IA is not currently recognized officially as a mental health disorder by official medical bodies such as the American Psychiatric Association or the World Health Organization. In fact, many researchers have argued that IA is a misnomer and an inadequate construct that should be abandoned altogether due to current lack of knowledge about its nature (Starcevic 2013). For example, Starcevic and Aboujaoude (2016) argued that the role of internet as a medium (to fuel other online addictions) does not exist, despite the fact that the internet can play an important role in eliciting addictive behaviors.

However, it is possible that the concept of IA is too heterogeneous because it includes a wide range of different online behaviors (Starcevic and Aboujaoude 2016). This observation has lead to recommendations to replace the term IA with terms pertaining to specific online behaviors (e.g., gambling, SNS use, gaming, etc.). There is currently no consensus on the terminology that should be used to describe IA, with research referring to this phenomenon as 'problematic internet use' (Shapira et al. 2000), 'pathological internet use' (Suler 1999), 'compulsive internet use' (Greenfield 1999), and 'internet use disorder' (Geisel et al. 2013) even though they may not necessarily be synonyms (Starcevic and Billieux 2017). The next subsection (see Sect. 5.1.4: *Conceptual frameworks*) will provide an overview of the latest advances regarding the definition, theory, and conceptualization of IA.

5.1.4 Conceptual Frameworks

Although several definitions exist for IA, these have been extensively debated over the last 25 years. Nevertheless, all existing definitions and theoretical frameworks exploring this issue refer to IA as a behavioral addiction that is associated with serious functional and health-related impairments (Pontes et al. 2015). Early theoretical approaches suggested that IA was akin to pathological gambling, further defining it as an impulse-control disorder that does not involve the ingestion of psychoactive intoxicants (Young 1998b). Similarly, Kuss and Pontes (2019) defined

IA as a behavioral pattern of internet use marked by dysfunctional craving underpinning unregulated and excessive usage that can lead to significant psychosocial and functional impairments not accounted for by any other disorder, these may include jeopardizing social, academic, and professional activities, and the development of comorbidities such as depression and axiety.

Alternate definitions have been suggested by other scholars. For example, IA has been defined in terms of poorly controlled cognitive preoccupation, urges, and behaviors related to excessive internet use leading to clinical impairment and distress (Weinstein et al. 2014). Block (2008) defined IA as a compulsive-impulsive disorder associated with excessive patterns of computer use, experience of withdrawal, tolerance, and other deleterious outcomes. Griffiths (1995) has also suggested that IA is a type of 'technological addiction', operationally defined as a nonchemical (behavioral) addiction involving excessive human–machine interactions. Within this framework, technological addictions (e.g., IA) include the experience of six key components that are common to all addictive behaviors which are salience, mood modification, tolerance, withdrawal, conflict, and relapse (Griffiths 2005).

In terms of theories and models that have been developed to explain the nature of IA, several important contributions have been made thus far. Below, two prominent conceptual frameworks will be explored, the early Cognitive-Behavioral Model (Davis 2001) and the more recent Interaction of Person-Affect-Cognition-Execution Model (Brand et al. 2019, 2016). We summarize three prevalent frameworks in Table 5.1.

5.1.4.1 The Cognitive-Behavioral Model

The Cognitive-Behavioral Model (Davis 2001) refers to IA as 'pathological internet use' and makes a key contribution by distinguishing between specific pathological internet use (SPIU) and generalized pathological internet use (GPIU). This important distinction emerged from the idea that the internet can serve different purposes. SPIU is defined as a type of pathological use of the internet whereby individuals become addicted to a specific online function/application. Conversely, GPIU is defined as a general, multidimensional behavioral pattern of pathological overuse. In this model, maladaptive cognitions play a major role in the development and maintenance of pathological use of the internet.

This theory adopts the concept of distal and proximal contributory causes to explain pathological internet use and illustrate its etiological process. Here, distal causes refer to existing psychiatric conditions (such as depression, social anxiety, etc.) and the behavioral reinforcement provided by the internet when individuals engage with new functionalities alongside the situational cues that contribute to conditioned responses. In contrast, proximal causes are related to maladaptive cognitions that are understood to be a sufficient condition for pathological use (in terms of both GPIU and SPIU) and the subsequent impairments that emerge due to pathological use of the internet (Davis 2001).

Table 5.1 Framework comparison

Model/conceptual framework	Definition of IA	Conceptual explanation	Symptoms	Framework for assessment and therapy
The cognitive-behavioral model	Pathological internet use, which is either specific, or generalized	Distal (existing conditions e.g., depression) and proximal contributory causes (maladaptive cognitions)	Mainly cognitive, including obsessive thoughts	Yes, more specific CBT therapy
The interaction of person-affect-cognition-execution model (I-PACE)	Focus on specific types of IA	Based on biological predisposition and affective/cognitive responses to stimuli, executive and inhibitory control as a process, and the role of conditioning	Varied, including coping responses, mood regulation etc.	Yes, focus on mediating and moderating factors to treat with pharmacological and psychological treatment
Biological bases	Neurobiology as investigated by neuroimaging	Explanations linked to decreased/altered gray matter, and increased activity in other brain areas	Impulsivity, craving, reductions in dopamine activity	Not explicitly evident, but points to pharmacological approaches

In this model, Davis (2001) proposed that GPIU is expressed at the behavioral level through the expenditure of excessive amounts of time using the internet with no direct purpose. This is why procrastination is thought to play an important role in the development and maintenance of GPIU. Interestingly, recent evidence supports a robust link between IA and procrastination within different contexts of internet use, even beyond generalized use (Aznar-Díaz et al. 2020; Geng et al. 2018; Müller et al. 2020; You et al. 2020).

Furthermore, this model argues that symptoms related to pathological internet use are primarily due to the experience of maladaptive cognitions (Davis 2001). This is because these symptoms are primarily experienced at the cognitive level, and encompass obsessive thoughts about internet use, reduced ability to control impulses, an inability to stop internet use, and the experience of a subjective and generalized feeling that the internet is the only place where the individuals feel good about themselves (Davis 2001). Additional symptoms can include thinking about the internet when not online, anticipating future time online, reduced interest in other previously enjoyed activities, and experiencing social isolation (Davis 2001).

In conclusion, the Cognitive-Behavioral Model provides a robust understanding of IA, including a framework for its assessment, despite being a theoretical model and not a comprehensive theory of IA in itself. This model is prominent due to the extensive support received from previous research (e.g., Caplan 2002, 2010) and its applicability to other internet-related problems such as Internet Gaming Disorder (IGD) (Haagsma et al. 2013). Additionally, this theory has inspired subsequent treatment approaches for IA (see Sect. 5.6*: Treatment Approaches*), including the cognitive behavior therapy for IA (CBT-IA) (Young 2011, 2013).

5.1.4.2 The Interaction of Person-Affect-Cognition-Execution Model (I-PACE)

Brand et al. (2016) developed the Interaction of Person-Affect-Cognition-Execution (I-PACE) model and recently revised it (see Brand et al. 2019). The I-PACE model highlights the process in which internet use disorders intertwine with person-affect-cognition-execution. This model expanded upon the Neuropsychology-Based Model of Internet Addiction (see Brand et al. 2014) by focusing on specific types of IA, while also conveying the idea that individuals have a first choice use, similarly to the idea of a first-choice drug in substance use disorders.

The I-PACE model aims to explain at the theoretical level, the main processes implicated in the development and maintenance of specific IA to certain contents and features people experience online (Brand et al. 2016). The model was devised according to the following components: predisposing factors, affective and cognitive responses to internal or external stimuli, executive and inhibitory control, decision-making behavior resulting in the use of specific online content, and consequences regarding the use of the internet and websites of choice.

When explaining the nature of IA, the I-PACE model can be thought of as a process model arguing that specific forms of IA are a consequence of the interactions between predisposing factors (e.g., neurobiological and psychological constitutions), moderators (e.g., individual coping styles and cognitive biases), and mediators (e.g., affective and cognitive responses to situational triggers in combination with reduced executive functioning). Moreover, this model suggests that conditioning processes can reinforce these associations as they occur within an addiction process (Brand et al. 2016).

The I-PACE model includes key components illustrating this process. The first component (the P component) is related to the person's biopsychological makeup, psychopathological characteristics, personality, social cognitions, and internet use motives. The second component (the A and C components) pertain to the affective and cognitive responses individuals display in response to external or internal stimuli. These allude to an individual's coping ability, internet-related cognitive biases, cue reactivity and craving, urge for mood regulation, and attentional biases. The third component (the E component) relates to the person's executive functions, inhibitory control, and decisions to use certain specific online features or websites. These denote impaired executive functions, inhibitory control, and decision making (Brand et al. 2016).

According to Brand et al. (2016), the decision to engage with specific online content or websites may lead to short-term positive experiences and gratifications, particularly in the early stages of the addiction process. The I-PACE model also suggests that the use of specific online content or websites, alongside the gratification experienced leads to greater cue reactivity and craving responses to certain stimuli as a result of both Pavlovian and instrumental-conditioning processes (Brand et al. 2016).

The I-PACE model proposes that some predisposing factors of IA are not plastic (e.g., genetic factors, early childhood experiences), whereas others may be difficult to change (e.g., psychopathological vulnerability factors, personality). One of the main implications of this model is that treatment approaches for IA should focus on addressing the moderating and mediating factors that can be targeted and modified across pharmacological and psychological treatments. As a result of the main propositions of this model, when treating individuals presenting with IA, consideration of key predisposing factors and systematic assessment of cognitive functions (e.g., attentional biases, implicit and explicit cognitions, executive functions, and inhibitory control capacities) should be prioritized.

Our current understanding of the nature of IA can thus be derived from different conceptual standpoints. Although two prominent models have been considered in this chapter, these are not the only theoretical frameworks developed so far. Therefore, interested readers are encouraged to expand their knowledge of the theoretical processes involved in IA through the examination of other important conceptual frameworks such as the Syndrome Model of Addiction (Shaffer et al. 2004), the Components Model of Addiction (Griffiths 2005), and/or the Neuropsychology-Based Model (Brand et al. 2014).

5.2 Biological Bases

Over the course of the past decade, researchers have begun unravelling the neurobiology of behavioral addictions, particularly in regard to IA (see Cerniglia et al. 2017; Park et al. 2017; Sharifat et al. 2018; Tereshchenko and Kasparov 2019). This section will briefly discuss some the key findings regarding the neurobiology of IA.

In the investigation of the biological underpinnings of IA, scholars have utilized a wide range of neuroimaging techniques in order to identify regions and processes in the human brain implicated in IA. Some of the imaging techniques researchers have adopted with high spatial resolution include structural magnetic resonance imaging (MRI) (e.g., voxel-based morphometry [VBM], diffusion tensor imaging [DTI]), functional magnetic resonance imaging (fMRI), and nuclear imaging (e.g., positron emission tomography [PET] and single photon emission computed tomography [SPECT]). Further studies have also been conducted using electroencephalography (EEG) to enhance our knowledge of the neurobiology of the temporal dynamics of neural activity underlying IA. For the purpose of conciseness, this section will not review EEG findings (see Burleigh et al. 2020 for a recent review).

When we consider the existing neurobiological evidence concerning IA, several important conclusions can be made. In terms of MRI findings, the current literature suggests that IA is commonly linked to decreased gray matter in a number of cortical regions (Tereshchenko and Kasparov 2019). Through the use of VBM, researchers have found that adolescents with IA present with lower gray matter density in the anterior cingulate cortex (ACC), left insula, left lingual gyrus, and left posterior cingulate cortex (PCC) (Zhou et al. 2011). Additional evidence has shown decreased gray matter density in the right orbitofrontal cortex (OFC), right supplementary motor area, and bilateral insula in patients diagnosed with IA (Weng et al. 2013). Another study of adolescents presenting with IA similarly observed reductions in gray matter density in the bilateral dorsolateral prefrontal cortex (DLPFC), OFC, supplementary motor area, cerebellum, and left rostral ACC (Yuan et al. 2011). Although these studies reported relatively similar results regarding decreased gray matter density among individuals with IA, the specific brain regions showing reduced gray matter were not entirely consistent across these studies, potentially reflecting methodological discrepancies related to assessment and recruitment of participants.

It is worth noting that several brain regions that have been reported to be altered in individuals with IA have been previously associated with functions related to the development of other addictive behaviors. For instance, damage to the prefrontal cortex (PFC) is commonly observed in addictive behaviors, and among IA individuals, gray matter atrophy of the PFC has been associated with loss of control related to internet use, which is a key characteristic of IA (Park et al. 2017). Interestingly, the OFC regulates impulse control and decision-making, while the dorsolateral PFC and rostral ACC are typically responsible for cognitive control (Krawczyk 2002).

When examining the neurobiology of IA, researchers have also employed DTI paradigms to quantify the status of white matter tracts via fractional anisotropy (FA), which measures the diffusivity of water molecules within the brain (Park et al. 2017).

Accordingly, previous research has found that FA was increased in the thalamus and left PCC of individuals presenting with specific types of IA (i.e., IGD) in comparison to healthy controls (Dong et al. 2012a). Further research has demonstrated that FA of IA individuals was lower in a number of brain regions (e.g., the orbitofrontal white matter, corpus callosum), and that no brain region presented with higher FA levels when compared to controls (Lin et al. 2012). Yuan et al. (2011) also found that FA of white matter in the right parahippocampal gyrus was decreased, and that FA was increased in the left posterior limb of the internal capsule among IA individuals.

In a more recent study, Cheng and Liu (2020) used resting-state fMRI and DTI techniques to investigate the functional and structural connectivity of the amygdala in individuals with IA. This study found that individuals with IA presented with decreased negative functional connectivity (FC) between the amygdala and the DLPFC, and increased negative FC between the amygdala and precuneus and superior occipital gyrus (SOG). Furthermore, Cheng and Liu (2020) reported that IA individuals in their study showed decreased positive FC between the amygdala and ACC and had increased positive FC between the amygdala and thalamus. Interestingly, the authors also reported that the duration of IA was linked with the FC between the left amygdala and right DLPFC. Overall, Cheng and Liu (2020) concluded that amygdala connectivity was altered in IA individuals, and that altered FC between the amygdala and DLPFC was correlated with the duration of IA. Taken together, this emerging body of evidence suggests that IA can lead to significant neurobiological impairments in both white and gray matter, and that observed changes to these structures may be associated with greater risk of developing IA. Nevertheless, most of the existing evidence is not causal in nature and therefore caution when interpreting such results is warranted.

Studies using fMRI techniques have led to the suggestion that increased activity in brain areas such as PFC and ACC is associated with impulsivity and craving responses in IA individuals (Park et al. 2017). For example, using an event-related fMRI Stroop color-word paradigm to assess inhibitory control, Dong et al. (2012b) found that male IA individuals exhibited greater blood-oxygen level-dependent (BOLD) signaling in the ACC and dorsal PCC in the incongruent Stroop trials, when compared to a control group (Dong et al. 2012b).

Current evidence also suggests that increased dorsal PCC activity is associated with incomplete disengagement of the default mode network, and impairment in the optimization of tasks related to attentional resources among IA individuals (Park et al. 2017). In another fMRI study, Dong et al. (2011) examined the differences in reward and punishment processing among male individuals presenting with IA as compared to healthy controls. In this study, participants presenting with IA showed significant increases in OFC activity in gain trials, and decreases in ACC activity in loss trials (Dong et al. 2011). This finding suggests that there is an increase in reward sensitivity and a decrease in loss sensitivity associated with IA since the OFC has been previously shown to be activated by reward (Gallagher et al. 1999), while the ACC is activated by losses (Petrovic et al. 2008).

Researchers have also conducted nuclear imaging research to further understand IA. Particular attention has been paid to research employing both SPECT and PET nuclear imaging techniques. Overall, the current evidence stemming from PET research supports the occurrence of impairments within the dopaminergic system in the brains of individuals experiencing IA (Park et al. 2017). In one key PET study (Kim et al. 2011), researchers assessed the D2 dopamine receptor availability of individuals presenting with IA using the radioligand 11C-raclopride. Results showed that participants with IA presented with reduced dopamine D2 receptor availability in the bilateral caudate and left putamen in comparison to participants in the control group, suggesting that reductions in dopamine activity are associated with IA (Kim et al. 2011). Interestingly, hypodopaminergic function (i.e., reduced dopamine D2 receptor) has been implicated in obesity in human and animals, and reward deficiency leading to increased reward seeking behaviors (e.g., compulsive eating) to compensate for the diminished dopamine activity (Beeler et al. 2016).

The dopamine transporter is a plasma membrane protein that is involved in the transport of dopamine from extracellular space into presynaptic neurons (Torres et al. 2003). Interestingly, in one study (Hou et al. 2012), researchers assessed the levels of dopamine transporter in individuals experiencing IA using 99mTc-TRODAT-1SPECT and reported that levels of striatal dopamine transporters decreased among those experiencing IA as compared to non-IA individuals. In a similar vein, comparable findings have been widely reported among individuals experiencing substance use disorders (Verma 2015; Zahniser and Sorkin 2004). In summary, the current evidence suggests that IA may lead to impairments in the dopaminergic systems in the brain indicating evidence for long term consequences for the nervous system, further impacting behaviors, mental health, and well-being.

Analysis of the current evidence suggests that IA is associated with both structural of functional impairments in the OFC, dorsolateral PFC, ACC, and PCC regions of the brain. These areas are implicated in many higher-level functions, such as the processing of reward, motivation, memory, and cognitive control abilities (Park et al. 2017). Furthermore, IA has also been associated with impairments of dopamine D2 receptor function, which is related to dysregulation of the OFC. Since these findings share important commonalities with those from the substance use disorder field, it is likely that IA and substance use disorders share specific underlying neurobiological mechanisms, although these may vary according to the behavior and substance being investigated (Park et al. 2017).

5.3 Comorbidities

For many years, researchers have been particularly interested in expanding our understanding about how IA may co-occur with other mental health disorders. The current literature supports consistent associations between IA and a wide range of psychiatric disorders, including but not limited to mood disorders (e.g., major depressive disorder, bipolar disorder), social anxiety, sleep but also neurodiverse conditions including attention-deficit/hyperactivity disorder (ADHD) disorders and autism spectrum disorder (ASD) (Carli et al. 2012; Karaca et al. 2017; Pluhar et al. 2019).

5.3.1 *Why Does IA Co-Occur with Other Psychopathologies?*

When we consider the dynamic relationship between IA and other comorbidities, previous theoretical models such as the I-PACE model (Brand et al. 2019, 2016) and the Cognitive-Behavioral Model (Davis 2001) suggested that existing psychopathologies and dysfunctional personality traits are key factors leading to the development of IA. Furthermore, Kuss and Pontes (2019) posited that the relationship between IA and other psychiatric disorders such as depression is bi-directional. In the case of mood disorders such as depression, this bi-directional association occurs as depressed individuals are more likely to become addicted to the internet because excessive internet use helps them manage daily stressors (e.g., low mood) (Morita et al. 2021).

This bi-directional process is justified by the fact that one of the key characteristics of behavioral addictions such as IA relates to its mood enhancing properties through the well-known mood modification capabilities of internet use (Griffiths 2005). Moreover, the deleterious effects, antecedents, and negative outcomes typically associated with IA (e.g., social isolation, loneliness, interpersonal conflict) (see Mamun et al. 2020; Tian et al. 2020; Zhang et al. 2017; Zhou et al. 2017) can further trigger and exacerbate ongoing depressive symptoms suggested by the social displacement hypothesis (Bessière et al. 2008; Kraut et al. 1998) which holds that time spent on the internet is to the detriment of high quality relationships.

5.3.2 *Mood Disorders in IA*

The existent literature suggests that individuals diagnosed with IA experience greater levels of depression when compared to non-IA individuals, with almost one in three individuals experiencing major depressive disorder also presenting with IA (Alpaslan et al. 2016). Furthermore, a study by Wölfling et al. (2015) reported that the overall prevalence of bipolar disorders among IA individuals was about 5.6%, with about 30.9% of those experiencing IA also having bipolar spectrum disorder diagnosis.

5.3.3 *Neurodiversity in IA*

A range of conditions such as autism, dyslexia, attention deficit hyperactivity disorder and Tourette's syndrome are now increasingly referred to as neurodiversity which was a movement originated by Judy Singer to frame human experience as a biological spectrum rather than collection of disorders. Our understanding of relevant conditions continues to advance, and in particular to two conditions linked to IA. ADHD was originally thought to be affecting young boys in particular, and refer to a range of difficulties linked to attention and impulsivity and has high prevalence rates. For example, ADHD affects over 10% of all children and adolescents in Australia and over a third of all these individuals will retain symptoms into adulthood (Visser et al. 2014). Additionally, ADHD has been reported to affect between 7.47% of children and adolescents in Africa (Ayano et al. 2020) and 6.26% of youth in China (Wang et al. 2017b). This condition is now recognized as a lifespan condition (Matheson et al. 2013) where it is common for ADHD-ers[1] to demonstrate hyperfocus on particular interests. Perhaps not surprisingly, previous research has also shown important links between IA and ADHD.

In a review of the relationship between IA and ADHD, Karaca et al. (2017) reported that the prevalence of ADHD among those experiencing IA ranged between 26.8 and 83.3%. Conversely, Karaca et al. (2017) also found that the co-occurrence of IA among ADHD-ers ranged from 15.7% to 71.8%.

Wang et al. (2017a) conducted a meta-analytic study reviewing the evidence from a total of 15 empirical studies investigating the relationship between IA and ADHD and found that IA was moderately associated with ADHD, and that those with IA were more likely to experience severe levels of ADHD (with high rates of inattention and impulsivity). In a more recent longitudinal study investigating the causal pathway between ADHD and IA, Zhou et al. (2020) found that ADHD levels measured at time 1 predicted IA levels at time 2 but not vice versa. Furthermore, the authors reported that those diagnosed with ADHD were more likely to experience IA than controls, suggesting that ADHD is a key risk factor for IA (Zhou et al. 2020). In summary, the present evidence provides a robust indication that IA and ADHD are highly comorbid.

5.3.4 ASD in IA

Similarly to ADHD, ASD is also associated with focus on particular interests, as well as restrictions in social communication and repetitive behaviors. Research on ASD and IA was prompted by early studies on internet use reporting that socially isolated individuals present with high levels of internet use (e.g., Sanders et al. 2000). In one of the first studies investigating this relationship, Romano et al. (2014) reported a correlation between the presence of autistic traits, IA and anxiety, where the relationship between IA and autistic traits is moderated by anxiety—where those with high anxiety use the internet less. Given the study design, further research is needed to disentangle any causal relationships however.

Research exploring the associations between ASD and IA has been conducted in children, adolescents, and adult populations both in terms of generalized and specific IA. More recently, So et al. (2019) recruited a clinical sample and conducted a two-year longitudinal study investigating the relationship between IA, ASD, and ADHD among Japanese adolescents that had been clinically diagnosed. According to the results, the prevalence of IA at baseline was about 8.9% ($n = 5$) among participants

[1] We deploy 'identity first language' consistent with current preferences of those affected.

classed as ASD only, 6.7% (n = 1) among participants classed as ADHD only, and about 22.22% (n = 4) among participants classed as both ASD and ADHD (So et al. 2019).

The relationship between ASD and IA has also been supported by clinical data. Engelhardt et al. (2017) conducted a study on a sample of 119 individuals with and without ASD to investigate the relationship between IGD and ASD. The authors found that participants with ASD spent significantly more time playing video games than non-ASD participants, and that the former group showed greater levels of IGD when compared to the latter group. These findings led Engelhardt et al. (2017) to conclude that the risk of specific IA related to online gaming is greater in adults with ASD than in non-ASD adults. Additional research is needed to investigate long-term effects of prolonged and sustained internet use and whether any effects are different to other sustained interests also present in ASD adults.

Further evidence exists supporting the link between IA and ASD, particularly among younger internet users. In one study, Kawabe et al. (2019) recruited 55 Japanese children and adolescents that had been diagnosed with ASD. Based on the clinical and psychometric assessments performed, the authors found that IA was prevalent in nearly half of all ASD participants (45%, n = 25). Kawabe et al. (2019) also concluded that hyperactivity symptoms in ASD adolescents might be an important factor explaining the onset of IA. Therefore, the evidence regarding co-occurrence for ASD generalized and specific IA in adolescents and adults, especially among those presenting with IGD warrants future investigation.

5.3.5 Social Anxiety in IA

The link between IA and social anxiety, which comprises fears about meeting other people and their judgment, has also been researched as developmental stages are known to play an important role in both social anxiety and internet use (Prizant-Passal et al. 2016). Notwithstanding this, the current evidence suggests that the relationship between IA and social anxiety is complex. In their review study, Prizant-Passal et al. (2016) found that although social anxiety was not associated with email use or instant messaging, it was linked with online gaming. Moreover, Prizant-Passal et al. (2016) found a significant association between IA and social anxiety, suggesting that social anxiety may be particularly prominent among high-severity IA individuals.

In a longitudinal study investigating a total of 2,293 adolescents, Ko et al. (2009) found that depression, ADHD, social anxiety, and hostility predicted IA in a two-year follow-up period, with hostility and ADHD being the most significant predictors of IA. Furthermore, in a large-scale study involving a sample of 1,460 students aged between 11 and 15 years, Yayan et al. (2016) found that about 13.7% of their sample presented with IA, and that there was a positive association between IA and social anxiety. Moreover, the authors noted that although IA was mostly associated with online gaming, dating, and browsing the internet, social anxiety was specifically related to homework, online gaming, and browsing the internet (Yayan et al. 2016).

Finally, Peterka-Bonetta et al. (2019) found that social anxiety and impulsivity also presented positive associations with higher levels of IA and substance use disorder, further cementing the current body of evidence supporting the relationship between IA and social anxiety across different age groups.

5.3.6 Sleep Disorders in IA

Another important phenomenon co-occurring with IA is related to sleep disorders. The relationship between IA and sleep disorders is multidimensional in nature, with the current evidence suggesting that IA may be implicated in different types of sleep disorders since both frequent internet use (Li et al. 2010) and excessive exposure to blue light electronic devices can negatively affect sleep due to its interferences with melatonin secretion (Chellappa et al. 2013). Previous research has demonstrated important links between IA and sleep disturbances such as short sleep duration (Guo et al. 2018), reduced sleep quality (AlAmer et al. 2020), insomnia (Tsumura et al. 2018), increased fatigue (Bener 2017), and daytime drowsiness (Alimoradi et al. 2019).

In a recent meta-analysis, Alimoradi et al. (2019) reviewed a total of 23 empirical studies examining the association between IA and sleep disturbances (i.e., sleep problems and duration). The authors found that individuals with IA were approximately 2.20 times more likely to experience sleep problems when compared to non-IA individuals. Additionally, Alimoradi et al. (2019) found that IA individuals presented with reduced overall sleep duration when compared to those without IA (overall pooled standard mean difference = −0.24 h). In a similar vein, with a sample of 1976 adolescents Wang et al. (2021) found that those with IA presented with significantly higher risk of developing behavioral and emotional problems than those without IA, with sleep disorders partially mediating this effect.

Further epidemiological evidence from review studies have shown a positive association between IA and specific sleep disorders, including insomnia, short sleep duration, and suboptimal sleeping quality (e.g., Lam 2014). Additionally, a recent study conducted among 4750 school-based adolescents found that IA was associated with greater risk of sleep disturbance with older adolescents having higher risk of experiencing sleep disturbances (Yang et al. 2018).

In terms of the association between IA and insomnia, several studies have been conducted using different samples. In a study by Cheung and Wong (2011), the authors found that about 51.7% of adolescents with IA were also identified as insomniacs and about 58.9% were depressed. The authors also found that both insomnia and IA were significantly associated with depressive symptoms, implying that a complex underlying mechanism exists between insomnia, IA, and depression (Cheung and Wong 2011). Similarly, another study reported that adolescents presenting with IA were significantly more likely to exhibit higher levels of insomnia, stress, anxiety, depression, and low self-esteem (Younes et al. 2016).

In addition to the studies discussed above, recent research has shown that IA affects other sleep-related factors that are important to maintaining adequate sleep health. For example, in a sample of emerging adults, Jahan et al. (2019) found that IA was associated with poor sleep quality. Furthermore, Jahan et al. (2019) found that those experiencing moderate and severe levels of IA were 75% and 95% less likely to have good sleep quality, respectively. Similarly, Karimy et al. (2020) found that young adults with higher levels of IA presented with poorer sleep quality compared to those without IA. Taken together, the current evidence suggests important links between IA and sleep disorders through excessive and dysregulated internet use.

5.4 Prevalence Rates

Although IA is not yet an officially recognized mental health disorder, several epidemiological studies have been conducted to estimate the different levels of prevalence rates of IA across many cultural contexts. This line of research is paramount to helping estimate the current demand for consulting, treatments, and preventative measures since a large number of empirical reports have found that IA affects a small minority of internet users (Griffiths et al. 2016b; Pontes et al. 2015).

Several review studies have been conducted to clarify the extent of problems caused by IA in different parts of the globe. In one of the most recent review and meta-analytic studies conducted to date examining the epidemiological evidence on IA from 113 studies (published from 1996 to 2018) that included a total of 693,306 individuals from 31 countries, the authors found a prevalence rate of 7.02% for IA and 2.47% for IGD (Pan et al. 2020). Furthermore, among healthcare professionals, a review study conducted by Buneviciene and Bunevicius (2020) found a pooled prevalence rate of IA of 9.7% based on a sample of 1,818 healthcare professionals, with IA being strongly associated with greater mental health symptoms and fatigue among healthcare workers.

In an earlier systematic review study conducted by Kuss et al. (2014) to assess the evidence from a total of 68 epidemiological studies on IA (published after the year of 2000) that included at least 1,000 individuals, the authors found prevalence rates of IA ranging from 0.8% in Italy to 26.7% in Hong Kong. Furthermore, the authors also noted that no gold standard of IA classification was found, with 21 different assessment tools used to assess the construct in these studies (Kuss et al. 2014). Moreover, Kuss et al. (2014) reported that these tools were based on the substance use disorder and/or pathological gambling clinical criteria, and that the majority of them had no (or few) relevant criteria for behavioral addiction diagnosis, time spent on the internet, or functional impairment.

In a meta-analytic study conducted by Cheng and Li (2014) assessing the overall prevalence of IA based on a total of 80 eligible epidemiological studies conducted across 31 nations from seven world regions, a worldwide prevalence rate of IA of 6%, with the highest rates observed in the Middle East (10.9%) and lowest rates in

Northern and Western Europe (2.6%). Interestingly, Cheng and Li (2014) also found that decreased quality of life was linked to higher prevalence rates of IA.

In another similar study, Pontes et al. (2015) reviewed the prevalence rates of IA reported from 12 robust empirical studies (published between 2014 to 2015) using nationally representative data from different countries. According to this review, prevalence rates of IA ranged from a minimum of 0% in Iran to a maximum of 18.7% in Taiwan (Pontes et al. 2015). Although all these studies were culturally diverse, all studies presented with several limitations as they adopted a cross-sectional design, had substantial heterogeneity in the way assessment of IA was conducted, and showed arbitrariness in terms of the cut-off points used to estimate prevalence rates of IA, even when studies adopted the same assessment tool for IA (Pontes et al. 2015).

Further epidemiological research has been conducted to assess the prevalence rates of IA in different countries using nationally representative data. Among adolescents, an overall prevalence rate of IA of 26.5% has been reported by Xin et al. (2018), with severe addiction levels affecting about 0.96% of individuals (based on a large nationally representative sample of Chinese adolescents [N = 6,648, age range = 10–18 years]). In Korea, Kim et al. (2020) found an overall prevalence rate of IA of 5.2% (7.7% among males and 3.8% among females) based on 22,542 adolescents (age range = 12–18 years). In Slovenia, (Macur et al. 2016) IA has been reported to affect 3.1% of the adult population and 20% of adolescents in the 8th grade (Pontes and Macur 2021). Similarly, Lewczuk et al. (2020) found, in a nationally representative sample of Polish adults (N = 1,036, age range = 18–69 years), that the prevalence of self-perceived IA (i.e., not assessed through a standardized diagnostic tool) was 23% and 4.2% for social media addiction.

In summary, based on the current epidemiological evidence on IA, it can be concluded that robust studies using large and nationally representative samples assessing IA on different cohorts are still scarce. However, this type of research is important to broaden our current understanding of how IA may impact different populations. Furthermore, discrepancies in the prevalence rates observed in previous research seem to be primarily due to research limitations (e.g., assessment strategy, study design, sampling technique) (Pontes et al. 2015).

5.5 Assessment Approaches

In terms of potential assessment approaches, several psychometric and clinical tools have been developed to assist the assessment of IA (Pontes et al. 2015). Traditionally, a wide range of assessment tools have been employed to assess IA, and this section will cover some of the main assessment tools that interested researchers and clinicians can adopt in their daily work.

One of the first assessment tools used in empirical research to assess and diagnose IA is the eight-item Diagnostic Questionnaire (DQ) (also known as Young's Diagnostic Questionnaire [YDQ]) (Young 1998b), which uses a dichotomous scoring system (i.e., 'yes' or 'no'). This tool was adapted from the criteria for pathological gambling as defined in the DSM-IV (American Psychiatric Association 1994). An example of an item includes: "*Do you feel restless, moody, depressed, or irritable when attempting to cut down or stop internet use?*" (see Young 1998b for a list of all items). In terms of diagnostic capabilities, this tool specifies that responding 'yes' to at least five out of the eight questions indicates IA.

The DQ represents an important early effort towards facilitating assessment of IA and inspired follow-up research. However, there is very little empirical and clinical evidence supporting the suggested cutoff points initially proposed. For example, Wartberg et al. (2017) found that the DQ presented with low levels of reliability, and concluded that assessing IA with the DQ may not be accurate as measurements across time may differ. Despite this, parental assessment of IA (especially in children and adolescent samples) is also possible using the DQ since Wartberg et al. (2016) developed a parental version of this tool with a nationally representative sample from Germany (N = 1,000 parents of adolescents aged 12–17 years). The parental version of the DQ adopts the same diagnostic principles of the original DQ, but all eight items must be responded to by parents or caretakers. All relevant information (e.g., items, scoring) about the parental DQ can be found within its original study (see Wartberg et al. 2016).

The most popular assessment tool for IA is the Internet Addiction Test (IAT) (Young 1998a). The IAT contains a total of 20 items measuring characteristics and behaviors associated with compulsive use of the internet and its impairments. Once completed, a score ranging from 0 to 100 can be obtained, with higher scores indicating greater severity of IA, and the following cutoff points being proposed: 'none' (0–30), 'mild' (31–49), 'moderate' (50–79), and 'severe' (80–100) (Young 1998a). The IAT has received substantial cross-cultural scrutiny, and has been developed in several languages, such as Pakistani (Waqas et al. 2017), Portuguese (Pontes et al. 2014), Turkish (Kaya et al. 2016), Polish (Hawi et al. 2015), and other languages.

The psychometric properties of the IAT have been well documented in the literature, through several studies. More specifically, a systematic review and meta-analysis conducted by Moon et al. (2018) reviewing the evidence from 25 studies using the IAT, concluded that the test presented with excellent levels of internal consistency (Pooled Cronbach's alpha ranging from 0.83 to 0.93), with the authors suggesting that the IAT has acceptable test–retest reliability, and convergent validity in specific samples. However, its clinical validity has been questioned by research reporting that the IAT presents with low diagnostic accuracy in clinical samples since it is only able to detect 42% of individuals who have already been clinically diagnosed with IA (Kim et al. 2013).

More recent assessment approaches have focused on the development of assessment tools for IA based on the DSM-5 criteria for IGD (American Psychiatric Association 2013). One such example is the Internet Disorder Scale–Short Form (IDS9-SF) (Pontes and Griffiths 2016), which includes nine items that can be used to assess broad symptoms of IA. According to the authors, endorsement of at least five out of the nine items may indicate IA. The IDS9-SF has been developed in other languages, such as Italian (Soraci et al. 2020) and Bangla (Saiful Islam et al. 2020). Overall, these studies report adequate construct validity and satisfactory levels of internal consistency.

Although there are several psychometric tests available to assess IA, previous review studies have criticized some of the existing assessment tools. A review by Király et al. (2014) of the nine most used IA psychometric tools reported different types of limitations that underscore inconsistencies across the instruments in terms of (i) underlying theoretical bases, (ii) factor structures, and (iii) psychometric properties. Moreover, Lortie and Guitton (2013) reviewed a total of 14 IA assessment tools and reported that they varied considerably on several important aspects. Specifically, the following three dimensions were present across all tools: compulsive use of the internet, negative outcomes, and salience. However, less commonly assessed dimensions among these tools related to using the internet to relieve adverse moods and withdrawal symptoms (Lortie and Guitton 2013).

In summary, although the field has developed several assessment tools for IA that have been widely adopted across numerous countries, more progress needs to be achieved to establish a sound assessment framework for IA since current tools appear to lack clinical validity and there is no established gold standard for diagnosing IA (Kuss et al. 2014). This means that diagnosis and identification of IA is largely left to clinical judgment and therefore the respective practitioners' awareness and training.

5.6 Treatment Approaches

Despite the relatively low prevalence of IA, affected individuals often seek professional help for their internet-related problems, with some countries providing psychiatric facilities entirely dedicated to treating IA (Zajac et al. 2017). Therefore, evaluating the existing treatment approaches is important to provide evidence supporting specific treatments.

To date, treatment for IA has been mainly conducted through psychological and pharmacological approaches. However, regardless of the chosen treatment approach, complete abstinence from the internet should not be the goal of any specific intervention, instead, regulated use of the internet should be achieved (Pontes et al. 2015). In terms of treatments, a meta-analysis conducted by Winkler et al. (2013) concluded that in relation to the efficacy of treatments for IA, effect sizes were high, robust, and maintained over a follow-up period.

In terms of pharmacotherapy, treatment methods have adopted the use of antidepressants such as escitalopram (Dell'Osso et al. 2008) and bupropion (for IGD) (Bae et al. 2018), opioid receptor antagonists such as naltrexone (for internet sex addiction) (Bostwick and Bucci 2008), and antipsychotics such as quetiapine (combined with citalopram) (Atmaca 2007). Furthermore, psychotherapy-based approaches for IA have demonstrated promising results employing Cognitive Behavioral Therapy (CBT) over 15 (Wölfling et al. 2014) and 12 (Young 2013) sessions, family therapy over six (Liu et al. 2015) and 14 sessions (Zhong et al. 2011), and using positive psychology interventions over 10 sessions (Khazaei et al. 2017).

In a systematic review study conducted by Przepiorka et al. (2014) on the existing evidence for CBT and pharmacological treatments of IA, the authors recommended that therapists should adopt both treatment approaches since a combined strategy has been found to be the most effective method for treating IA.

To summarize, despite the relatively high amount of treatment studies on IA published to date, previous review studies have suggested a paucity of well-designed treatment research studies of IA (Kuss and Lopez-Fernandez 2016; Winkler et al. 2013). Moreover, pharmacological treatments for IA are still in development as very little is known about their efficacy. Despite the promising results so far, psychotherapy treatments studies have also been hindered by weak methodological standards, leading some authors to conclude that there are currently no treatments for IA that fully meet the criteria for being considered evidence-based or even a possibly efficacious intervention (Zajac et al. 2017).

5.7 Final Conclusions

The present chapter reviewed the current literature on IA and attempted to provide a comprehensive understanding of several aspects pertaining to this issue, namely its: potential positive and negative effects, main conceptual frameworks, biological bases, comorbidity factors, prevalence rates, assessment methodologies, and treatment approaches employed thus far.

Despite the large amount of research that has been conducted on IA, current approaches and theoretical understanding of this phenomenon suggest that the term IA in itself is problematic when used to understand the issue as a broad mental health condition. Instead, researchers seem to favor the adoption of more precise and specific terminology to describe different forms of IA linked to unique online activities, implying that IA is better conceptualized as a spectrum than a single entity (Starcevic and Billieux 2017).

Due to the existing conceptual and theoretical challenges hindering the recognition of IA as a *bona fide* addictive disorder, it is likely that official medical bodies (e.g., World Health Organization and American Psychiatric Association) will focus more on the future merits of specific forms of IA based on the addictive potential and subsequent functional impairments of specific online activities when revising their diagnostic manuals. Notwithstanding the current conceptual challenges and lack of medical recognition, we believe that prematurely abandoning IA may be may lead to serious unintended harmful psychosocial consequences. This is because the existing scientific conundrum should not be used to generate further stigma by invalidating peoples' own experiences of distress in relation to dysregulated digital technology use. There are groups of adolescents and adults who engage in internet use to an extent which is harmful to their wellbeing. The onus is on researchers to collaborate with practitioners, including clinicians, psychologists and therapists to devise accessible and evidence-based treatment approaches. This is pramount given also the increased vulnerability of individuals with mental health conditions and neurodiverse people.

References

AlAmer M, Shdaifat E, Alshowkan A, Eldeen AG, Jamama A (2020) Exploring associations between internet addiction, depressive symptoms, and sleep disturbance among Saudi nursing students the open. Nurs J 14:29–36. https://doi.org/10.2174/1874434602014010029

Alimoradi Z et al (2019) Internet addiction and sleep problems: a systematic review and meta-analysis. Sleep Med Rev 47:51–61. https://doi.org/10.1016/j.smrv.2019.06.004

Alpaslan AH, Soylu N, Kocak U, Guzel HI (2016) Problematic internet use was more common in Turkish adolescents with major depressive disorders than controls. Acta Paediatr 105:695–700. https://doi.org/10.1111/apa.13355

American Psychiatric Association (1994) Diagnostic and statistical manual of mental disorders, 4th edn. Author, Washington, DC

American Psychiatric Association (2013) Diagnostic and statistical manual of mental disorders (5th ed.). https://doi.org/10.1176/appi.books.9780890425596

Atmaca M (2007) A case of problematic internet use successfully treated with an SSRI-antipsychotic combination. Prog Neuropsychopharmacol Biol Psychiatry 31:961–962. https://doi.org/10.1016/j.pnpbp.2007.01.003

Ayano G, Yohannes K, Abraha M (2020) Epidemiology of attention-deficit/hyperactivity disorder (ADHD) in children and adolescents in Africa: a systematic review and meta-analysis. Ann General Psychiatry 19:21. https://doi.org/10.1186/s12991-020-00271-w

Aznar-Díaz I, Romero-Rodríguez JM, García-González A, Ramírez-Montoya MS (2020) Mexican and Spanish university students' internet addiction and academic procrastination: correlation and potential factors. PLoS ONE 15:e0233655. https://doi.org/10.1371/journal.pone.0233655

Bae S, Hong JS, Kim SM, Han DH (2018) Bupropion shows different effects on brain functional connectivity in patients with Internet-based gambling disorder and Internet gaming disorder. Front Psychiatry 9. doi:https://doi.org/10.3389/fpsyt.2018.00130

Beeler JA, Faust RP, Turkson S, Ye H, Zhuang X (2016) Low dopamine D_2 receptor increases vulnerability to obesity via reduced physical activity. Not Increased Appet Motivat Biol Psychiatry 79:887–897. https://doi.org/10.1016/j.biopsych.2015.07.009

Bener A (2017) Lifestyle factors and internet addiction among school children. Eur Psychiatry 41:S431–S431. https://doi.org/10.1016/j.eurpsy.2017.01.414

Bessière K, Kiesler S, Kraut R, Boneva BS (2008) Effects of internet use and social resources on changes in depression information. Commun Soc 11:47–70. https://doi.org/10.1080/13691180701858851

Block JJ (2008) Issues for DSM-V: internet addiction. Am J Psychiatry 165:306–307. https://doi.org/10.1176/appi.ajp.2007.07101556

Bostwick JM, Bucci JA (2008) Internet sex addiction treated with naltrexone. Mayo Clin Proc 83:226–230. https://doi.org/10.4065/83.2.226

Brand M, Wegmann E, Stark R, Müller A, Wölfling K, Robbins TW, Potenza MN (2019) The interaction of person-affect-cognition-execution (I-PACE) model for addictive behaviors: update, generalization to addictive behaviors beyond internet-use disorders, and specification of the process character of addictive behaviors. Neurosci Biobehav Rev 104:1–10. https://doi.org/10.1016/j.neubiorev.2019.06.032

Brand M, Young K, Laier C, Wölfling K, Potenza MN (2016) Integrating psychological and neurobiological considerations regarding the development and maintenance of specific Internet-use disorders: an interaction of person-affect-cognition-execution (I-PACE) model. Neurosci Biobehav Rev 71:252–266. https://doi.org/10.1016/j.neubiorev.2016.08.033

Brand M, Young KS, Laier C (2014) Prefrontal control and Internet addiction: a theoretical model and review of neuropsychological and neuroimaging findings. Front Hum Neurosci 8. https://doi.org/10.3389/fnhum.2014.00375

Buneviciene I, Bunevicius A (2020) Prevalence of internet addiction in healthcare professionals: systematic review and meta-analysis. Int J Soc Psychiatry. https://doi.org/10.1177/0020764020959093

Burleigh TL, Griffiths MD, Sumich A, Wang GY, Kuss DJ (2020) Gaming disorder and internet addiction: a systematic review of resting-state EEG studies. Addict Behav 107:106429. https://doi.org/10.1016/j.addbeh.2020.106429

Caplan SE (2002) Problematic internet use and psychosocial well-being: development of a theory-based cognitive–behavioral measurement instrument. Comput Hum Behav 18:553–575. https://doi.org/10.1016/S0747-5632(02)00004-3

Caplan SE (2010) Theory and measurement of generalized problematic Internet use: a two-step approach. Comput Hum Behav 26:1089–1097. https://doi.org/10.1016/j.chb.2010.03.012

Carli V et al (2012) The association between pathological Internet use and comorbid psychopathology: a systematic review. Psychopathology 46:1–13

Cerniglia L, Zoratto F, Cimino S, Laviola G, Ammaniti M, Adriani W (2017) Internet addiction in adolescence: neurobiological, psychosocial and clinical issues. Neurosci Biobehav Rev 76(Part A):174–184. https://doi.org/10.1016/j.neubiorev.2016.12.024

Chellappa SL, Steiner R, Oelhafen P, Lang D, Götz T, Krebs J, Cajochen C (2013) Acute exposure to evening blue-enriched light impacts on human sleep. J Sleep Res 22:573–580. https://doi.org/10.1111/jsr.12050

Chen YRR, Schulz PJ (2016) The effect of information communication technology interventions on reducing social isolation in the elderly: a systematic review. J Med Internet Res 18:e18. https://doi.org/10.2196/jmir.4596

Cheng C, Li AYl (2014) Internet addiction prevalence and quality of (real) life: a meta-analysis of 31 nations across seven world regions. Cyberpsychol Behav Soc Network 17:755–760. https://doi.org/10.1089/cyber.2014.0317

Cheng H, Liu J (2020) Alterations in amygdala connectivity in internet addiction disorder. Sci Rep 10:2370. https://doi.org/10.1038/s41598-020-59195-w

Cheung LM, Wong WS (2011) The effects of insomnia and internet addiction on depression in Hong Kong Chinese adolescents: an exploratory cross-sectional analysis. J Sleep Res 20:311–317. https://doi.org/10.1111/j.1365-2869.2010.00883.x

Choi N, DiNitto DM (2013) Internet use among older adults: association with health needs, psychological capital, and social capital. J Med Internet Res 15:e97. https://doi.org/10.2196/jmir.2333

Coyne SM, Stockdale L (2020) Growing up with grand theft auto: a 10-year study of longitudinal growth of violent video game play in adolescents cyberpsychology. Behav Soc Network 24:11–16. https://doi.org/10.1089/cyber.2020.0049

Davis RA (2001) A cognitive-behavioral model of pathological internet use. Comput Hum Behav 17:187–195. https://doi.org/10.1016/S0747-5632(00)00041-8

Dell'Osso B, Hadley S, Allen A, Baker B, Chaplin WF, Hollander E (2008) Escitalopram in the treatment of impulsive-compulsive internet usage disorder: an open-label trial followed by a double-blind discontinuation phase. J Clin Psychiatry 69:452–456. https://doi.org/10.4088/JCP.v69n0316

Dong G, DeVito E, Huang J, Du X (2012a) Diffusion tensor imaging reveals thalamus and posterior cingulate cortex abnormalities in internet gaming addicts. J Psychiatr Res 46:1212–1216

Dong G, DeVito EE, Du X, Cui Z (2012b) Impaired inhibitory control in 'Internet addiction disorder': a functional magnetic resonance imaging study. Psychiatry Res: Neuroimaging 203:153–158. https://doi.org/10.1016/j.pscychresns.2012.02.001

Dong G, Huang J, Du X (2011) Enhanced reward sensitivity and decreased loss sensitivity in Internet addicts: an fMRI study during a guessing task. J Psychiatr Res 45:1525–1529. https://doi.org/10.1016/j.jpsychires.2011.06.017

Engelhardt CR, Mazurek MO, Hilgard J (2017) Pathological game use in adults with and without autism spectrum disorder. PeerJ 5:e3393. https://doi.org/10.7717/peerj.3393

Gallagher M, McMahan RW, Schoenbaum G (1999) Orbitofrontal cortex and representation of incentive value in associative learning. J Neurosci 19:6610. https://doi.org/10.1523/JNEUROSCI.19-15-06610.1999

Geisel O, Banas R, Schneider M, Hellweg R, Müller CA (2013) Serum levels of brain-derived neurotrophic factor in patients with Internet use disorder. Psychiatry Res 209:525–528. https://doi.org/10.1016/j.psychres.2012.12.020

Geng J, Han L, Gao F, Jou M, Huang CC (2018) Internet addiction and procrastination among Chinese young adults: a moderated mediation model. Comput Hum Behav 84:320–333. https://doi.org/10.1016/j.chb.2018.03.013

Greenfield DN (1999) Psychological characteristics of compulsive internet use: a preliminary analysis. Cyber Psychol Behav 2:403–412. https://doi.org/10.1089/cpb.1999.2.403

Griffiths MD (1995) Technological addictions. Clin Psychol Forum 76:14–19

Griffiths MD (1996) Internet "addiction": an issue for clinical psychology? Clin Psychol Forum 97:32–36

Griffiths MD (2005) A 'components' model of addiction within a biopsychosocial framework. J Substance Use 10:191–197. https://doi.org/10.1080/14659890500114359

Griffiths MD, Kuss DJ, Billieux J, Pontes HM (2016a) The evolution of internet addiction: a global perspective. Addict Behav 53:193–195. https://doi.org/10.1016/j.addbeh.2015.11.001

Griffiths MD, Pontes HM, Kuss DJ (2016b) Online addictions: conceptualizations, debates, and controversies Addicta. Turkish J Addict 3:1–20. https://doi.org/10.15805/addicta.2016b.3.0101

Guo L et al (2018) Association between problematic internet use, sleep disturbance, and suicidal behavior in Chinese adolescents. J Behav Addict 7:965–975. https://doi.org/10.1556/2006.7.2018.115

Haagsma MC, Caplan SE, Peters O, Pieterse ME (2013) A cognitive-behavioral model of problematic online gaming in adolescents aged 12–22 years. Comput Hum Behav 29:202–209. https://doi.org/10.1016/j.chb.2012.08.006

Hawi NS, Blachnio A, Przepiorka A (2015) Polish validation of the internet addiction test. Comput Hum Behav 48:548–553. https://doi.org/10.1016/j.chb.2015.01.058

Hou H, Jia S, Hu S, Fan R, Sun W, Sun T, Zhang H (2012) Reduced striatal dopamine transporters in people with Internet addiction disorder. Biomed Res Int 2012:5. https://doi.org/10.1155/2012/854524

Jahan SM, Hossain SR, Sayeed UB, Wahab A, Rahman T, Hossain A (2019) Association between internet addiction and sleep quality among students: a cross-sectional study in Bangladesh. Sleep Biol Rhythms. https://doi.org/10.1007/s41105-019-00219-y

Karaca S, Saleh A, Canan F, Potenza MN (2017) Comorbidity between behavioral addictions and attention deficit/hyperactivity disorder: a systematic review. Int J Ment Heal Addict 15:701–724. https://doi.org/10.1007/s11469-016-9660-8

Karimy M, Parvizi F, Rouhani MR, Griffiths MD, Armoon B, Fattah Moghaddam L (2020) The association between internet addiction, sleep quality, and health-related quality of life among Iranian medical students. J Addict Dis 1–9. https://doi.org/10.1080/10550887.2020.1762826

Kawabe K, Horiuchi F, Miyama T, Jogamoto T, Aibara K, Ishii E, Ueno S (2019) Internet addiction and attention-deficit/hyperactivity disorder symptoms in adolescents with autism spectrum disorder. Res Dev Disabil 89:22–28. https://doi.org/10.1016/j.ridd.2019.03.002

Kaya F, Delen E, Young KS (2016) Psychometric properties of the internet addiction test in Turkish. J Behav Addict 5:130–134. https://doi.org/10.1556/2006.4.2015.042

Khazaei F, Khazaei O, Ghanbari-H B (2017) Positive psychology interventions for internet addiction treatment. Comput Hum Behav 72:304–311. https://doi.org/10.1016/j.chb.2017.02.065

Kim KM, Kim H, Choi JW, Kim SY, Kim JW (2020) What Types of internet services make adolescents addicted? Corr Prob Internet Use Neuropsychiatric Dis Treatment 16:1031–1041. https://doi.org/10.2147/NDT.S247292

Kim SH, Baik SH, Park CS, Kim SJ, Choi SW, Kim SE (2011) Reduced striatal dopamine D2 receptors in people with Internet addiction. NeuroReport 22:407–411. https://doi.org/10.1097/WNR.0b013e328346e16e

Kim SJ, Park DH, Ryu SH, Yu J, Ha JH (2013) Usefulness of young's internet addiction test for clinical populations. Nord J Psychiatry 67:393–399. https://doi.org/10.3109/08039488.2012.748826

Király O, Nagygyörgy K, Koronczai B, Griffiths MD, Demetrovics Z (2014) Assessment of problematic Internet use and online video gaming. In: Starcevic V, Aboujaoude E (eds) Mental health in the digital age: grave dangers, great promise. Oxford University Press, Oxford, pp 46–68. https://doi.org/10.1093/med/9780199380183.001.0001

Ko C, Yen J, Chen C, Yeh Y, Yen C (2009) Predictive values of psychiatric symptoms for internet addiction in adolescents: a 2-year prospective study. Arch Pediatrics Adolescent Med 163:937–943. https://doi.org/10.1001/archpediatrics.2009.159

Kraut R, Patterson M, Lundmark V, Kiesler S, Mukophadhyay T, Scherlis W (1998) Internet paradox: a social technology that reduces social involvement and psychological well-being? Am Psychol 53:1017

Krawczyk DC (2002) Contributions of the prefrontal cortex to the neural basis of human decision making. Neurosci Biobehav Rev 26:631–664. https://doi.org/10.1016/S0149-7634(02)00021-0

Kuss DJ, Griffiths MD, Karila L, Billieux J (2014) Internet addiction: a systematic review of epidemiological research for the last decade. Curr Pharm Des 20:4026–4052. https://doi.org/10.2174/13816128113199990617

Kuss DJ, Lopez-Fernandez O (2016) Internet addiction and problematic Internet use: a systematic review of clinical research. World J Psychiatry 6:143–176. https://doi.org/10.5498/wjp.v6.i1.143

Kuss DJ, Pontes HM (2019) Internet addiction vol 41. Advances in psychotherapy – evidence-based practice. Hogrefe Publishing Corporation, Boston, MA

Lam LT (2014) Internet gaming addiction, problematic use of the Internet, and sleep problems: a systematic review. Curr Psychiatry Rep 16:444. https://doi.org/10.1007/s11920-014-0444-1

Lewczuk K, Nowakowska I, Lewandowska K, Potenza MN, Gola M (2020) Frequency of use, moral incongruence and religiosity and their relationships with self-perceived addiction to pornography, internet use, social networking and online gaming. Addiction https://doi.org/10.1111/add.15272

Li S, Zhu S, Jin X, Yan C, Wu S, Jiang F, Shen X (2010) Risk factors associated with short sleep duration among Chinese school-aged children. Sleep Med 11:907–916. https://doi.org/10.1016/j.sleep.2010.03.018

Lin F, Zhou Y, Du Y, Qin L, Zhao Z, Xu J, Lei H (2012) Abnormal white matter integrity in adolescents with internet addiction disorder: a tract-based spatial statistics study. PLoS ONE 7:e30253. https://doi.org/10.1371/journal.pone.0030253

Lissitsa S, Chachashvili-Bolotin S (2016) Life satisfaction in the internet age – changes in the past decade. Comput Hum Behav 54:197–206. https://doi.org/10.1016/j.chb.2015.08.001

Liu QX, Fang XY, Yan N, Zhou ZK, Yuan XJ, Lan J, Liu CY (2015) Multi-family group therapy for adolescent Internet addiction: exploring the underlying mechanisms. Addict Behav 42:1–8. https://doi.org/10.1016/j.addbeh.2014.10.021

Lortie CL, Guitton MJ (2013) Internet addiction assessment tools: dimensional structure and methodological status. Addiction 108:1207–1216

Macur M, Király O, Maraz A, Nagygyörgy K, Demetrovics Z (2016) Prevalence of problematic internet use in Slovenia Slovenian. J Public Health 55:1–10. https://doi.org/10.1515/sjph-2016-0026

Matheson SL, Shepherd AM, Pinchbeck RM, Laurens KR, Carr VJ (2013) Childhood adversity in schizophrenia: a systematic meta-analysis. Psychological medicine 43(2):225–238. https://doi.org/10.1017/S0033291712000785

Mamun MA, Hossain MS, Moonajilin MS, Masud MT, Misti JM, Griffiths MD (2020) Does loneliness, self-esteem and psychological distress correlate with problematic internet use? A Bangladeshi survey study. Asia-Pacific Psychiatry n/a:e12386. https://doi.org/10.1111/appy.12386

Marin MG, Nuñez X, de Almeida RMM (2021) Internet Addiction and attention in adolescents: a systematic review cyberpsychology. Behav Soc Network 24:237–249. https://doi.org/10.1089/cyber.2019.0698

Moon SJ, Hwang JS, Kim JY, Shin AL, Bae SM, Kim JW (2018) Psychometric properties of the internet addiction test: a systematic review and meta-analysis cyberpsychology. Behav Soc Network 21:473–484. https://doi.org/10.1089/cyber.2018.0154

Morita M et al (2021) Bidirectional relationship of problematic Internet use with hyperactivity/inattention and depressive symptoms in adolescents: a population-based cohort study. Eur Child Adolesc Psychiatry. https://doi.org/10.1007/s00787-021-01808-4

Müller SM, Wegmann E, Stolze D, Brand M (2020) Maximizing social outcomes? Social zapping and fear of missing out mediate the effects of maximization and procrastination on problematic social networks use. Comput Hum Behav 107:106296. https://doi.org/10.1016/j.chb.2020.106296

Näsi M, Räsänen P, Sarpila O (2012) ICT activity in later life: Internet use and leisure activities amongst senior citizens in Finland. Eur J Ageing 9:169–176. https://doi.org/10.1007/s10433-011-0210-8

Pan YC, Chiu YC, Lin YH (2020) Systematic review and meta-analysis of epidemiology of internet addiction. Neurosci Biobehav Rev 118:612–622. https://doi.org/10.1016/j.neubiorev.2020.08.013

Park B, Han DH, Roh S (2017) Neurobiological findings related to internet use disorders. Psychiatry Clin Neurosci 71:467–478. https://doi.org/10.1111/pcn.12422

Peterka-Bonetta J, Sindermann C, Elhai JD, Montag C (2019) Personality associations with smartphone and internet use disorder: a comparison study including links to impulsivity and social anxiety. Front Public Health 7. https://doi.org/10.3389/fpubh.2019.00127

Petrovic P, Pleger B, Seymour B, Klöppel S, De Martino B, Critchley H, Dolan RJ (2008) Blocking central opiate function modulates hedonic impact and anterior cingulate response to rewards and losses. J Neurosci 28:10509. https://doi.org/10.1523/JNEUROSCI.2807-08.2008

Pluhar E, Kavanaugh JR, Levinson JA, Rich M (2019) Problematic interactive meda use in teens: comorbidities, assessment, and treatment. Psychol Res Behav Manage 12:447–455. https://doi.org/10.2147/PRBM.S208968

Pontes HM, Griffiths MD (2016) The development and psychometric properties of the internet disorder scale–short form (IDS9-SF) addicta. Turkish J Addict 3:303–318. https://doi.org/10.15805/addicta.2016.3.0102

Pontes HM, Kuss DJ, Griffiths MD (2015) Clinical psychology of internet addiction: a review of its conceptualization, prevalence, neuronal processes, and implications for treatment. Neurosci Neuroecon 4:11–23. https://doi.org/10.2147/NAN.S60982

Pontes HM, Macur M (2021) Problematic internet use profiles and psychosocial risk among adolescents. PLoS ONE 16:e0257329. https://doi.org/10.1371/journal.pone.0257329

Pontes HM, Patrão IM, Griffiths MD (2014) Portuguese validation of the internet addiction test: an empirical study. J Behav Addict 3:107–114. https://doi.org/10.1556/JBA.3.2014.2.4

Prizant-Passal S, Shechner T, Aderka IM (2016) Social anxiety and Internet use – a meta-analysis: what do we know? What are we missing? Comput Hum Behav 62:221–229. https://doi.org/10.1016/j.chb.2016.04.003

Przepiorka AM, Blachnio A, Miziak B, Czuczwar SJ (2014) Clinical approaches to treatment of Internet addiction. Pharmacol Rep 66:187–191. https://doi.org/10.1016/j.pharep.2013.10.001

Quittschalle J, Stein J, Luppa M, Pabst A, Löbner M, Koenig HH, Riedel-Heller SG (2020) Internet use in old age: results of a german population-representative survey. J Med Internet Res 22:e15543. https://doi.org/10.2196/15543

Rayan A, Dadoul AM, Jabareen H, Sulieman Z, Alzayyat A, Baker O (2017) Internet use among university students in south West Bank: prevalence, advantages and disadvantages, and association with psychological health. Int J Mental Health Addict 15:118–129. https://doi.org/10.1007/s11469-016-9658-2

Romano M, Truzoli R, Osborne LA, Reed P (2014) The relationship between autism quotient, anxiety, and internet addiction. Res Autism Spectrum Disorders 8:1521–1526. https://doi.org/10.1016/j.rasd.2014.08.002

Saiful Islam M, Rahman ME, Moonajilin MS, Griffiths MD (2020) Validation and evaluation of the psychometric properties of Bangla nine-item internet disorder scale-short form. J Addict Dis 38:540–549. https://doi.org/10.1080/10550887.2020.1799134

Sanders CE, Field TM, Diego M, Kaplan M (2000) The relationship of Internet use to depression and social isolation among adolescents. Adolescence 35:237–242

Shaffer HJ, LaPlante DA, LaBrie RA, Kidman RC, Donato AN, Stanton MV (2004) Toward a syndrome model of addiction: multiple expressions, common etiology. Harvard Rev Psychiatry 12:367–374. https://doi.org/10.1080/10673220490905705

Shapira NA, Goldsmith TD, Keck PE Jr, Khosla UM, McElroy SL (2000) Psychiatric features of individuals with problematic Internet use. J Affect Disord 57:267–272. https://doi.org/10.1016/S0165-0327(99)00107-X

Sharifat H, Rashid AA, Suppiah S (2018) Systematic review of the utility of functional MRI to investigate internet addiction disorder: recent updates on resting state and task-based fMRI Malaysian. J Med Health Sci 14:21–33

So R et al (2019) The 2-year course of internet addiction among a Japanese adolescent psychiatric clinic sample with autism spectrum disorder and/or attention-deficit hyperactivity disorder. J Autism Dev Disorders. https://doi.org/10.1007/s10803-019-04169-9

Soraci P, Ferrari A, Bonanno E, Rosanna DP, Repice E, Griffiths MD (2020) Psychometric validation of the internet disorder scale-short form in an Italian adult sample. Int J Ment Heal Addict. https://doi.org/10.1007/s11469-020-00279-z

Starcevic V (2013) Is Internet addiction a useful concept? Aust N Z J Psychiatry 47:16–19. https://doi.org/10.1177/0004867412461693

Starcevic V, Aboujaoude E (2016) Internet addiction: reappraisal of an increasingly inadequate concept. CNS Spectr 22:7–13. https://doi.org/10.1017/S1092852915000863

Starcevic V, Billieux J (2017) Does the construct of internet addiction reflect a single entity or a spectrum of disorders? Clin Neuropsychiatry 14:5–10

Statista (2021) Worldwide digital population as of January 2021. https://www.statista.com/statistics/617136/digital-population-worldwide. Accessed 23 Sept 2021

Suler JR (1999) To get what you need: healthy and pathological internet use. Cyber Psychol Behav 2:385–393. https://doi.org/10.1089/cpb.1999.2.385

Tereshchenko S, Kasparov E (2019) Neurobiological risk factors for the development of internet addiction in adolescents. Behav Sci 9:62

Tian Y, Qin N, Cao S, Gao F (2020) Reciprocal associations between shyness, self-esteem, loneliness, depression and Internet addiction in Chinese adolescents. Addict Res Theory 1–13. https://doi.org/10.1080/16066359.2020.1755657

Torres GE, Gainetdinov RR, Caron MG (2003) Plasma membrane monoamine transporters: structure, regulation and function. Nat Rev Neurosci 4:13–25. https://doi.org/10.1038/nrn1008

Tsumura H, Kanda H, Sugaya N, Tsuboi S, Fukuda M, Takahashi K (2018) Problematic internet use and its relationship with psychological distress, insomnia, and alcoholism among schoolteachers in Japan Cyberpsychology. Behav Soc Network 21:788–796. https://doi.org/10.1089/cyber.2018.0233

Verheijen GP, Burk WJ, Stoltz SEMJ, van den Berg YHM, Cillessen AHN (2021) A longitudinal social network perspective on adolescents' exposure to violent video games and aggression cyberpsychology. Behav Soc Network 24:24–31. https://doi.org/10.1089/cyber.2019.0776

Verma V (2015) Classic studies on the interaction of cocaine and the dopamine transporter. Clin Psychopharmacol Neurosci 13:227–238. https://doi.org/10.9758/cpn.2015.13.3.227

Visser SN et al (2014) Trends in the parent-report of health care provider-diagnosed and medicated attention-deficit/hyperactivity disorder: United States, 2003–2011. J Am Acad Child Adolescent Psychiatry 53:34–46.e32 https://doi.org/10.1016/j.jaac.2013.09.001

Wang BQ, Yao NQ, Zhou X, Liu J, Lv ZT (2017a) The association between attention deficit/hyperactivity disorder and Internet addiction: a systematic review and meta-analysis. BMC Psychiatry 17:260. https://doi.org/10.1186/s12888-017-1408-x

Wang T, Liu K, Li Z, Xu Y, Liu Y, Shi W, Chen L (2017b) Prevalence of attention deficit/hyperactivity disorder among children and adolescents in China: a systematic review and meta-analysis. BMC Psychiatry 17:32. https://doi.org/10.1186/s12888-016-1187-9

Wang P, Wang X, Zhao M, Wu Y, Wang Y, Lei L (2019) Can social networking sites alleviate depression? The relation between authentic online self-presentation and adolescent depression: a

mediation model of perceived social support and rumination. Curr Psychol 38:1515–1521. https://doi.org/10.1007/s12144-017-9711-8
Wang W, Du X, Guo Y, Li W, Zhang S, Guo L, Lu C (2021) Association between problematic internet use and behavioral/emotional problems among Chinese adolescents: the mediating role of sleep Disorders. PeerJ 9:e10839. https://doi.org/10.7717/peerj.10839
Waqas A et al (2017) Validation of the internet addiction test in students at a Pakistani medical and dental school. Psychiatr Q. https://doi.org/10.1007/s11126-017-9528-5
Wartberg L et al (2017) Psychometric properties of a German version of the Young Diagnostic Questionnaire (YDQ) in two independent samples of adolescents. Int J Mental Health Addict 15:182–190. https://doi.org/10.1007/s11469-016-9654-6
Wartberg L, Kriston L, Kegel K, Thomasius R (2016) Adaptation and psychometric evaluation of the Young Diagnostic Questionnaire (YDQ) for parental assessment of adolescent problematic internet use. J Behav Addict 5:311–317. https://doi.org/10.1556/2006.5.2016.049
Weinstein A, Feder LC, Rosenberg KP, Dannon P (2014) Internet addiction disorder: overview and controversies. In: Rosenberg KP, Feder LC (eds) Behavioral addictions. Academic Press, San Diego, CA, pp 99–117. https://doi.org/10.1016/B978-0-12-407724-9.00005-7
Weng CB, Qian RB, Fu XM, Lin B, Han XP, Niu CS, Wang YH (2013) Gray matter and white matter abnormalities in online game addiction. Eur J Radiol 82:1308–1312. https://doi.org/10.1016/j.ejrad.2013.01.031
Winkler A, Dörsing B, Rief W, Shen Y, Glombiewski JA (2013) Treatment of internet addiction: a meta-analysis. Clin Psychol Rev 33:317–329. https://doi.org/10.1016/j.cpr.2012.12.005
Wölfling K, Beutel ME, Dreier M, Müller KW (2014) Treatment outcomes in patients with internet addiction: a clinical pilot study on the effects of a cognitive-behavioral therapy program. Biomed Res Int 2014:425924. https://doi.org/10.1155/2014/425924
Wölfling K, Beutel ME, Dreier M, Müller KW (2015) Bipolar spectrum disorders in a clinical sample of patients with Internet addiction: hidden comorbidity or differential diagnosis? J Behav Addict 4:101–105. https://doi.org/10.1556/2006.4.2015.011
Xin M, Xing J, Pengfei W, Houru L, Mengcheng W, Hong Z (2018) Online activities, prevalence of Internet addiction and risk factors related to family and school among adolescents in China addictive. Behav Rep 7:14–18. https://doi.org/10.1016/j.abrep.2017.10.003
Yang J et al. (2018) Association between problematic internet use and sleep disturbance among adolescents: the role of the child's sex. Int J Environ Res Public Health 15. https://doi.org/10.3390/ijerph15122682
Yayan EH, Arikan D, Saban F, N GB, Özcan Ö (2016) Examination of the correlation between Internet addiction and social phobia in adolescents. Western J Nurs Res 1–15. https://doi.org/10.1177/0193945916665820
You Z, Mei W, Ye N, Zhang L, Andrasik F (2020) Mediating effects of rumination and bedtime procrastination on the relationship between internet addiction and poor sleep quality. J Behav Addict. https://doi.org/10.1556/2006.2020.00104
Younes F, Halawi G, Jabbour H, El Osta N, Karam L, Hajj A, Rabbaa Khabbaz L (2016) Internet addiction and relationships with insomnia, anxiety, depression, stress and self-esteem in university students: a cross-sectional designed study. PLOS ONE 11:e0161126. https://doi.org/10.1371/journal.pone.0161126
Young KS (1996) Psychology of computer use: XL addictive use of the internet: a case that breaks the stereotype. Psychol Rep 79:899–902. https://doi.org/10.2466/pr0.1996.79.3.899
Young KS (1998a) Caught in the net: How to recognize the signs of Internet addiction - and a winning strategy for recovering. Wiley, New York, NY
Young KS (1998b) Internet addiction: the emergence of a new clinical disorder. Cyber Psychol Behav 1:237–244. https://doi.org/10.1089/cpb.1998.1.237
Young KS (2011) CBT-IA: the first treatment model for internet addiction. J Cogn Psychother 25:304–312. https://doi.org/10.1891/0889-8391.25.4.304
Young KS (2013) Treatment outcomes using CBT-IA with internet-addicted patients. J Behav Addict 2:209–215. https://doi.org/10.1556/JBA.2.2013.4.3

Yuan K et al (2011) Microstructure abnormalities in adolescents with internet addiction disorder. PLoS ONE 6:e20708. https://doi.org/10.1371/journal.pone.0020708

Zahniser NR, Sorkin A (2004) Rapid regulation of the dopamine transporter: role in stimulant addiction? Neuropharmacology 47:80–91. https://doi.org/10.1016/j.neuropharm.2004.07.010

Zajac K, Ginley MK, Chang R, Petry NM (2017) Treatments for internet gaming disorder and internet addiction: a systematic review. Psychol Addict Behav 31:979–994. https://doi.org/10.1037/adb0000315

Zhang H, Spinrad TL, Eisenberg N, Luo Y, Wang Z (2017) Young adults' Internet addiction: prediction by the interaction of parental marital conflict and respiratory sinus arrhythmia. Int J Psychophysiol 120:148–156. https://doi.org/10.1016/j.ijpsycho.2017.08.002

Zhong X et al (2011) The effect of a family-based intervention model on internet-addicted Chinese adolescents. Int J Soc Behav Person 39:1021–1034. https://doi.org/10.2224/sbp.2011.39.8.1021

Zhou B, Zhang W, Li Y, Xue J, Zhang-James Y (2020) Motivational but not executive dysfunction in attention deficit/hyperactivity disorder predicts internet addiction: evidence from a longitudinal study. Psychiatry Res 285:112814. https://doi.org/10.1016/j.psychres.2020.112814

Zhou Y, Li D, Jia J, Li X, Zhao L, Sun W, Wang Y (2017) Interparental conflict and adolescent Internet addiction: the mediating role of emotional insecurity and the moderating role of big five personality traits. Comput Hum Behav 73:470–478. https://doi.org/10.1016/j.chb.2017.04.012

Zhou Y, Lin FC, Du YS, Qin LD, Zhao ZM, Xu JR, Lei H (2011) Gray matter abnormalities in Internet addiction: a voxel-based morphometry study. Eur J Radiol 79:92–95. https://doi.org/10.1016/j.ejrad.2009.10.025

Chapter 6
Love Addiction

Taya Bockmann and Marsal Sanches

Abstract Love addiction (LA), also termed as pathological love, is characterized by a maladaptive relationship behavior encompassing excessive longing for one or several romantic partners. Even though LA is not currently recognized by 5th Diagnostic and Statistical Manual of Mental Disorders (DSM-5) as a specific diagnostic category, it seems to represent a specific type of behavioral addcition. In this chapter, we critically analyze available literature and data regarding LA as a diagnostic construct, its putative biological basis, and its diagnosis and treatment.

Keywords Love · Addiction · Behavioral addictions · Pharmacology · Neurobiology

6.1 Introduction (Definition; Theoretical Model)

For most of us, the feeling of falling in love with someone is one of the most intense and pleasurable experiences. We meet this one particular person, our heart races when we are with them, and when we are not, we can think of little else than the next encounter. If everything goes well, this lasts a couple of months, and without us noticing, the obsession slowly passes and smoothly transitions into a different kind of bond. Now we can entirely focus on our work again, on friendships, on time alone. However, what if it does not pass, if this rush never stops or, in contrast, we miss it to a point we do not want it to ever cease of?

The concept of love addiction (LA), also called pathological love, is characterized by a maladaptive relationship behavior that includes excessive longing for one or several romantic partners. The affected person experiences this longing as pervasive and uncontrollable, which can have detrimental results. They can find themselves losing other interests, existing hobbies, friendships, and experiencing a negative state of emotion whenever they cannot spend time with the subject of their romantic appeal. They might show a decline in their work performance and risk ruining their

T. Bockmann (✉) · M. Sanches
UT Center of Excellence On Mood Disorders, Department of Psychiatry and Behavioral Sciences, UT Houston Medical School, 1541 East Road, Houston, TX 77054, US
e-mail: Taya.Bockmann@uth.tmc.edu

H. M. Pontes (ed.), *Behavioral Addictions*, Studies in Neuroscience, Psychology and Behavioral Economics, https://doi.org/10.1007/978-3-031-04772-5_6

financial well-being in their pursuit of love. An addiction to love, or to the rush one experiences when falling in love, can severely impact a person's well-being in various ways (Sanches and John 2019). Although we currently lack a clear definition of LA, researchers have described its concept from different perspectives.

For example, Earp et al. compare LA from two different approaches, namely the narrow and the broad view. The former would include only the most extreme cases of love related behaviors as potential instances of addiction. This perspective would assume that addictive behaviors result from abnormal brain processes that do not exist in non-addicted people's brains. Thus, natural rewards like relationships or even sweet foods could trigger different brain processes in addicted persons' brains, but not in healthy individuals. According to this perspective, if these abnormal brain processes are present, and the person engages in extremely maladaptive relationship behaviors, one could indeed be addicted to love. This person might tolerate toxic or abusive relationships that threaten their own mental and physical well-being and their safety, or they might engage in actions that lead to severe social or legal costs for the sake of the relationship(s).

The broad view, on the other hand, describes addictions as a spectrum of motivation. Here, addiction relies on neurobiological processes that are present in the brains of healthy people as well. Researchers assume that rewards of any kind can trigger this brain response. A person feels the temporary satisfaction of a specific need, like drinking or trying their luck on a slot machine. When these needs rise to a level of urgency, distraction, and uncontrollability, a person might be addicted. Both love and addiction seem to have the capacity to be strong enough forces to lead a person to make irrational choices with sometimes detrimental outcomes.

To illustrate the parallels between addiction to a chemical substance and LA, we can compare extreme romantic behavior to the criteria for Substance Use Disorder from the latest 5th edition of the Diagnostic and Statistical Manual for Mental Disorders (DSM-5) (American Psychiatric Association 2013). The behavioral similarities illustrated in the table below suggest that addiction to a substance and LA might draw on similar biological substrates.

DSM-5 criteria for substance use disorder	Analog to love addiction
Taking the substance in larger amounts or for longer than you are meant to	Spending more time with the partner(s) or with thought content about the partner(s) then you are meant to
Wanting to cut down or stop using the substance but not managing to	Wanting to end the relationship(s) or spend less time with the partner(s) or with thoughts about them, but not managing to
Spending excessive time getting, using, or recovering from use of the substance	Spending much time planning the next encounter with the partner(s) or thinking about the last encounter
Cravings and urges to use the substance	Urges to spend time with, talk to, or be intimate with your partner(s)

(continued)

(continued)

DSM-5 criteria for substance use disorder	Analog to love addiction
Not managing to do what you should at work, home, or school because of substance use	Not managing what you should at work, home, or school because you are either spending time with your partner(s) or are very distracted by thoughts about your partner(s)
Continuing to use, even when it causes problems in relationships	Continuing the relationship(s) against well-intended advice from family and close friends and even though the relationship causes problems with family and friends
Giving up important social, occupational, or recreational activities because of substance use	Giving up important social, occupational, or recreational activities because of the partner(s)
Using substances again and again, even when it puts you in danger	Continuing spending time with the partner(s), even when this puts you in danger
Continuing to use, even when you know you have a physical or psychological problem that could have been caused or made worse by the substance	Continuing the relationship, even when you know you have physical or psychological problems that could have been caused or made worse by the relationship(s) or your partner(s)
Needing more of the substance to get the effect you want (tolerance)	Needing more time with the partner or more intimacy with the partner to get the effect you want
Development of withdrawal symptoms, which can be relieved by taking more of the substance	Growing feelings of longing and desperation when not around the partner(s), which can be relieved by spending time with or talking to your partner(s)

The above table shows that compulsive romantic behavior can be very similar to those observed in patients addicted to substances. This comparison can be extended. People and laboratory animals can relapse to drug-taking, which can be triggered by different factors. Three major contributors to relapse are stressful life events, environmental stimuli that the person associates with the drug, and re-exposure to the substance (Anderson and Pierce 2005). Similar factors can also contribute to maladaptive romantic behavior. Someone who has gotten out of a dysfunctional relationship and thinks they are now past the danger of reinstating the same relationship might find themselves returning to that partner after experiencing stressful events, being reminded of their former partner by external stimuli, learning about someone else's engagement or marriaged, or after being reexposed to the former partner. The same circumstances can trigger other types of maladaptive romantic behavior. For example, stressful life events can drive a person to ever new romantic relationships, even though they might have just decided to stay away from dating for a while.

Even though LA is not currently recognized by DSM-5 as an official mental health disorder, clinical experience and literature data suggest it corresponds to a specific clinical condition. The next sessions of this chapter will explore potential biological bases that might contribute to maladaptive romantic behavior, potential comorbidities of this particular addiction, and approaches for diagnosis and treatment.

6.2 Biological Bases

In romantic love, elevated levels of dopamine, norepinephrine, endogenous opioids, and neuropeptides such as oxytocin (OT) and arginine vasopressin (AVP) have been recorded, while serotonin levels seem to decrease (Sussman 2010). Similar responses are found during drug use behavior. Animals like monogamous prairie voles have been used to explore the biological bases of social and drug-related behavior and have shed light on exciting similarities between the brain's neurochemical processes that are involved in both (Anderson and Pierce 2005; Arias-Carrión et al. 2010; Burkett and Young 2012; Caldwell 2012; Dreyer et al. 2010; McGregor and Bowen 2012; Ross and Young 2009).

6.2.1 Dopamine

Engaging in sexual activity can feel very similar to the experiences associated with consuming drugs like cocaine. Both sex and drugs trigger activity in the so-called dopaminergic reward system in the human brain. For example, both drugs and sex lead to a release of dopamine in the nucleus accumbens, possibly contributing to the experienced similarities between a cocaine rush and sexual behavior (Burkett and Young 2012). Dopamine is a neurotransmitter in the central nervous system that has long been established as an essential participant in movement control. In the more recent past, though, it has been shown that dopamine is involved in various processes, like several psychiatric diseases. It is also essential for the reward system and thus involved in the abuse of and addiction to different drugs (Arias-Carrión et al. 2010).

In the human adult brain, most dopaminergic neurons are found in the ventral part of the mesencephalon, mainly the substantia nigra (SNc) and the ventral tegmental area (VTA). On the one hand, axons of these neurons project mainly to the nucleus accumbens (NAc) and the olfactory tubercle. On the other hand, they lead to the prefrontal, cingulate, and perirhinal cortices (Arias-Carrión et al. 2010).

The activity of dopamine is mediated by five distinct receptors, belonging to two subfamilies, namely D1-like (D1R) and D2-like receptors (D2R) (Arias-Carrión et al. 2010). D1R are excitatory receptors with a low affinity for dopamine, thus responding to a high concentration. D2R, on the other hand, is a class of inhibitory receptors that respond to low concentrations of dopamine due to their high affinity for the neurotransmitter. High concentrations of dopamine are mainly reached during phasic firing of dopaminergic neurons, while low concentrations are found during tonic firing (Burkett and Young 2012). Animal studies have shown a relationship between tonic and phasic dopamine release, but the exact relationship between this, the effect on the postsynaptic receptors, and behavior have yet to be explored (Dreyer et al. 2010).

Evidence gathered on studies on prairie voles suggests that D1R and D2R have distinct influences on pair bonding. Prairie voles are studied to learn more about social decision-making because they show social behaviors that are quite similar to that of humans in many ways. After mating, these mammals bond with their partner and show a monogamous social structure, which is very unconventional in mammals. Male prairie voles show high involvement in the upbringing of their pups. Pair bonding behavior in prairie voles is so pronounced that animals often do not accept a new mate upon losing their partner (Aragona and Wang 2009). When researchers artificially blocked D2R in the NAC shell, this bonding was not observed, while artificial activation of the same receptors induced pair-bonding even without mating. When D1R was artificially activated, no pair-bonding behavior was observed, not influenced by the activation state of D2R. Blocking D1R, on the other hand, does not influence pair-bonding behavior in either direction. Researchers highlighted this interaction by introducing different doses of apomorphine into the NAC shell. Apomorphine bonds to both D1R and D2R. As the receptor types have a different affinity towards the neurotransmitter, it depends on the concentration of apomorphine, which class it preferentially binds to. In low concentrations, apomorphine is expected to bind to D2R, as these receptors have a higher affinity than D1R, while at high concentrations, it is expected to bind to both receptor classes. When researchers injected low doses of apomorphine into the CAN shell of their prairie voles, they observed that the animals showed more pair-bonding behavior than before the injection. At high doses, pair-bonding behavior was not increased (Burkett and Young 2012).

Studies on prairie voles have also shown exciting parallels between the plastic and functional changes occurring during pair bonding and drug addiction. Pair bonding leads to an upregulation of D1R in the NAC, which corresponds to higher aggressive behavior towards unfamiliar females in male prairie voles, stabilizing the monogamous partnership. This same upregulation of D1R is also observed when administering amphetamines for several days. Moreover, when sexually naive male prairie voles receive this treatment, they react in the same way pair-bonded male voles do. They reject potential females and do not engage in a partnership.

The effects of cocaine, which inhibits the reuptake of dopamine, serotonin, and norepinephrine, are mainly mediated by the high transmission of dopamine within the mesocorticolimbic paths. Even though the exact physiological effects of prolonged cocaine exposure remain to be explored in more detail and seem to depend upon a variety of factors, it has been established that one effect is the enhanced inhibition within the NAC, mediated by increased sensitivity of D1R. Above, we have described the reinstatement of cocaine-seeking behavior. In animals who have formerly shown drug self-administration, and in which self-administering behavior has been extinguished, administration of either dopamine reuptake inhibitors or D2R agonists leads to reinstatement of cocaine-seeking (Anderson and Pierce 2005).

All these results strongly suggest a significant involvement of dopamine and the expression of dopaminergic neurons in romantic and sexual behavior, and interestingly also in drug addiction, in a noticeably similar way. Neurological alterations might thus lead people to display maladaptive relationship behavior implicated in LA.

6.2.2 Opioids

The endogenous opioid system plays a vital role in addiction, as it is one of the modulators of the rewarding and reinforcing effects of all known drugs of abuse, food, sex, and other rewarding activities. Animal studies have shown that it is also involved in maternal and pair-bonding behavior. Studies on the role of the opioid system in pair-bonding behavior have revealed interactions between the opioid and the dopamine system (Burkett and Young 2012).

Three types of inhibitory receptors comprise the opioid system: μ (MOR), δ (DOR), and κ (KOR). All three receptor types mediate the reward response in the human brain, where MOR and DOR are involved in positive motivation and KOR in mediating negative motivation. Researchers have conducted animal studies to specify the involvement of these receptors in the reinforcing effect shared by drugs of abuse and natural rewards such as food and sex. These studies suggest that the rewarding effect is mediated mainly by MOR in the NAC, the ventral pallidum (VP) in the basal ganglia, and VTA, while also DOR in the NAC is playing a role in these processes. The rewarding effect of drugs is not observed if MOR antagonists are injected into the NAC, VP, or VTA. In humans, opioid antagonists can reduce cravings in opioid dependence, alcoholism, and obesity. KOR seems to be involved in addiction through upregulation of its ligand, which inhibits dopamine release. This inhibition might play a role in withdrawal, as the decreased dopamine levels could contribute to negative affect and relapse (Burkett and Young 2012).

The opioid system is involved in the social behavior of prairie voles as well. It has been shown that the administration of non-selective opioid antagonists prevents pair bonding and reduces mating when administered peripherally. When a MOR-selective antagonist is administered into the caudate-putamen, it prevents pair-bonding without affecting sexual behavior.

In humans, the functions of MOR and the CP seem to be analogous to these animals. A genetic variant of the MOR gene, A118G, is correlated with altered reinforcement learning and increased risk for alcohol dependence in humans. This same variation also increases a person's sensitivity to rejection and possibly signals an altered attachment style. These symptoms are typical for what is considered an addiction to love. The CP of humans is activated when looking at pictures of loved ones, and this activation correlates with romantic love and passion scores (Burkett and Young 2012).

Studies conducted on humans to examine the relationship between opioids and social connection suggest that opioids modulate mostly the reward of social interaction with people in already established relationships. In contrast, the rush-like experience of falling in love or forming a new relationship is modulated to a higher degree by dopamine and OT. This would suggest that opioids are involved in the maintenance of a relationship. Researchers have shown in different experiments that opioid antagonists reduce the positive feelings induced by social experiences. For example, naltrexone, which blocks opioid receptors, reduced the felt connection while reading loving messages (Inagaki 2018).

6.2.3 Corticotropin-Releasing Factor (CRF)

While dopamine and OT are involved in most or all aspects of drug-seeking behavior, CRF seems to be involved only in withdrawal (Burkett and Young 2012). This protein, also called corticotropin-releasing hormone, is part of the CRF system known for coordinating an organism's stress response via the hypothalamic–pituitary–adrenal (HPA) axis. The distribution of the receptors within the CRF system shows that the CRF system also functions outside the HPA axis. Although the exact location of the CRF-producing neurons modulating social behavior is not yet fully understood, research strongly suggests the amygdala as the primary location (Hostetler and Ryabinin 2013).

The CRF system consists of two receptor types, CRF-R1 and CRF-R2. Studies on rats have shown that a non-selective CRF antagonist and a CRF-R1 antagonist similarly block excessive drug consumption in dependent rats, but not in non-dependent rats. Furthermore, the effect of CRF-R2 antagonists depends on the location of administration. In formerly drug-dependent rats that have been subjected to stress, increased CRF release in the NAC shell and VTA has been observed, which then led to drug reinstatement. This effect may partly be dopamine-dependent, as dopamine levels are also influenced by stress (Burkett and Young 2012).

Involving the human brain's stress response when discussing LA is paramount since separation from a loved one can induce strong negative affect and stress. When a person has to be separated from someone they have just fallen in love with, they might feel a growing obsession with thinking about that other person, imagining the next encounter, and making more desperate plans about achieving this reunion (Burkett and Young 2012).

To examine if the stress response to separation from a mate is dependent on CRF, researchers have again used prairie voles. They observed that, after four days of separation from their mate, male prairie voles showed a depressive-like increase in passive coping, while separation from a male sibling did not induce this behavior. These depressive-like symptoms could be reversed by injecting CRF-R1 and CRF-R2 antagonists into a part of the extended amygdala, namely the bed nucleus of the stria terminalis. The experience of negative affect and stress upon separation from the mate might make the male vole return to their partner, which stabilized the pair bond and ensures the species' survival. Researchers found that upon injection of CRF into the rats' NAC shell, the animals showed depressive-like behavior, which is reversed by the injection of a CRF-R1 antagonist. Based on these results, it can be hypothesized that CRF released in the NAC shell during social separation may induce negative affect, motivating the affected animal to return to the partner (Burkett and Young 2012). The CRF-system is involved in pair-bonding, which has been demonstrated by the administration of CRF directly into the NAC leads male prairie voles to show increased partner preference behavior, even at low doses of CRF (Hostetler and Ryabinin 2013).

6.2.4 Oxytocin (OT)

OT is synthesized in the hypothalamus and is released into the peripheral nervous system from the posterior pituitary gland. It is also released into the brain, where it modulates social behavior. OT has only one receptor, present in both the periphery and central nervous system, OTR (Burkett and Young 2012).

In addition to the well-established role OT plays in the mother-pub bond in animals, it also influences the pair bonding in adult prairie voles. In females, the evidence regarding this is much clearer than in male prairie voles. Females sometimes show pair-bonding with males even when mating did not occur. This bonding behavior can be induced artificially by an intracerebral infusion of OT during cohabitation. In females who mated with a partner during this cohabitation, the development of a pair bond can be inhibited by infusion of an OT antagonist, even 24 h after cohabitation. In male animals, these artificial manipulations of pair-bonding behavior by OT or OT antagonists do not work in the same way. Nonetheless, researchers did find that a high dose of OT via a central infusion was inducing partner preference behavior in male voles, while an OT antagonist or a vasopressin antagonist blocked this OT-induced behavior. In male prairie voles, vasopressin seems to be the more significant contributor to partner preference behavior (Ross and Young 2009).

In human drug use behavior, OT is best known for mediating the prosocial effect of common party drugs like 3,4-methylenedioxymethamphetamine (MDMA, Ecstasy). People who consumed this drug report feeling love and closeness and greater openness for others' feelings. These prosocial actions of the drug are also replicated in animal studies. For example, MDMA reduces aggressive behavior towards unfamiliar individuals in rats and increases OT production in the hypothalamus. The prosocial effects induced by MDMA interestingly more closely correlate with the plasma OT levels than with the plasma MDMA levels. The increased OT release is caused by a cascade of functions, in which MDMA leads to an increase in serotonin release, which then finally causes increased OT release (McGregor and Bowen 2012).

The above-explained evidence shows that OT is involved in MDMA's prosocial effects, even though it is unclear whether OT causes the reward response directly or indirectly. Scientific evidence suggests that OT is less directly rewarding itself than priming the brain for social rewards. OT has been shown to interact with dopamine when modulating processes like pair-bonding and drug reward. How exactly OT modulates peoples' social experience and how it interacts with other neurotransmitters needs to be examined further before final conclusions can be drawn (McGregor and Bowen 2012).

6.2.5 Arginine Vasopressin (AVP)

Like OT, AVP is synthesized in the hypothalamus and released into the peripheral circulation and the brain. Vasopressin has two receptors in the brain: V1aR and

V1bR. Of these, V1aR is the primary receptor modulating social processes (Burkett and Young 2012). AVP plays a role in modulating aggression and affiliation, both mechanisms necessary for a species' survival. When researchers administered an AVP1aR antagonist to prairie voles before mating, this inhibited partner preference in male voles, while administering AVP facilitated said behavior. The role of AVP in female animals is much less understood. As OT, also AVP might interact with the dopamine system to modulate social behavior (Caldwell 2012).

6.3 Specific Comorbidities

The development of an addiction during a person's lifetime has been shown to be highly influenced by their genetic predisposition. Prescott and Kendler (1999) used data from the Virginia Twin Registry to determine if someone is more likely to suffer from alcoholism if they have an identical twin than if they have a non-identical twin. The results of this suggested that about 48–58% of the variation in liability was attributable to genetic factors (Prescott and Kendler 1999).

The specific expression of a person's genome, which then indirectly shapes the brain's functional organization, depends on a multitude of environmental and developmental factors. Thus it is important to consider that stressful or even traumatic experiences during childhood and adolescence, both critical periods of brain development, can lead a person to develop maladaptive behaviors and disorders. This correlation has been scientifically established numerous times. There is an overwhelming body of evidence showing that negative childhood stressors are positively correlated with various outcomes (e.g., substance abuse and depressive disorders) (Parsaik et al. 2017). Moreover, early exposition to traumatic experiences can increase the genetic expression of CRF in the hypothalamus. As explored above, an increase in CRF in animals leads to increased depressive-like and partner preference behavior. Thus an early trauma could contribute to maladaptive relationship behavior as well (Anda et al. 2006).

A leading hypothesis is that dysfunctioning of the monoamine neurotransmitter systems (e.g., dopamine), which helps organize and orchestrate complex neural functions, contributes to many neuropsychiatric syndromes, including but not limited to suicidality, substance use disorder and dependencies, and interpersonal problems. The person's childhood experiences influence the way these monoamine systems function in the adult body. Early abuse survivors tend to show multiple comorbidities. The current body of evidence suggests that co- and cross-occurring disorders are common in sex and LA, including various other addictions, substance use, depression, and other mental health issues (Swendsen and Merikangas 2000; Thege et al. 2016).

Multiple types of addictive disorders often go along with so-called cross-addictions or co-occurring addictions. People who suffer from cross-addictions switch to a secondary addiction if they cannot or do not want to engage in their primary addiction. In the context of LA, we could be talking about a hypothetical individual who lives away from his spouse and child most of the year to work and engages in short and intense relationships during this time. During the brief periods when the individual is reunited with their family, they are able to refrain from these behaviors but, to cope with the negative emotions, the individual in question increased their consumption of alcohol and finds themselves showing signs of alcohol abuse or dependence.

Co-occurring addiction, on the other hand, means that the person engages in several addictions simultaneously. For instance, we could imagine a hypothetical couple who struggles with a dysfunctional relationship, with frequent arguments, break-ups, and reconciliations. While together, they display other behavioral addictions, such as gaming disorder. During the brief periods when they are apart, the addiction to gaming subsides, just to resume when they reconcile and are again back together.

In the search for comorbidities of LA, the results are blog posts, homepages about drug abuse, and studies about compulsive sexual behavior. There is no evidence suggesting that compulsive sexual behavior and LA are closely related. Therefore, conclusions about comorbidities drawn from studies about sexual behavior should be interpreted with caution.

6.4 Prevalence Rates

As LA is not a recognized disorder according to the DSM-5 and is a quite controversial topic, there is not much information about prevalence rates at this point in time. Overall prevalence rates ranging from 5 to 10% have been previously reported, while other researchers found a relatively low rate of 3% in the general US population and rates as high as 25% in young college students (Cook 1987; Sussman et al. 2011; Timmreck 1990).

There does not seem to be a significant gender difference in the prevalence of LA. Gender seems to have a more significant effect on specific symptomatology than on prevalence, which might call for gender-specific diagnostics (Feeney and Noller 1990; Sussman 2010).

6.5 Assessment Approaches

The symptomatology of LA has not yet been sufficiently explored or described. Therefore, there is not yet an evidence-based instrument that can be used to diagnose this disorder or to assess specific symptoms.

In the 1970s, the three primary love styles were first introduced: eros, ludus, and storge. Eros describes romantic and passionate love, ludus a game-playing love style that involves the avoidance of attachment, and storge a kind of love that is very focused on friendship and trust. With these primary love styles came three secondary ones: mania, pragma, and agape. Mania involves possession and dependence on the partner, pragma is focused on a relationship based on logic, and agape describes a self-less and very giving kind of love (Lee 1973).

A few years later, it was proposed that this differentiation was not without flaw and could be reduced to resemble the three attachment styles proposed by attachment theory. Attachment theory proposes that a child develops a particular attachment style based on the parents' responsiveness to the child's signals in the first few years of its life. The three styles of attachment are secure, avoidant, and anxious-ambivalent attachment. Secure attachment results from appropriately responsive parents, avoidance from too little responsiveness. Anxious-ambivalent attachment results from fluctuating responsiveness, leaving the child unable to predict the parent's response to its signals. It has been proposed that secure attachment results in eros and a less extreme form of agape, anxious-ambivalent attachment resembles the mania love style, and avoidant attachment in the equivalent of the ludus love style (Ainsworth et al. 2015; Shaver et al. 1988).

Based on these early works, it has been hypothesized that an anxious-ambivalent attachment style could predict a relationship behavior that could be described as LA. This hypothesis was examined in a study in 1990, which found that adults who showed the anxious-ambivalent attachment style had a more extreme approach to love than the other attachment styles. They show a high degree of self-sacrifice, obsessive preoccupation, and emotional reliance on the partner. Additionally, they report lower levels of self-esteem than securely attached participants.

These results and the above-explained similarities between LA and substance use disorder might be a good base for developing tools that specifically assess the symptomatology of LA.

6.6 Treatment Approaches

Because LA is not yet recognized as a formal mental health disorder, there is little information on treatment approaches and their efficacy in this specific behavioral pattern.

The most common and popular intervention for LA is self-help groups, such as Sex and Love Addicts Anonymous (SLAA), the majority being 12-step-based groups. As the name of the SLAA suggests, most of these are not solely focusing on LA but include the treatment of sex addiction. Regarding self-help groups, to the best of the authors' knowledge there are no controlled studies published assessing their efficacy in the treatment of LA. Due to the inherently inclusive nature of these groups, admission criteria to self-help groups are often not very strict or specific. The symptoms of LA overlap to a certain degree with the symptoms of various other diagnoses that affect romantic behavior (e.g., borderline personality disorder or sex addiction), and this overlap in symptoms makes it difficult to determine the actual diagnosis of group members (Sanches and John 2019).

LA seems to involve dysfunctional thought processes that suggest cognitive-behavioral therapy (CBT) as potentially beneficial in treating this disorder. It has been suggested that LA might best be treated using rational self-counseling. This approach involves the improvement of the separation between feelings/thoughts and facts regarding the relationship. This approach might help alleviate LA symptoms, though unfortunately, no studies assessing CBT as a treatment of LA could be found (Sanches and John 2019).

People who suffer from LA often report attachment difficulties in early childhood. Therefore a psychodynamic treatment approach might sound promising. However, no studies examining this could be found. This treatment approach might prove difficult for this particular disorder as the patients suffer from attachment-related problems and might be prone to transferential and counter-transferential reactions (Sanches and John 2019).

Group therapy is another psychosocial intervention that can be considered for the treatment of LA. Unfortunately, only one study assessing this treatment approach could be found in the literature. In this particular study, psychodrama group therapy was used. The study participants showed some improvement on a scale assessing a relationship's healthiness, but these results should be interpreted carefully as no control group was present in this study, and five out of the eight subjects also met the criteria for Major Depressive Disorder (Lorena et al. 2008).

Regarding the psychopharmacological treatment of LA, no data is available of studies assessing any pharmacological agents' safety and efficacy in treating LA. To assess these, LA would first need to be clearly defined, and the bioethical issues involved in the pharmacological treatment of this addiction would need to be fully considered. It would not be possible to target only the symptoms a person experiences in one specific relationship as a pharmacological treatment would potentially influence that person's social behavior and experience as a whole. This kind of treatment would only be ethically considerable if it happened voluntarily (Earp et al. 2017; Sanches and John 2019). Due to the potential benefits patients with this addiction could gain from pharmacological treatment, it is nevertheless worth exploring potential options.

Some pharmacological medications can be hypothesized to be beneficial in the treatment of LA although, to date, there is no clinical evidence supporting their efficacy. Treatment with certain antidepressants, specifically selective serotonin reuptake inhibitors (SSRI), might be a promising approach as these can be efficient in treating obsessive/compulsive disorder, which shows some similarities with LA. SSRIs seem to reduce dopaminergic transmission in the VTA, which has been discussed above as potentially involved in LA's symptomatology. People who suffer from LA tend to show impulsive behavior, and as such, treatment with mood stabilizers like lithium could also be considered. As antipsychotics can act as dopamine antagonists, they might also show beneficial effects in the treatment of LA. Finally, assessing the efficacy of treating LA with neuropeptides such as OT and vasopressin can be considered as animal research has shown a link between these and binding behavior (Sanches and John 2019).

Because of a lack of clinical trials assessing the safety and efficacy of pharmacological agents in the treatment of LA, pharmacological treatment of this disorder can, at this point, not be recommended.

6.7 Final Conclusions

Despite some controversies and a lack of consensus as for its nosological status, available evidence suggests that LA does exist as an independent diagnostic construct. Given the potential implications associated with its proper delineation and concerns about the risk of "medicalization" of falling in love, a behavior considered universal and, to some extent, normal, large scale studies aiming at a better characterization of LA from a phenomenological standpoint and, maybe more important, of its diagnostic boundaries.

It is likely that LA should be seen, on the one hand, from a spectrum perspective, with a continuum (without a clear limit) between what would be a so called "normal" love experience and pathological love or LA. On the other hand, research aiming at clarifying the limits between LA and other conditions that involve dysfunctional affective relationships, particularly borderline personality disorder and dependent personality disorder, is crucial for the establishment of putative diagnostic criteria for LA and their subsequent validation. That, on its turn, will enable systematic investigations addressing the efficacy of therapeutic interventions targeting this condition.

References

Ainsworth MDS, Blehar MC, Waters E, Wall SN (2015) Patterns of attachment: a psychological study of the strange situation. Psychology Press and Routledge Classic Editions. Taylor & Francis Group, New York

Anda RF et al (2006) The enduring effects of abuse and related adverse experiences in childhood. Eur Arch Psychiatry Clin Neurosci 256:174–186. https://doi.org/10.1007/s00406-005-0624-4

Anderson SM, Pierce RC (2005) Cocaine-induced alterations in dopamine receptor signaling: implications for reinforcement and reinstatement. Pharmacol Ther 106:389–403. https://doi.org/10.1016/j.pharmthera.2004.12.004

American Psychiatric Association (2013) Diagnostic and statistical manual of mental disorders (5th ed.). https://doi.org/10.1176/appi.books.9780890425596

Aragona B, Wang Z (2009) Dopamine regulation of social choice in a monogamous rodent species. Front Behav Neurosci 3. https://doi.org/10.3389/neuro.08.015.2009

Arias-Carrión O, Stamelou M, Murillo-Rodríguez E, Menéndez-González M, Pöppel E (2010) Dopaminergic reward system: a short integrative review. Int Arch Med 3:24. https://doi.org/10.1186/1755-7682-3-24

Burkett JP, Young LJ (2012) The behavioral, anatomical and pharmacological parallels between social attachment, love and addiction. Psychopharmacology 224:1–26. https://doi.org/10.1007/s00213-012-2794-x

Caldwell HK (2012) Neurobiology of sociability. In: López-Larrea C (ed) Sensing in nature. Springer US, New York, NY, pp 187–205. https://doi.org/10.1007/978-1-4614-1704-0_12

Cook DR (1987) Self-identified addictions and emotional disturbances in a sample of college students. Psych Addict Behav 1:55–61. https://doi.org/10.1037/h0080429

Dreyer JK, Herrik KF, Berg RW, Hounsgaard JD (2010) Influence of phasic and tonic dopamine release on receptor activation. J Neurosci 30:14273–14283. https://doi.org/10.1523/jneurosci.1894-10.2010

Earp BD, Wudarczyk OA, Foddy B, Savulescu J (2017) Addicted to love: what is love addiction and when should it be treated? Philos Psychiatry Psychol 24:77–92. https://doi.org/10.1353/ppp.2017.0011

Feeney JA, Noller P (1990) Attachment style as a predictor of adult romantic relationships. J Pers Soc Psychol 58:281–291. https://doi.org/10.1037/0022-3514.58.2.281

Hostetler C, Ryabinin A (2013) The CRF system and social behavior: a review. Front Neurosci 7. https://doi.org/10.3389/fnins.2013.00092

Inagaki TK (2018) Opioids and social connection. Curr Dir Psychol Sci 27:85–90. https://doi.org/10.1177/0963721417735531

Lee JA (1973) The colors of love: an exloration of the ways of loving. New Press, Ontario

Lorena A, Sophia EC, Mello C, Tavares H, Zilberman ML (2008) Group therapy for pathological love. Braz J Psychiatry 30:292–301. https://doi.org/10.1590/S1516-44462008000300019

McGregor IS, Bowen MT (2012) Breaking the loop: oxytocin as a potential treatment for drug addiction. Horm Behav 61:331–339. https://doi.org/10.1016/j.yhbeh.2011.12.001

Parsaik AK, Abdelgawad N, Chotalia JK, Lane SD, Pigott TA (2017) Early-life Trauma in hospitalized patients with mood disorders and its association with clinical outcomes. J Psychiatr Pract 23. https://doi.org/10.1097/PRA.0000000000000202

Prescott CA, Kendler KS (1999) Genetic and environmental contributions to alcohol abuse and dependence in a population-based sample of male twins American. J Psychiatry 156:34–40. https://doi.org/10.1176/ajp.156.1.34

Ross HE, Young LJ (2009) Oxytocin and the neural mechanisms regulating social cognition and affiliative behavior. Front Neuroendocrinol 30:534–547. https://doi.org/10.1016/j.yfrne.2009.05.004

Sanches M, John VP (2019) Treatment of love addiction: current status and perspectives. Eur J Psychiatry 33:38–44. https://doi.org/10.1016/j.ejpsy.2018.07.002

Shaver P, Hazan C, Bradshaw D (1988) Love as attachment: the integration of three behavioral systems. In: Sternberg RJ, Barnes ML (eds) The psychology of love. Yale University Press, New Haven, pp 68–99

Sussman S (2010) Love addiction: definition, etiology, treatment. Sex Addict Compuls 17:31–45. https://doi.org/10.1080/10720161003604095

Sussman S, Lisha N, Griffiths MD (2011) Prevalence of the addictions: a problem of the majority or the minority? Eval Health Prof 34:3–56. https://doi.org/10.1177/0163278710380124

Swendsen JD, Merikangas KR (2000) The comorbidity of depression and substance use disorders. Clin Psych Rev 20:173–189. https://doi.org/10.1016/S0272-7358(99)00026-4

Thege BK, Hodgins DC, Wild TC (2016) Co-occurring substance-related and behavioral addiction problems: a person-centered, lay epidemiology approach. J Behav Addict 5:614–622. https://doi.org/10.1556/2006.5.2016.079

Timmreck TC (1990) Overcoming the loss of a love: preventing love addiction and promoting positive emotional health. Psychol Rep 66:515–528. https://doi.org/10.2466/pr0.1990.66.2.515

Chapter 7
Sex Addiction

Manpreet Dhuffar-Pottiwal

Abstract Although empirical literature into sex addiction has flourished in the last two decades, the lack of universal agreement, and diagnostic criteria within such studies was reflected in the paucity of treatment provisions for those experiencing compulsive sexual behaviors. Since its ultimate inclusion in the International Classification of Diseases-11 (ICD-11), Compulsive Sexual Behavior Disorder (CSBD) provides a solid foundation in which gold-standard treatment interventions can be designed and implemented. This chapter will provide a theoretical overview of this issue, its biological basis, co-morbidities, prevalence rates, assessment approaches and treatment approaches reported in the literature. While research in CSBD has accelerated and there is evidence that examines the lasting consequences of the disorder, the field has been has subjected to a number of treatment barriers. The implications of these are further discussed.

Keywords Compulsive sexual behavior disorder · Compulsive sexual behavior · Hypersexuality · Sex addiction treatment · Treatment interventions · Behavioral addictions

7.1 Introduction

Over the last few decades in particular, sex addiction or compulsive sexual behaviors have not only become more accepted in society (e.g., the public admittance of famous figures) but it has also become a topic of scientific interest (Griffiths and Dhuffar 2014; Grubbs et al. 2020; Kafka 2010). With this evolution, researchers and clinicians in the field have heightened understanding about the concept of an excessive appetite pertaining to sexual behaviors. While, compulsive sexual behaviors have remained controversial for a number of years (Dhuffar 2015), hypersexuality has been discussed under difference labels including early categorizations such as nymphomania and satyriasis (Krafft-Ebbing 1886/1965) to more familiar

M. Dhuffar-Pottiwal (✉)
The University of the West of England, Faculty of Health and Applied Sciences, Frenchay Campus, Bristol, UK
e-mail: Manpreet.dhuffar-pottiwal@uwe.ac.uk

H. M. Pontes (ed.), *Behavioral Addictions*, Studies in Neuroscience, Psychology and Behavioral Economics, https://doi.org/10.1007/978-3-031-04772-5_7

terminology including sexual addiction (Carnes 1986, 1998, 2001; Carnes et al. 2005; Goodman 1993, 2001; Herkov 2006; Leedes 2001; Myers 1995; Ragan and Martin 2000; Schneider and Irons 2001; Schwartz and Brasted 1985), sexual compulsivity (Coleman 1987; Cooper and Lebo 2001; Herring 2001; Quadland 1985; Travin 1995; Weissberg and Levay 1986), compulsive sexual behavior (Black et al. 1997; Coleman 1991; Coleman et al. 2000; Raymond et al. 2003), sexual dependency (Wan et al. 2000), sexual impulsivity (Barth and Kinder 1987; Gabbard and Bennett 2005), and hypersexuality (Brandell and Nol 1992; Finlayson et al. 2001; Kafka 1997, 2001, 2003, 2010; Orford 1978; Reid 2007; Reid and Carpenter 2009; Rinehart and McCabe 1997, 1998; Salzman 1972). Although the debates concerning the addictive and compulsive conceptualizations of hypersexuality have continued, the empirical research in 2020 has outweighed the theoretical speculations outlined in earlier literature by Carnes in (1983) and Gold and Heffner (1998).

Since there was no universal agreement on the classification, diagnostic criteria, and—by default—treatment until recently, proponents of the field (i.e., Kafka 2010; Reid 2013) worked towards the development of Hypersexual Disorder introduced by Kafka (2010) for potential inclusion in the American Psychiatric Association's fifth edition of the *Diagnostics and Statistics Manual for Mental Health Disorders* (DSM-5). Kafka (2010) defined Hypersexual Disorder as a "sexual desire disorder characterized by an increased frequency and intensity of sexually motivated fantasies, arousal, urges and enacted behavior in association with an impulsivity component—a maladaptive behavioral response with adverse consequences" (p. 385). Furthermore, to meet the criteria for Hypersexual Disorder, sexual behaviors could not be a consequence of a substance. While field trials into Hypersexual Disorder were promising (Reid et al. 2012), Hypersexual Disorder was not included in the DSM-5 due to apprehensions relating to its diagnostic criteria, cultural limitations, and general lack of acceptance within psychiatry (Kafka 2014). As asserted by Grubbs et al. (2020), although Hypersexual Disorder was rejected from the DSM-5, research into this area has "flourished, ultimately culminating in diagnostic recognition of specific cases of out-of-control sexual behaviors" (p. 4). Advancements in technology now have also become an intrinsic part of hypersexuality, as they have the ability to influence the onset, progression, and maintenance of hypersexual behaviors.

It is also argued that modern ways of experiencing compulsive sexual behaviors do not mandate physical contact (Weiss 2013a). The addition of the COVID-19 pandemic and its social and economic have induced widespread anxiety, depression and an increase in other psychological difficulties, as well as the expected increase in addictive behavior, both substance and behavioral addictions (Qiu et al. 2020). Growing evidence has suggested that sex addiction is an important clinical reality with the potential of harmful consequences if left untreated (Dhuffar 2015; Dhuffar and Griffiths 2016, 2017; Kraus et al. 2018). Therefore, the setting of research pertaining to sex addiction has evolved in line with the proposal and subsequent inclusion of CSBD in the 11th edition of the *International Classification of Diseases* (ICD-11 2019). In parallel to Hypersexual Disorder, CSBD is defined as: *the persistence, repetitive engagement in sexual behaviors that result in impairment in one's life in addition to failed attempts to reduce or stop such behaviors* (Kraus et al. 2018).

Table 7.1 Diagnostic criteria for compulsive behavior disorder for ICD-11

Essential (required) features for compulsive sexual behavior disorder:
1. A persistent pattern of failure to control intense, repetitive sexual impulses or urges resulting in repetitive sexual behavior, must be manifested in one or more of the following: • 1a. Engaging in repetitive sexual activities has become a central focus of the person's life to the point of neglecting health and personal care or other interests, activities, and responsibilities (yes/no) • 1b. The person has made numerous unsuccessful efforts to control or significantly reduce repetitive sexual behavior (yes/no) • 1c. The person continues to engage in repetitive sexual behavior despite adverse consequences (e.g., repeated relationship disruption, occupational consequences, negative impact on health) (yes/no) • 1d. The person continues to engage in repetitive sexual behavior even when the individual derives little or no satisfaction from it (yes/no)
2. The pattern of failure to control intense, sexual impulses or urges and resulting in repetitive sexual behavior is manifested over an extended period (e.g., 6 months or more) (Must be met)
3. The pattern of repetitive sexual behavior causes marked distress or significant impairment in personal, family, social, educational, occupational, or other important areas of functioning (Must be met)

Note for rule out Distress that is entirely related to moral judgments and disapproval about sexual impulses, urges, or behaviors is not enough to meet this requirement

Although the CSBD has been included in the ICD-11, a conventional position has been recommended. Given that we do not yet have definitive information on whether the processes involved in the onset and maintenance of the disorder are equivalent to substance abuse disorders, gambling, and gaming (Kraus et al. 2016), CSBD is not included in the grouping of disorders due to substance and addictive behaviors, but rather in that of impulse control disorders. (Kraus et al. 2018). The diagnostic criteria for this new disorder are displayed in Table 7.1.

Due to being in its infancy, CSBD will be used interchangeably with sex addiction and Hypersexual Disorder throughout this chapter to provide an overview of its evolvement. Additionally, some scholars (i.e., Goodman 1992; Griffiths and Dhuffar 2014; Schneider 2000) argued that the disorder remains a clinical reality, regardless of what it is called.

7.2 Biological Basis of Sex Addiction

Neuroscientific studies of CSBD remain in their infancy and despite the considerable increase in studies, relatively little is known about the neural underpinnings of CSBD (Kowalewska et al. 2018). Given the lack of diagnostic acknowledgement until recent years, this is expected. In their review, Kowalewska et al. (2018) examined scientific literature pertaining to the neural mechanisms of CSBD from three standpoints: addictive, impulse-control, and obsessive–compulsive. They found some evidence to support the association between CSBD and increased sensitivity for erotic rewards or cues predicting these rewards (Voon et al. 2014; Gola et al. 2017). Studies included in the review also suggested that symptoms of CSBD are correlated with higher levels of anxiety (Gola et al. 2016).

Davis (2001; Davis et al. 2002) and more recently, Laier and Brand (2014) proposed a cognitive–behavioral theory of problematic internet use and cybersex addiction, in which they view the concepts as arising from a unique pattern of internet-related cognitions and behaviors. Reward circuits in the brain are typically activated by 'natural' positive reinforcers such as food, water, and sex—all of which are vital to survival. However, 'unnatural' reinforcers such as drugs, alcohol, gambling, and internet use can prove more powerful, causing people to neglect personal hygiene, work, food, and health. The reward-deficiency hypothesis suggests that those who achieve less satisfaction from natural rewards turn to substances to seek an enhanced stimulation of reward pathways (Blum et al. 1996). Certain activities carried out on the internet (e.g., pornography use, gambling, video gaming, social networking, and stock market trading) have the capacity to provide immediate reward with minimal delay, mimicking the stimulation provided by alcohol or drugs.

Preliminary evidence suggests that dopamine may contribute to CSBD (Kraus et al. 2016). In Parkinson's disease, dopamine replacement therapies have been associated with CSBD and other impulse-control disorders (Weintraub et al. 2010). Incentive motivational processes relate to sexual cue reactivity. CSBD vs non-CSBD men had greater sex cue-related activation of the anterior cingulate, ventral striatum, and amygdala (Voon et al. 2014). In CSBD participants, functional connectivity correlated with cue-related sexual desire, therefore echoing findings reported in substance use disorder research (Voon et al. 2014). Furthermore, CSBD men showed enhanced attentional bias to pornographic cues, implicating early attentional orienting responses as in substance use disorders (Mechelmans et al. 2014). An additional review by Grubbs et al. (2020) identified 21 studies that adopted neuroscientific methods to investigate CSBD, including structural magnetic resonance imaging, functional magnetic resonance imaging and electroencephalograms (Grubbs et al. 2020—see review for a comprehensive summary of studies). It was concluded that across neuroimaging studies, the research on CSBD is only just developing, without sufficient controls for confounding variables (e.g., comorbidity) and samples only comprising heterosexual men.

7.3 Specific Comorbidities in Sex Addiction

7.3.1 Substance Use and Comorbidities with Other Behavioral Addictions

Early literature (i.e., Carnes et al. 2005) has sought to understand addiction interaction and comorbidities. Specifically, Carnes et al. (2005) examined a sample of 1603 sex addicts and found that 69% of men and 79% of women reported a lifetime prevalence of other addictive and abusive behaviors. Furthermore, 40% of heterosexual men, 40% of heterosexual women, and 60% of homosexual men engaged in compulsive sexual behaviors simultaneously involved in other addictive behaviors such as substance abuse, gambling, or eating disorders.

Moreover, Sussman et al. (2011) examined the relationship between the co-occurrence of several addictions and estimated that 47% of sex addicts had a substance-related disorder. Similarly, Reid and Meyer (2016) conducted a review of the literature and highlighted an increased rate of substance use (i.e., alcohol) with CSBD. In a study of pathological gamblers, Grant and Steinberg (2005) found that 19.6% of their participants also met the criteria for sex addiction. 70% of the participants who met the criteria for both disorders reported that CSBD had preceded problematic gambling behaviors. Finally, Power (2007) outlined the co-occurrence and cross tolerance of food and sex addiction and argued that both processes are clinically significant.

7.3.2 Technological Factors

Cybersex addiction has been referred to as the "crack, cocaine of sexual compulsivity" given its ability for rapid escalation (Cooper et al. 1998). The advances of the internet have increasingly eliminated traditional offline methods of obtaining pornographic material, thus decreasing the sense of shame yet increasing isolation and the escapism into a fantasy world that to some extent replace real-life relationships (Dhuffar and Griffiths 2015a). This growth, in large part, can be attributed to greater access, affordability of, and anonymity while on the computer (Cooper 1998). Additionally, there are a variety of positive impacts that the use of the internet has had on the experiences of people's sexuality globally, especially among women (Leiblum 2001) and youth (Lou et al. 2006), as well as extremely marginalized populations, such as gay (McLelland 2002), lesbian (Correll 1995), bisexual (Koch and Schockman 1998), transgendered, and disabled individuals (Kaufman et al. 2007). The internet provides a 'safe' space for sexual exploration that poses less physical and social danger than offline activities (Turkle 1995; Weiss and Schneider 2006). However, such technological development has given rise to problematic internet-enabled sexual behavior and has led to a minority of individual's family relationships, work productivity, and academic success being negatively affected.

Sexual activities by females in the electronic age are often misunderstood, under-assessed, and under and/or misdiagnosed due to the wide array of various media and sexual outlets (Ferree et al. 2012; Laier and Brand 2014). The fast paced and evolving world of technology provides a broad arena for which sexual-based activities such as sexting, cybersex, photo exchanging, and streaming real-time video are now experienced (Weiss 2010). This has changed the landscape of how women define and experience intimacy, connection, and attachment, how they manage social skills, and how females deal with anxiety and self-esteem. There is no doubt that today's technology is a contributing factor to the etiology of sex and love addiction (SLA), especially among women who because of cultural and familial influences engage in anonymous pursuits without the presence of shame, thus increasing the risk of addiction (Ferree et al. 2012).

Corley and Hook (2012) supported this concept, they reviewed changes in use of internet by women over the past decade compared to previous decades. Data were collected from a sample of 525 self-identified female sex and love addicts (FSLA) and non-addicts. Data collected related to internet use revealed significant differences on hypersexual behavior and cybersex subscales between women who identify as FSLA having cybersex problems, FSLAs without cybersex problems, and women without cybersex or FLSA. Findings showed that 36% of the sample identified as solely a sex addict. Clinical observations in which significant differences were found included symptoms of depression, withdrawal, attempted suicide, childhood sexual abuse, and exposure to pornography as a child. This study was significant for a number of reasons as it reported subgroup differences where females were either love and sex addicts or solely sex addicts. It showed how cybersexual behaviors had changed in the last two decades and it reported clinical observations whereby there was an association to child sexual abuse and suicide attempts in adulthood and early exposure to pornography was a significant predictor of hypersexuality.

In addition to the above, Laier and brand (2014) conducted a study that investigated tendencies towards cybersex addiction in women by comparing Internet Pornography Use (IPU; $n = 51$) versus Non-Internet Pornography Use (NIPU; $n = 51$). Findings showed that female IPU rated internet pornographic pictures as more arousing and reported greater craving due to internet pornographic picture presentation compared with NIPU. Such findings provide greater insights to the various types of behaviors among male and female sex addicts.

7.3.3 Psychopathology

Numerous studies have examined psychopathological comorbidities associated with CSBD (e.g., Dhuffar et al. 2015; Kraus et al. 2015; Reid et al. 2012; Reid 2013; Rinehart and McCabe 1997).

Reid et al. (2012) investigated facets of personality (with the administration of the NEO-Personality Inventory) in a treatment-seeking sample of hypersexual women ($n = 31$) compared with hypersexual men ($n = 47$). Results indicated a number of striking parallels between the two groups including similar levels of impulsivity, emotional dysregulation and difficulties coping with stress. Hypersexual women exhibited higher levels of distrust towards others, lower levels of confidence and ambition, and a greater inclination for excitement and stimulation. In another study, Odlaug et al. (2013) examined sexual behaviors and their consequences in 1837 university students. The authors reported that 2% of the sample ($n = 36$) were classified as having sex addiction using the compulsive sexual behavior assessment tool of the Minnesota Impulsive Disorders Interview. The results also showed that university students with sex addiction reported more depressive and anxiety symptoms, higher levels of stress, poorer self-esteem, and higher rates of social anxiety disorder, attention-deficit/hyperactivity disorder, compulsive buying, pathological gambling, and kleptomania. Even with little research on hypersexual behaviors within various populations, some studies have indicated that women do engage in compulsive high-risk sexual behaviors accompanied by a variety of consequences such as emotional dysregulation (e.g., Arnold et al. 2002; Dodge et al. 2004; Gaither and Sellbom 2003).

The literature on personality in hypersexual populations to date has been limited to male samples. Although, one exception has examined whether women diagnosed with borderline personality disorder exhibited greater sexual impulsivity, but the researchers were unsuccessful in finding support for this hypothesis (Sansone and Muennich 2008) despite anecdotal reports of such an association (e.g., Rickards and Laaser 1999). Studies among hypersexual men using the Five Factor Model of personality (Digman 1990) have found traits (e.g., extroversion) related to emotional dysregulation, stress vulnerability, distrust, perceived competence, societal conformity, and impulsivity (Reid et al. 2011, 2009) to be implicated in CSBD.

Moreover, Boredom proneness has also been linked to increased solitary sexual behavior in adults (Gana et al. 2001) as well as greater tendencies to engage in obsessive thinking (Rinehart and McCabe 1998; Reid and Carpenter 2009). Similarly, Kastner and Sellbom (2012) examined the role of pathology (i.e., fearless-dominance and impulsive-antisociality) in hypersexuality (e.g., sensation seeking and compulsivity) in a sample of 393 college students. Results indicated that both facets of psychopathy were strong predictors of hypersexual behaviors. Furthermore, Dhuffar et al. (2015) explored the correlates of hypersexual behaviors and investigated whether hypersexual among university students ($n = 165$) can be predicted by variables relating to negative mood states (i.e., emotional dysregulation, loneliness, shame and life satisfaction) in a British sample. It was found that emotional dysregulation was a significant predictor of hypersexuality and that 19.4% of the sample met the criteria for sex addiction.

The terms affect and/or emotional dysregulation have also gained increased recognition and use in the sex addiction literature. In order to understand more entirely the significance of affect regulation, researchers (e.g., Batson et al. 1992; Bradley 2000; Goodman 1998; Nuttall 2000) have attempted to distinguish between these concepts either based on how they are formed for an individual or through their relative function for the individual. Affect is fundamentally apparent when we consider that it is responsible for awareness of internal states, as it focuses attention on the stimuli that provokes it. Similarly, van der Kolk and Fisler (1994) stressed the significance of the ability to regulate internal states as being central to self-definition and an individual's attitude towards their surrounding. For example, a maladaptive strategy like sex addiction helps the addict to escape from troublesome affects that then activate an addictive process. The solution in turn becomes the problem. Moskowitz and Roloff (2007) suggest that sex addiction is better explained by the depression/mood model, as some individuals, when in depressive states, become unconcerned about the consequences of their behaviors, in turn, engaging in more risky sexual activities (e.g., Wright 2012). Sexual behaviors and/or encounters may then serve as an instrument to recover psychological stability, since depressive states activate a search for satisfying the need (Bancroft and Vukadinovic 2004).

According to Goodman (1998), '*the addictive process originates in a disorder of self-regulation*' (p.197). The system he proposes is made up of three primary functions: (i) *Affect regulation functions:* the ability to avoid becoming overwhelmed by strong affective states with the use of self-soothing, self-enlivening, and arousal-balancing skills; (ii) *Self-care functions:* one's ability to provide protection and nurturance to themselves. The ability to recognize high-risk or dangerous situations and to respond appropriately is a self-protective skill; recognizing needs and setting priorities to meet them are linked to self-nurture; and (iii) *Self-governance functions:* internal beliefs, values, and standards that contribute to the experience of appropriate self-esteem and an interconnected sense of self.

Shame is a self-conscious emotion that was recently acknowledged as having a unique contribution to psychopathology (Candea and Szentagotai 2013). Furthermore, shame has been understood as an intrinsic contributor to sex addiction as noted by a number of researchers (e.g., Carnes 1983; Dhuffar and Griffiths 2016), therefore, it is not surprising that shame is used to manage and anaesthetize against hypersexuality. Birchard (2004) referred to this process as the oscillating cycle of control and release. Some studies (i.e., Dhuffar and Griffiths 2014; Gilliland et al. 2011; Reid et al. 2009; Reid et al. 2011, 2014; see Dhuffar and Griffiths 2016 for a review) explain the role of shame among those exhibiting hypersexual behaviors. The majority of studies reported that shame is a salient feature of hypersexuality, even within an increasingly technological society where sexual behavior is often carried out via internet-enabled applications. Despite the considerable theoretical debate in the last three decades, it is evident that shame being empirically assessed as a predictor of hypersexuality has only begun to be investigated in the last 10 years.

7.4 Prevalence Rates of Sex Addiction

To date, there have been no epidemiological studies, only a limited number of studies into CSB (Grubbs et al. 2020). Kafka (2010) noted that "*any operational definition for hypersexuality should first be derived from large non-clinical community samples where a normative range of sexual behaviors can be ascertained for*" (p. 379). Among population samples, all are limited to Western countries and cultures. In addition to this, the absence of consistent operationalization of CSBD has led to inaccurate prevalence rates despite its recent inclusion in the ICD-11. Previous studies (using various terms) that have made attempts to establish a baseline prevalence are outlined below.

Although some researchers in the field have estimated that about 3% to 6% of the general population in the United States may be sexually addicted, no nationally representative published studies have been carried out to support this statistic (Reid 2013). For example, Coleman estimated that 5% of the population met criteria for 'sexual compulsivity' Carnes et al. (1991) estimated a figure of 3–6% in the population of the United States suffer from sex addiction. Exactly how these conclusions were reached is not entirely clear. Kuzma and Black (2008) pointed out the difficulty in obtaining proper data as problematic sexual behavior is obviously of a private nature and continues to be stigmatized. Underreporting is highly likely and may also be due to a lack of knowledge or denial of the problem (Kuzma and Black 2008), but empirical studies in this regard are lacking (Ragan and Martin 2000).

Outside the United States, Langstrom and Hanson (2006) have reported prevalence rates of such behaviors in a Swedish sample based on frequency of masturbation, pornography consumption, and sexual infidelity. Findings from this study indicated that 12.1% of men (n = 1144) and 7.0% of women (n = 1142) met the criteria for out-of-control sexual behaviors.

Another study by Skegg et al. (2010) assessed a non-clinical sample of men (n = 474) and women (n = 466) in New Zealand with regard to their "out of control" sexual fantasies and urges over a 12-month period. While 12.7% of men and 6.7% reported having lack of control over their sexual fantasies and urges, these rates dropped significantly when actual sexual behavior was analyzed (0.8% and 0.6% respectively). Moreover, Winters et al. (2010) conducted an online study in a sample of 5834 men and 7251 women to investigate differences between sexually dysregulated behavior and levels of sexual desire. They found that 1.83% of men and 0.95% of women had significantly elevated scores on the Sexual Compulsivity Scale and had a history of seeking treatment for sexual compulsivity, addiction, and/or impulsivity.

More recently, beyond sex addiction, Lewczuk et al. (2020) examined pornography use in a Polish sample comprising 1,036 participants. Approximately 6% of lifetime pornography users (n = 880, 55.1% male) indicated concurrence with the statement "*I am addicted to pornography*", while only 3% of women concurred. The study also found that religion and moral incongruence were positive predictors of self-reported problems with pornography and self-perceived addiction. Relatedly, in a nationally representative sample of the Australian population (n = 20,094), among lifetime pornography users, 4.4% of men and 1.2% of women reported that they agreed with the statement "*You feel that you are addicted to pornography*" (Rissel et al. 2017). Importantly, these rates were only among those who reported ever viewing pornography. Subsequent analyses indicated that being male, between the age intervals of 16 to 19 years or 30 to 39 years, or living in a rural area were all associated with an increased likelihood of reporting addiction to sex.

While these studies relate to both male and female prevalence rates of hypersexuality, it is also worth mentioning that figures with regards to female-only data are also vague. According to Carnes, 3% of the American population are sexually addicted. In other words, of all-American sex addicts, 37.5% are female. Carnes' research also indicated that approximately 20% of those seeking help are female. Following on from the work of Carnes, Weiss (2011) also suggested that 8% to 12% of those seeking CSBD treatments are women. Dhuffar (2015) concluded that although females did make up for a portion of treatment-seekers in the United Kingdom, it was apparent that sought help through anonymized and private clinics, therefore, a representative statistic is warranted.

More recently, the 2018 United States national probability survey ($N = 2325$), recruited by Knowledge Panel (GfK Research), found that 10.3% of men and 7.0% of women reported at least occasional subjective feelings of distress over difficulty controlling sexual behaviors, impulses, or desires (Dickenson et al. 2018). Dickenson et al. (2018) found that both sexual and ethnic minorities were significantly more likely to report such concerns and were disproportionately represented among positive screens. Although prevalence rates (online and offline) remain unclear among the general population, pornography viewing online is presumed to be the most widespread form of cybersex, at least for males (Laier and Brand 2014; Brand et al. 2016). As noted by Grubbs et al. (2020) and Reid (2013), while research in this area has accelerated over the last two decades, the lack of theoretical integration and rigorous research design in the study of sex addiction prevents accurate estimates of the prevalence rates. And like other forms of sex research, under-reporting is not an unusual happenstance.

7.5 Assessment Approaches

7.5.1 Measurements of CSBD

Given the lack of consistency in definition, discrepancies in diagnostic criteria, and the general absence of theoretical unison, it is not surprising there is diversity in the measurement of CSBD. Currently (as of October 2020), there are about 40 self-report measures that have been formulated and identified in the empirical literature, 39 of these have been outlined in a prominent review undertaken by Grubbs et al. (2020). Additional (systematic) literature on methodological evaluations of assessments are outlined by Fernandez and Griffiths (2019); Grubbs et al. (2017), Grubbs et al. (2020); Hook et al. (2010); Marshall and Miller (2019) and; Womack et al. (2013). The most recent self-report measure was developed in line with the ICD-11 based screening, the *Compulsive Sexual Behavior Disorder Scale-19* (CSBD-19; Bothe et al. 2020). The construct validity and reliability of CSBD-19 were cross-validated in three languages across four independent samples from the United States, Hungary, and Germany. The CSBD-19 is therefore a sound measure of CSBD based on the latest ICD-11 diagnostic guidelines. Bothe et al. (2020) concluded that the CSBD-19 can be used in large-scale, cross cultural, multi-language studies, and can distinguish between those at increased risk of CSBD.

7.5.2 *Non-Clinical Populations*

7.5.2.1 Gender

Historically women have been under investigated in comparison with men and have generally not been targeted in treatment (e.g., Brady et al. 2009; Dhuffar 2015; Dhuffar and Griffiths 2016; Kaplan and Krueger 2010). While addiction research in the last two decades has identified significant differences between genders in relation to alcohol and drug use and treatment (Brady and Ashley 2005; Brady et al. 2009), gender differences that encompass biological, psychological, social, and cultural factors in CSBD remain scarce and often rely on theoretical speculations of 'potential' differences that lack empirical support.

Carnes et al. (1991) examined gender differences in sex addiction using the Sexual Behavior Inventory (SBI; Carnes et al. 1991). They investigated 752 males and 177 females in the United States. The participants completed the SBI that examined sexual behaviors, thoughts, and feelings. The main findings of this study showed that women tended to engage in sexual behaviors that distorted power by gaining control over another or by being a victim and the participation of behaviors that drew attention to them. Men were more excessive in sexual behaviors that objectified the partner and required little emotional involvement. The frequency of behaviors and perceptions of loss of control by sexually addicted women differed significantly to those experienced by male counterparts. Female sex addicts as a group scored the highest on fantasy sex, seductive role sex, trading sex, and pain exchange while the male group scored higher in fantasy sex, voyeuristic sex, exhibitionism, and anonymous sex. Females in this study did not intensify trends that existed on a socially 'acceptable' level. Rather, it was apparent that they were breaking social conventions about women (Carnes et al. 1991).

Although all individuals who experience sex addiction exhibit some similarities [i.e., decreased control, preoccupation, continued engagement despite negative consequences, and resulting shame (Schneider 1991)] that can be identified at the surface level, women often present with significant differences to men. Moreover, Dhuffar and Griffiths (2015a, b) qualitatively examined the conceptualizations of female sex addiction and recovery. Three women were interviewed in depth, and findings showed an escalation in risk, emotional dysregulation, and an apparent generation gap between the participants. For example, they ranged from opportunity-induced whereby the internet served as a mediator to cruise for sex to hunting for sexual conquests in person.

Cohen (2008) has argued that research into risky sexual behaviors among university students has been understudied and even more so among females. Cohen sought to expand the literature by investigating the role of sexual sensation seeking and sexual compulsivity on high-risk sexual behavior among heterosexual female university students. It was found that the combination of sexual sensation seeking and sexual compulsivity was the strongest predictor of risky sexual behaviors (Cohen 2008). While this study provides greater insight into risky sexual behaviors on the university campus, it was limited as it focused on: (i) heterosexual females, and (ii) risky sexual behaviors in regards to Sexually transmitted diseases (STDs) and human immunodeficiency virus (HIV), without accounting for other risks such as unwanted pregnancies and abortion. This was supported in a recent study by Klein et al. (2014) who examined which sexual behavioral patterns were associated with risky sexual behaviors and hypersexual behaviors using the Hypersexual Behavior Inventory (HBI) in a sample of 988 females. Findings indicated that increased frequency of masturbation, number of sexual partners, and pornography was associated with increased hypersexual behaviors. It was also reported that these findings do not support the typical characteristics of female hypersexuality (i.e., passive behaviors), instead, hypersexuality in women was characterized by impersonal sexual activity.

The physiological make-up of a woman can potentially intensify the vulnerability to SLA. The menstrual cycle alters levels of neuro-hormones that shape the brain structure and chemistry may regulate arousal (Cosgrove et al. 2007). Therefore, heightening the possibility to commit infidelity when ovulating (Dixit 2010). Since infidelity in a digital world does not necessitate physical contact (Weiss 2013b), such neuro-hormones may potentially lead women to engage in online adultery.

7.5.2.2 Age

While the symptoms of sex addiction may become more prevalent in adulthood, most sex and love addicts trace their initial acting out to adolescence or before (Griffin-Shelley 2002). Although age is recognized as a contributing factor to the onset of hypersexual behaviors, research into this particular factor is anecdotal, as the majority of evidence relies heavily on childhood trauma and/or sexual abuse as the root of sex addiction, without accounting for the evolving profile of a typical sex addict (Hall 2013). Therefore, it is also worth mentioning that sex addiction can be opportunity-induced, with reference to the availability of sexual information through various media that can be accessed by many adolescents. For example, in a survey of 350 individuals, 44% reported no history of sexual abuse or trauma, and 26.5% reported never having an attachment-related issue (Hall 2012). While these results lack empirical grounding, the findings from this survey shed light to the diversity in the onset and maintenance of related sexual activities that can potentially be escalated on the internet. Furthermore, Weiss (2013a) suggested that age plays a significant role in which we now engage in sexual experiences. Due to the technological shifts in society and generation gaps between 'digital natives' and/or 'screenagers' (i.e., those individuals under 30 years) and 'digital migrants' (i.e., those individuals over 30 years), the level of shame experienced online is often reduced as a result of anonymity.

However, Stulhofer et al. (2008), described a retrospective study to assess the relationship between early exposure to pornography and sexual compulsivity but administering a number of measures online among a sample of Croatian young adults. The sample comprised 1528 heterosexual adults (99 females) with a mean age of 22 years. No significant gender differences were observed in age of first sexual intercourse or lifetime number of sexual partners. In comparison to men, women reported being first exposed to pornography at slightly older age. On average, the first male exposure occurred at the age of 10 years while the first female exposure occurred a year later. Over one-third of women and almost a half of men reported the first contact with pornography by the age of 10 or younger. As anticipated by the authors, there was a significant gender difference in sexual compulsivity, with men scoring significantly higher on the four-item Sexual Compulsivity Scale than women.

Seegers (2003) investigated sex addiction symptoms among a sample of 240 college students, aged between 17 and 51 years. This study aimed to explore definition, categories, and prevalence rates, using Carnes' Sexual Addiction Screen Test (SAST) and the Woman Sexual Addiction Screening Test (W-SAST). It was hypothesized that there would be equal scores among both males and females that suggested sex addiction. Findings showed that a 32.2% of female participants fell in the category of needing to seek further evaluation and treatment while 9.6% were at risk. However, Seegers (2003) argued that further research should be conducted concerning the validity and reliability of the W-SAST. Though the measure may not be reliable, this study entailed a larger number of females than males and highlighted that there is an existence of female sex addiction.

7.5.3 Clinical Populations

In a study conducted by Schwartz and Southern (2000), results indicated that 76.2% of females within a clinical population of cybersex abusers showed a history of sexual abuse, and 54.4% met the criteria for Post-Traumatic Stress Disorder (PTSD). These findings coincided with those of Carnes et al. (1991), who found that 63% of sexually addicted women reported such abuse. Schwartz (2004) concurred that 'child abuse and neglect are common factors in the histories of individuals who manifest hypo- and hyper-sexuality (p. 572, Schwartz 2004). Wan et al. (2000) sought to determine recovery-related outcomes for 59 patients who undertook a residential program for sex addiction. Preliminary analysis of variables was associated with patient demographics, clinical profiles, recovery data, and patient evaluations of the program in order to identify possible factors underlying abstinence and relapse. The main findings of the study suggested that 71% of the sample reported relapse in some (or all) of their identified sex addiction behaviors at follow-up. While this study provides rich data for those in treatment, it did not separate gender and how they differed at recovery stages. Winters et al. (2010) made an attempt to assess dysregulated sexuality and differentiate it from high sexual desire among a sample of 14,396 participants, some of whom had sought treatment for sexual compulsivity, addiction, or impulsivity. The participants completed an online survey comprised of various sexuality measures. This was the first time that scores on the Sexual Compulsivity Scale have been reported for individuals who have sought treatment for sexual compulsivity, addiction, or impulsivity. Treatment seeking females scored higher on psychological symptoms and negative affects while their male counterparts and non-treatment females did not.

7.6 Treatment Approaches

Collectively, researchers (e.g., Carnes 1983; Kafka 2010; Reid et al. 2012) in this area tend to agree that CSBD stems from a desire to eliminate negative or dysphoric mood states (i.e., anxiety, pain, loneliness, worthlessness, powerlessness, shame, etc.) and unlike substance addiction treatment, whose goal is abstinence from all psychoactive substances, the therapeutic goal in CSBD is the abstinence only from compulsive, self-destructive, and self-defeating sexual behaviors (Schneider and Irons 2001). However, Leahy (2011) argues that if psychotherapy is to be taken seriously, it must rely on empirical research. Despite its frequent discussion in related literature, treatments for CSBD are limited and varied.

Grubbs et al. (2020) assert that although previous reviews (i.e., Dhuffar and Griffiths 2015a, b; Hook et al. 2014) have concluded that there are some promising treatments for CSBD, the absence of quality treatment studies are potentially due to the lack of formal recognition of such a behavior. However, the lack of gold-standard approaches to treatment such as a randomized control trial are also striking given that CSBD has been studied within the psychiatric literature over the last four decades (Dhuffar 2015; Grubbs et al. 2020). The following sections outline both psychological and pharmacological treatments available for CSBD respectively.

7.6.1 Psychological Treatments

Exploratory evidence indicates that Cognitive Behavioral Therapy (CBT; Hallberg et al. 2019), acceptance and commitment therapy, and mindfulness-based interventions (Reid et al. 2014) have the potential to reduce the intensity of hypersexual behaviors. Empirical literature pertaining to psychological interventions are discussed below.

Orzack and colleagues reported the first known empirically based outcome study in the United States highlighting that cognitive-behavioral and motivational interview techniques in a group format were effective forms of treatment and increased significantly the participants' quality of life and diminished the severity of depressive symptoms. In a similar study, Sadiza et al. (2011) evaluated the existence of sexually compulsive behaviors; emotional states (i.e., anxiety and depression) and the effectiveness of CBT in a group format among 12 men (mean age 28 years). Utilizing both self-report measures and semi-structured interviews, it was found that CBT was effective in stabilizing individual emotional states of depression, anxiety, and learning of useful coping management skills. Though group therapy addressed compulsive sexual behavior, the study was limited to men.

CBT has been suggested as the preferred mode of treatment for sexually compulsive disorders (e.g., Young, 2007) and from a cognitive perspective, thought processes of an addict will for no logical reason feel apprehensive when anticipating disaster (Hall and Parsons 2001). While addicts are not the only people who worry and anticipate negative happenings, they tend to do this more often than other people. Young (1998) first suggested that this type of catastrophic thinking might contribute to compulsive internet use in providing a psychological escape mechanism to avoid real or perceived problems. Later studies (e.g., Caplan 2002; Davis 2001; LaRose et al. 2001) hypothesized that other maladaptive cognitions such as overgeneralizing or catastrophizing and negative core beliefs also contribute to compulsive use of the internet. Those who suffer from negative thinking often suffer from low self-esteem and maintain pessimistic attitudes. They may be the ones drawn to the most anonymous interactive capabilities of the internet in order to overcome their personal feelings of low self-esteem and worth (Young 2007, 2011). Therefore, the cognitive model enables us to better understand why hypersexual and cybersexual addicts develop a habit or compulsive use and how negative self-thoughts maintain patterns of compulsive behavior (Weiss 2013b).

In addition to this, Weiss (2013b) argued that behavioral problems require behavioral forms of intervention and treatment. Therefore, a thorough assessment that focuses on antecedents (i.e., trauma and abuse) of sex addiction should not be sought in the initial stages of therapy. He further asserted that problematic sexual behaviors must first be contained using the CBT model (for at least three months) to structure treatment (i.e., psychoeducation, agenda and boundary setting, contracting, and reformulating to ensure rigor in sessions) before beginning to implement any deep-rooted psychodynamic and/or trauma therapy.

While Reid et al. (2020) have noted emotional dysregulation, stress proneness, and impulsivity as potential treatment foci for individuals with CSBD, culture and ethnicity are important themes that are often overlooked in the presentation of such behaviors.

7.6.2 Pharmacological Treatments

Proponents of pharmacotherapy have argued that it can be effective in the treatment of CSBD, however clinical awareness of the side effects can potentially help or hinder treatment seekers. Others have argued that psychotherapy preferably CBT should be used in association with pharmacotherapy and treatments of comorbidities (Malandain et al. 2020).

In a review of pharmacological treatments of CSBD, selective serotonin reuptake inhibitors (SSRIs) appear to be the first-line treatment for CSBD (Malandian et al. 2020) given that they reduce desire, arousal and orgasm (Kafka 2010).

In addition to SSRIs, naltrexone (commonly used for substance addictions) demonstrated efficacy for problem gamblers and could potentially alleviate symptoms of CSBD (Fong 2006). In a retrospective review of 19 adult males treated with naltrexone for compulsive sexual behavior, 89% indicated a reduction in CSBD (Raymond et al. 2010). Naltrexone, in general, decreases the hedonic experience of orgasm, and thereby may be helpful in reducing addictive behavior yet hurtful for healthy sexual relations (Holloway 2012). The psychiatrist should also be aware of the fact that, theoretically, opiate antagonists are known to increase sex hormones such as testosterone and can thereby increase sexual urges.

Anti-androgens that can directly diminish all phases of the sexual response cycle present in extreme cases of sexual acting out (e.g., sentenced sex offenders). While this does not relate to the intensity of sexual preoccupation, urges or behaviors, either deviant or non-deviant, lowering testosterone levels will significantly reduce sexual arousal in men. Guay (2009) proposed combination treatment of SSRIs and anti-androgenic treatments for extremely distressed patients. Other data pertaining to pharmacotherapy has been reported in cases studies. For example, Elmore (2005) administered psychotropic medication for non-paraphilic sex addiction. The participant was treated with venlafaxine, sodium valproate, and risperidone that helped normalize the capacity for sexual relations and orgasm, while paroxetine diminished their sexual drive and symptoms of depression. Further, Gulsun et al. (2007) administered clomipramine but compulsive masturbation persisted, and two weeks later valproic acid was added and the combination of the two different medications for a two-week period resulted in a decrease of masturbation, depressive symptoms, and psychological distress.

7.6.3 Barriers to Treatment

While treatment for CSBD emerges in contemporary society in line with the new ICD-11. It is evident that many National Health Services (in particular the United Kingdom, Griffiths and Dhuffar 2014; Dhuffar-Pottiwal, in preparation) do not have the provision for such a disorder. Consequently, there are a number of treatment-related challenges that can potentially be overlooked. The additional layer of the unprecedented global impact of the Novel Coronavirus pandemic upon daily activities and a way of life have further impacted the ways in which mental health care provisions are provided. This "black swan" moment (an unforeseen event that changes everything; Wind et al. 2020), has led to significant changes in the role of psychological processes as well as the potential of further harm on top of the pandemic (WHO 2019). In their study, Dhuffar and Griffiths (2016) examined barriers for women seeking CSB treatment, i.e., (i) innate barriers, (ii) social barriers, (iii) research and screening tool barriers, and; (iv) treatment barriers. The following section outlines these barriers and extends them to men, young adults as well as the unforeseen barriers due to the pandemic.

7.6.3.1 Innate Barriers

This section examines innate barriers to seeking treatment. Such barriers are operationally defined as those that are directly linked to the non-treatment seeking motivations of women themselves (Dhuffar and Griffiths 2016).

- *Denial*: Those with sex addiction may fail or refuse to acknowledge there is a problem concerning their sexual behaviors, especially if they feel they have gained control through a period of abstaining from all sexual-related activities. Carnes has suggested that sex addiction lies at one end of the spectrum, the other end being characterized by the lack of any form of sexual activity or involvement. As he considers this as a form of sexual 'self-starvation', the term 'sexual anorexia' was used to describe it and refers to it as an obsessive state in which physical, mental, and emotional task of avoiding sex dominates one's life. In this, the themes of disconnection, rejection, and fear of intimacy are significantly more apparent.
- *Not wanting to seek treatment*: Some individuals may know that their sexual behaviors are spiraling out-of-control but may choose not to seek treatment for it. Ferree highlighted that women are more inclined to experience ambivalence towards treatment seeking due to the strong message of guilt, shame, and religious restrictions that portray sex as sinful or dirty.
- *Occurrence of spontaneous remission*: It may be that some individuals have problems but are able to overcome them without the need to resort to professional help.
- *Committing suicide prior to seeking treatment*: Some individuals may commit suicide because of the shame associated with their sexual behaviors before they are able to seek treatment. Hall (2012) conducted a survey of 350 respondents and it was found that 19.4% reported had a serious desire to commit suicide. However, actual suicide rate as a consequence of sexual behaviors remains unknown.
- *Excess in youth is commonplace and may not be pathologized*: Youth is a time of general excess and young people may not see such behavior as problematic. Consequently, the behaviors may not be perceived as warranting treatment given the normative data (actual or perceived).
- *Not seeking treatment in general*: Individuals (particularly young persons) may be resistant to treatment for any problem, not just sex addiction.
- *Seeking other forms of mental health treatment before getting treatment for sex addiction*: Individuals may be more likely to seek help for common mental health problems (i.e., depression, anxiety, and trauma) that are 'socially acceptable' before seeking treatment for sex addiction due to potential stigma.
- *Treatment of other underlying problems that indirectly diminish hypersexual behaviors*: Given that women are more likely to seek help for an underlying mental health problem that maintains their hypersexual behaviors, such a treatment could indirectly help diminish their problematic sexual behaviors. However, treatment for another problem (e.g., depression) may help diminish hypersexual

behaviors if the sexual behavior is used as a coping strategy [therefore the potential of sex addiction being an Axis I disorder rather than an Axis II disorder as described by Kafka and Prentky].
- *Negative consequences not unique to hypersexuality*: The negative consequences of a number of sexual behaviors share similar traits to sexual addiction (i.e., STDs, unwanted pregnancies, loss of relationship, etc.); therefore, hypersexual behaviors may not be treated as an underlying factor in problem behavior especially among young persons (see Cohen 2008).

7.6.3.2 Social Barriers

This section examines social barriers to seeking treatment. Such barriers are operationally defined as those that directly impact the individual on a micro- (e.g., interpersonal environment, family and friends), meso- (i.e., organizations and community) and macro-system (i.e., public policy) level.

- *Family bailout*: Turner and Lieu highlighted that young people are most likely to seek help when the consequences of their behavior are most severe. However, if family are supportive and willing to assist with an unwanted pregnancy or the process of an abortion, then help is less likely to be sought.
- *Parents*: According to Griffin-Shelley (2002), the main obstacle for young person's seeking treatment is the parent. Firstly, most carry shame about sexual, relational, and cultural issues. Second, parents may fear being seen as 'bad parents' and also apprehensive to look at their own behaviors that potentially may have influenced the behaviors of their child (Griffin-Shelley, 2002).
- *Socially undesirable*: In women and young persons, a lack of sexual behaviors is viewed as desirable, that are associated with shame in ethnic minorities residing in Western societies, making seeking treatment less likely.
- *Peers*: Griffin-Shelley (2002) asserted that peers could be a barrier to help and treatment out of their own shame, embarrassment, or lack of understanding about sex addiction (i.e., lack of exposure to the concept in comparison to substance-related addictions).
- *Socio-economic status*: Attrition in mental health treatment is more prominent among those who are economically disadvantaged (Barrett et al. 2008).
- *Family systems*: The COVID-19 pandemic has led to limited privacy, thus hindering further the potential to access treatment. The living circumstances and the lack of a confidential space may lead to relapse if in treatment. Especially since there is the potential of three-four generations residing in one home.

7.6.4 Research and Screening Tool Barriers

This section examines research and screening tool barriers to seeking treatment. Such barriers are operationally defined as those that are directly linked to resources currently available to assess and/or formulate SA among females.

- *Lack of universal agreement*: Until recently, the absence of a diagnostic criteria has presented difficulties for medical and clinical professionals. While CSBD has now been included in the ICD-11, treatment interventions are yet to be implemented given that the provision of treatment status remains unchanged since data were collected in 2014 (Griffiths and Dhuffar 2014)) and again more recently in 2021 (Dhuffar-Pottiwal, in preparation).
- *Lack of research*: Given that sex addiction has only recently become a topic of interest in mainstream society (see Dhuffar 2015), treatment-related, empirically driven research remains understudied and—to an extent—outdated whereby randomized control trials and outcome studies are yet to be conducted.
- *Lying or distortion on self-report measures*: Individuals may be malingering and/or minimizing the effects of symptoms and consequences.
- *Invalid screening instruments for measuring compulsive behaviors in women*: Another reason for skewed prevalence rates may be the tools used to measure sexual behaviors. However, some self-report measures are gender-specific (i.e., Sexual Addiction Screening Test; Carnes et al. 1991).
- *Misinterpretation of questions*: The overuse of formal language and information about sex addiction may lead some women to endorse items they should not, and only doing so because they do not understand the measure.
- *Screening measures being used incorrectly*: The inconsistency in methodology, definitions, measures, and cut-off points may also account for the current heterogeneity in prevalence rates of female sex addiction.
- *Exaggeration of the problem by professionals to enhance career needs*: Chevalier and Griffiths (2004) highlighted that an explanation for increased rates of adolescent gambling may be that researchers who rely on funding in this area may overstate such rates for personal career gains. While it is essential to point this out, this does not appear to be the case with sex addiction, as any governing body in the United Kingdom does not fund it. However, it may explain elevated rates of females experiencing sex addiction in the United States.
- *Lack of funding*: As sex addiction was not officially recognized as an official mental health disorder, this results in minimal (or lack of) clinically-based research studies of different forms of treatment that can be subjected to direct comparison of outcome, or methodological flaws such as small sample sizes.

Therapeutic Barriers

This section examines specific treatment barriers to seeking psychological treatment. Such barriers are operationally defined as the lack of clinical treatments and/or psychotherapies currently available that potentially demotivate women to seek treatment within the National Health Service (NHS) or within the private sector in the United Kingdom.

- *Sex addiction may not viewed as a "clinical reality"*: sex addiction is still a concept that is not listed within diagnostic manuals (e.g., DSM-5), therefore, it is not only a challenge to diagnose but treatment is generally not funded by the Department of Health nor does it appear in the guidelines of *The National Institute for Clinical Excellence (NICE)* whereby treatment-seekers could potentially receive psychotherapy under national health care systems (e.g., NHS) (see Griffiths and Dhuffar 2014).
- *Lack of trained professionals*: Only a small number of therapists are trained to treat sex addiction, and a majority of them are in private practice.
- *Sex addiction requires a specialist service*: The misconception that sex addiction services are specialist services was a significant finding in Griffiths and Dhuffar's (2014), a follow up by Dhuffar-Pottiwal (in preparation) in 2020 indicated similar findings study, therefore may lead a treatment seeker to go to a more established treatment location in another country.
- *Digital natives vs. digital migrants*: If treatment is accessed, the therapist or counsellor may have limited understanding of some of the online behaviors that are involved within online sex addiction (Weiss 2013). Dhuffar and Griffiths (2015b) found that there was a significant difference in acting-out behaviors among women who were considered traditional sex addicts (i.e., engaging with sexual activity offline) versus contemporary sex addicts (i.e., where online sex was a central part in meeting potential sex partners offline).
- *Lack of women-only treatment programs*: The lack of availability of treatment services for sex addiction is reflected in the small number of those who seek help. A majority of treatment settings are designed for men (Hall 2013).
- *Treatment programs being unsuitable*: The treatment programs that are available may cover other aspects of sexuality (i.e., hyposexual disorder and/or other mental health disorders whereby hypersexual behavior is a secondary disorder).
- *Fear of failure ('failure neurosis')*: The potential fear of relapsing may prevent a sex addict from seeking treatment. The prevalence rates of 'failure neurosis' (coined by Laforgue 1941) among females (significantly more than males) has been documented in the psychodynamic literature (i.e., Freud 1957; Kanefield 1985) whereby women are more inclined to experience low self-esteem and perceive themselves as worthless.
- *Ambivalence in client-therapist relationship*: A therapist can only go with what the client brings to the therapeutic alliance. Clients may themselves create a barrier by concealing certain behaviors they typically engage in out of the therapy sessions (Miller and Rollnick 1991).
- *Lack of engagement when in treatment*: Although some individuals tend to seek therapeutic treatment in mental health or primary care settings, they are less likely to engage in specialist treatment programs.

- *Stigma towards treatment seeking*: While all addictions encompass an element of shame, shame is more prevalent in sex addiction than other addictions (Adams and Robinson 2001; Carnes 1983). This may lead individuals to attend self-help programs (such as *Sex Addicts Anonymous, Sex and Love Addicts Anonymous, and SMART Recovery*) that are anonymous in nature.
- *Length of referral process*: The length of time between initial referral, initial contact, and the beginning of regular sessions may impact on therapy adherence (e.g., the longer the time period between an initial referral and the onset of treatment, the worse the adherence; Godden and Pollock 2007).
- *Religious and/or cultural barriers*: Therapists and counsellors may have little understanding of sexual 'norms' in other religions, cultures and ethnic minorities.
- *Language barriers*: As with other addictions, some individuals may not get the treatment they need due to language barriers when attempting to access the limited treatment for sex addiction. While multi-lingual therapists and counsellors are readily available within the NHS in the United Kingdom, it must also be noted that some individuals may not want to see someone who is from a similar ethnic background to them due to the fear of being 'judged' and others in their community finding out. Additionally, the language of sex addiction may be difficult to translate as clients may potentially view this as derogatory. Therefore, increasing levels of shame and consequently, the patient declining treatment.
- *Shame from interpreters*: Language barriers may also lead therapists to seek the help of external interpreters who may have limited understanding of mental health in general, thus not being able to relay the information from the client to the therapist.
- *Forced transference to electronic therapy*: Although there are advantages to online therapy, the COVID-19 pandemic has resulted in no other option for some individuals to terminate therapy.

7.7 Conclusions

Despite the emergence of comprehensive empirical literature, it is evident treatment aspects of sex addiction remains in its infancy. Given the recent inclusion of CSBD in the ICD-11 and the ways in which this condition is experienced, it is anticipated that gold-standard treatments factoring in cultural differences; and the unprecedented changes of a global pandemic alongside other treatment targets mentioned above will be developed. Additionally, the idiosyncratic experience of sex addiction is shaped by what meaning treatment-seekers themselves place on their experiences of the process itself, rather than the content per se. Socio-cultural experiences play a significant role and often viewed as an extension of which sex addiction is formed and one that needs to mastered and understood.

References

Adams KM, Robinson DW (2001) Shame reduction, affect regulation, and sexual boundary development: essential building blocks of sexual addiction treatment. Sex Addict Compuls 8(1):23–44. https://doi.org/10.1080/10720160127559

Arnold P, Fletcher S, Farrow R (2002) Condom use and psychological sensation seeking by college students. Sex Relatsh Ther 17:355–366

Bancroft J, Vukadinovic Z (2004) Sexual addiction, sexual compulsivity, sexual impulsivity, or what? Toward a theoretical model. J Sex Res 41:225–234

Barrett MS, Chua WJ, Crits-Christoph P, Gibbons MB, Casiano D, Thompson D (2008) Early withdrawal from mental health treatment: implications for psychotherapy practice. Psychotherapy (chic) 45(2):247–267. https://doi.org/10.1037/0033-3204.45.2.247

Barth RJ, Kinder BN (1987) The mislabeling of sexual impulsivity. J Sex Marital Ther 13:15–23

Batson CD, Shaw LL, Oleson KC (1992) Differentiating affect, mood, and emotion: toward functionally based conceptual distinctions. In: Clark M (ed) Review of personality and social psychology, vol 13. Sage, Thousand Oaks, CA, pp 294–326

Birchard T (2004) The snake and the seraph: sexual addiction and religious behaviour. Couns Psychol Q 17:81–88

Black DW, Kehrberg LD, Flumerfelt DL, Schlosser SS (1997) Characteristics of 36 subjects reporting compulsive sexual behavior. Am J Psychiatry 154:243–249

Blum K, Cull JG, Braverman ER, Comings DE (1996) Reward deficiency syndrome. Am Sci 84:132–145

Bothe B, Potenza MN, Griffiths MD, Kraus, et al (2020) The development of the compulsive sexual behavior disorder scale (CSBD-19): an ICD-11 based screening measure across three languages. J Behav Addictions. https://doi.org/10.1556/2006.2020.00034

Bradley SJ (2000) Affect regulation and the development of psychopathology. Guildford, New York

Brady KT, Back SD, Greenfield SF (2009) Women and addiction: a comprehensive handbook. Guildford Press, New York

Brady T, Ashley O (2005) Women in substance abuse treatment: results from the alcohol and drug service study (ADSS). Substance abuse and mental health services administration, Rockville, MD. Retrieved September 20, 2020 from www.oas.samhsa.gov/WomenTXWomenTX.htm

Brand M, Snagowski J, Laier C, Maderwald S (2016) Ventral striatum activity when watching preferred pornographic pictures is correlated with symptoms of Internet pornography addiction. Neuroimage 129:224–232. https://doi.org/10.1016/j.neuroimage.2016.01.033

Brandell JR, Nol J (1992) Hypersexuality as a disorder of the self. Psychotherapy Patient 8:51–64

Candea DM, Szentagotai A (2013) Shame and psychopathology: from research to clinical practice. J Cogn Behav Psychother 1:97–109

Caplan SE (2002) Problematic Internet use and psychosocial well-being: development of a theory-based cognitive-behavioral measurement instrument. Comput Hum Behav 18:553–575

Carnes PJ (1998) The presidential diagnosis. Sex Addict Compuls 5:153–158

Carnes PJ (2001) Out of the shadows: understanding sexual addiction, 3rd edn. Hazelden, Center City, MN

Carnes PJ, Nonemaker D, Skilling N (1991) Gender differences in normal and sexually addicted populations. Am J Prev Psychiatry Neurol 3:16–23

Carnes PJ, Murray RE, Charpentier L (2005) Bargains with chaos: sex addicts and addiction interaction disorder. Sex Addict Compuls 12:79–120

Carnes P (1983) Out of the shadows: understanding sexual addiction. Minneapolis, MN: CompCare

Carnes PJ (1986) Progress in sexual addiction: an addiction perspective. SIECUS: Report 14:4–6

Chevalier S, Griffith MD (2004) Why don't adolescents turn up for gambling treatment (revisited)? J Gambl Issues 11. Available: http://www.camh.net/egambling/issue11/jgi_11_chevalier_griffiths.html. Last accessed 29 Sep 2020

Cohen JE (2008) Sexual sensation seeking and sexual compulsivity: predicting high risk sexual behaviour among female university students (Doctoral Dissertation, University of Memphis). Dissertation Abs Int 68(83-B):1562

Coleman E (1987) Sexual compulsivity: definition, etiology and treatment. Chem Dependency Intimacy Dysfunction 1:189–204

Coleman E (1991) Compulsive sexual behavior: new concepts and treatments. J Psychol Hum Sex 4:37–52

Coleman E, Gratzer T, Nesvacil L, Raymond NC (2000) Nefazodone and the treatment of nonparaphilic compulsive sexual behavior: a retrospective study. J Clin Psychiatry 61:282–284

Cooper A (1998) Sexuality and the Internet: surfing into the new millennium. CyberPsychol Behav 1:187–193

Cooper M, Lebo RA (2001) Assessment and treatment of sexual compulsivity: a multimodal perspective. J Soc Work Pract Addict 1:61–74

Corley MD, Hook J (2012) Women, female sex and love addicts, and the use of the Internet. Sex Addict Compuls 19:53–76

Correll S (1995) The ethnography of an electronic bar—the lesbian café. J Contemp Ethnogr 24:270–298

Cosgrove K, Mazure C, Staley J (2007) Evolving knowledge of sex differences in brain structure, function, and chemistry. Biol Psychiat 62:847–855

Davis RA (2001) A cognitive–behavioral model of pathological internet use. Comput Hum Behav 17:187–195

Davis RA, Flett GL, Besser A (2002) Validation of a new scale for measuring problematic internet us: implications for pre-employment screening. Cyberpsychol Behav 5:331–345

Dhuffar M, Griffiths M (2014) Understanding the role of shame and its consequences in female hypersexual behaviours: a pilot study. J Behav Addict 3(4):231–237

Dhuffar MK, Griffiths MD (2015a) A systematic review of online sex addiction and clinical treatments using CONSORT evaluation. Curr Addict Rep 2(2):163–174. https://doi.org/10.1007/s40429-015-0055-x

Dhuffar MK, Griffiths MD (2015b) Understanding conceptualisations of female sex addiction and recovery using interpretative phenomenological analysis. Psychol Res 5:585–603. https://doi.org/10.17265/2159-5542/2015.10.001

Dhuffar MK, Griffiths MD (2016) Barriers to female sex addiction treatment in the UK. J Behav Addict 5(4):562–567. https://doi.org/10.1556/2006.5.2016.072

Dhuffar MK, Pontes HM, Griffiths MD (2015) The role of negative mood states and consequences of hypersexual behaviours in predicting hypersexuality among university students. J Behav Addict 3:181–188. https://doi.org/10.1556/2006.4.2015.030

Dhuffar MK, Griffiths MD. The role of shame in sexual addiction. In: Birchard T, Benfield J (eds) The Routledge international handbook of sexual addiction. (Abingdon: Routledge, 13 Sep 2017), accessed 17 Oct 2020, Routledge Handbooks Online.

Dhuffar MK (2015) Female hypersexuality in the UK: a mixed-methods investigation and implications for contemporary clinical practice. PhD Thesis, Nottingham Trent University, Nottingham, U.K

Dhuffar-Pottiwal MK (in preparation) The treatment status of compulsive sexual behaviour disorder in the British NHS: a follow up study

Dickenson JA, Gleason N, Coleman E, Miner MH (2018) Prevalence of distress associated with difficulty controlling sexual urges, feelings, and behaviors in the united states. JAMA Netw Open 1(7):e184468. https://doi.org/10.1001/jamanetworkopen.2018.4468

Digman JM (1990) Personality structure: emergence of the five-factor model. Annual Review Psychol 41:417–440

Dixit J (2010) The cheat sheet. Psychology Today. Retrieved 26 Sep 2020 from https://www.psychologytoday.com/articles/201002/sex-ed-the-cheat-sheet?collection=36248

Dodge B, Reece M, Cole SL, Sandfort TGM (2004) Sexual compulsivity among heterosexual college students. J Sex Res 41:343–350

Elmore JL (2005) Psychotropic medication control of non-paraphilic sexual addiction in a female. Sex Relatsh Ther 20:211–213

Fernandez DP, Griffiths MD (2019) Psychometric instruments for problematic pornography use: a systematic review. Eval Health Prof 163278719861688. https://doi.org/10.1177/0163278719861688

Ferree M, Hudson L, Katehakis A, McDaniel K, Valenti-Anderson A (2012) Etiology of female sex and love addiction: a biopsychosocial perspective. Making Advances: A comprehensive guide for treating female sex and love addicts. Royston, GA: SASH, pp 44–66

Finlayson AJR, Sealy J, Martin PR (2001) The differential diagnosis of problematic hypersexuality. J Sex Addict Compulsivity 8:241–251

Fong TW (2006) Understanding and managing compulsive sexual behaviors. Psychiatry 3:51–58

Freud S (1957) The dynamics of transference. In: Strachey J (ed) (Trans.) The standard edition of the complete psychological works of Sigmund Freud, vol 12. London, UK: Hogarth Press, pp 97–108

Gabbard GO, Bennett TJ (2005) Dilemmas in the psychotherapy of sexually impulsive patients. Am J Psychiatry 162:859–865

Gaither GA, Sellbom M (2003) The sexual sensation seeking scale: reliability and validity within a heterosexual college student sample. J Pers Assess 81:157–167

Gana K, Trouillet R, Martin B (2001) The relationship between boredom proneness and solitary sexual behaviors in adults. Soc Behav Pers 29:385–389

Gilliland R, South M, Carpenter BN et al (2011) The roles of shame and guilt in hypersexual behaviour. Sex Addict Compuls 18:12–29

Godden S, Pollock A (2007) Waiting list and waiting time statistics in Britain: a critical review. Available: http://www.allysonpollock.com/wp-content/uploads/2013/04/CIPHP_2008_Godden_WaitingListWaitingTime.pdf. Last accessed 28 Sep 2020

Gola M, Lewczuk K, Skorko M (2016) What matters: quantity or quality of pornography use? Psychological and behavioral factors of seeking treatment for problematic pornography use. J Sex Med 13(5):815–824. https://doi.org/10.1016/j.jsxm.2016.02.169

Gola M, Wordecha M, Sescousse G, Lew-Starowicz M, Kossowski B, Wypych M, Makeig S, Potenza MN, Marchewka A (2017) Can pornography be addictive? An fMRI study of men seeking treatment for problematic pornography use. Neuropsychopharmacology 42(10):2021–2031. https://doi.org/10.1038/npp.2017.78

Gold SN, Heffner CL (1998) Sexual addiction: many concepts, minimal data. Clin Psychol Rev 18:367–381

Goodman A (1992) Sexual designation and treatment. J Sex Marital Ther 18:303–314

Goodman A (1993) The addictive process: a psychoanalytic understanding. J Am Acad Psychoanal 2:189–205

Goodman A (1998) A sexual addiction: an integrated approach. International Universities Press, Madison, CT

Goodman A (2001) What's in a name? Terminology for designating a syndrome of driven sexual behaviour. Sex Addict Compuls 8:191–213

Grant JE, Steinberg M (2005) Compulsive sexual behaviour and pathological gambling. Sex Addict Compuls 12:235–244

Griffin-Shelley E (2002) Adolescent sex and love addicts. In: Carnes PJ, Adams KM (eds) Clinical management of sex addiction. Routledge, New York, pp 343–360

Griffiths MD, Dhuffar MK (2014) Treatment of sexual addiction in the British national health service. Int J Ment Heal Addict 12(5):561–571. https://doi.org/10.1007/s11469-014-9485-2

Grubbs JB, Hook JN, Griffin BJ, Penberthy JK, Kraus SW (2017) Clinical assessment and diagnosis of sexual addiction. In: Birchard T, Benfield J (eds) Routledge international handbook of sexual addiction, 1st edn. Routledge

Grubbs JB, Grant JT, Lee BN, Hoagland KC, Davison P, Reid RC, Kraus SW (2020) Sexual addiction 25 years on: a systematic and methodological review of empirical literature and a an agenda for future research

Guay DR (2009) Drug treatment of paraphilic and nonparaphilic sexual disorders. Clin Ther 31(1):1–31

Gulsun M, Gulcaf Z, Aydin H (2007) Treatment of compulsive sexual behaviour with clomipramine and valproic acid. Clin Drug Invest 3:219–233

Hall P (2013) Understanding and treating sex addiction. Routledge, New York

Hall AS, Parsons J (2001) Internet addiction: college students case study using best practices in behaviour therapy. J Ment Health Couns 23:312–322

Hall P (2012) Survey of 350 UK people who identify as suffering with sex addiction, conducted via survey monkey. Unpublished manuscript

Hallberg J, Kaldo V, Arver S, Dhejne C, Jokinen J, Oberg KG (2019) A randomized controlled study of group-administered cognitive behavioral therapy for hypersexual disorder in men. J Sex Med 16(5):733–745. https://doi.org/10.1016/j.jsxm.2019.03.005

Herkov M (2006) What is sexual addiction? Retreived 12 Sep 2020, from http://psychcentral.com/lib/2006/what-is-sexual-addiction

Herring B (2001) Ethical guidelines in the treatment of compulsive sexual behavior. Sex Addict Compuls 8:13–22

Holloway KS (2012) Opioid mediation of learned sexual behavior. Socioaffect Neurosci Psychol 2:14874

Hook JN, Hook JP, Davis DE, Worthington ELJ, Penberthy JK (2010) Measuring sexual addiction and compulsivity: a critical review of instruments. J Sex Marital Ther 36(3):227–260. https://doi.org/10.1080/00926231003719673

Hook JN, Reid RC, Penberthy JK et al (2014) Methodological review of treatments for nonparaphilic hypersexual behavior. J Sex Marital Ther 40:294–308

Kafka MP (1997) Hypersexual desire in males: an operational definition and clinical implications for males with paraphilias and paraphilia-related disorders. Arch Sex Behav 26:505–526

Kafka MP (2001) The paraphilia-related disorders: a proposal for a unified classification of nonparaphilic hypersexuality disorders. Sex Addict Compuls 8:227–239

Kafka MP (2003) Sex offending and sexual appetite: the clinical and theoretical relevance of hypersexual desire. Int J Offender Ther Comp Criminol 47:439–451

Kafka MP (2010) Hypersexual disorder: a proposed diagnosis for DSM-V. Arch Sex Behav 39:377–400

Kanefield L (1985) Psychoanalytic constructions of female development and women's conflicts about achievement. Part I. J Am Acad Psychoanal 13:347–366

Kaplan MS, Krueger RB (2010) Diagnosis, assessment and treatment of hypersexuality. J Sex Res 47:181–198

Kastner RM, Sellbom M (2012) Hypersexuality in college students: the role of psychopathy. Personality Individ Differ 53:644–649

Kaufman M, Silverberg C, Odette F (2007) The ultimate guide to sex and disability. Cleis, San Francisco, CA

Koch NS, Schockman HE (1998) Democratising internet access in the lesbian, gay and bisexual communities. In: Ebo B (ed) Cyberghetto or cybertopia? Race, class and gender on the internet. Praeger, Westport, CT, pp 171–184

Kowalewska E, Grubbs JB, Potenza MN et al (2018) Neurocognitive mechanisms in compulsive sexual behavior disorder. Curr Sex Health Rep 10:255–264. https://doi.org/10.1007/s11930-018-0176-z

von Krafft-Ebbing R (1965) Psychopathia sexualis. (trans: Klaf F). Stein and Day, New York

Kraus SW, Rosenberg H, Tompsett CJ (2015) Assessment of self-efficacy to employ self-initiated pornography use-reduction strategies. Addict Behav 40:115–118. https://doi.org/10.1016/j.addbeh.2014.09.012

Kraus SW, Voon V, Potenza MN (2016) Should compulsive sexual behavior be considered an addiction? Addiction 111(12):2097–2106. https://doi.org/10.1111/add.13297

Kraus SW, Krueger RB, Briken P, First MB, Stein DJ, Kaplan MS, Voon V, Abdo CHN, Grant JE, Atalla E, Reed GM (2018) Compulsive sexual behaviour disorder in the ICD-11. World Psychiatry 17(1):109–110. https://doi.org/10.1002/wps.20499101

Kuzma JM, Black DW (2008) Epidemiology, prevalence, and natural history of compulsive sexual behaviour. Psychiatr Clin North Am 31:603–611

Laforgue R (1941) Psychopathologie de l'echec [Psychopathology of failure]. Marseilles, France: Less Cahier du Sud

Laier C, Brand M (2014) Empirical evidence and theoretical considerations on factors contributing to cybersex addiction from a cognitive-behavioral view. Sex Addict Compuls 21:305–321

Langstrom N, Hanson RK (2006) Hight rates of sexual behavior in the general population: correlates and predictors. Arch Sex Behav 35:37–52

LaRose R, Mastro D, Eastin MS (2001) Understanding Internet usage: a social-cognitive approach to uses and gratifications. Soc Sci Comput Rev 19:395–413

Leahy RL (2011) Cognitive-behavioral therapy: proven effectiveness. Psychology today. Retrieved Oct 2020 from: http://www.psychologytoday.com/blog/anxiety-files/202009/cognitive-behavioral-therapy-proven-effectiveness

Leedes R (2001) The three most important criteria in diagnosing sexual addictions: obsession, obsession, and obsession. Sex Addict Compuls 8:215–226

Leiblum SR (2001) Women, sex and the internet. Sex Relatsh Ther 16:389–405

Lewczuk K, Glica A, Nowakowska I, Gola M, Grubbs JB (2020) Evaluating pornography problems due to moral incongruence model. J Sex Med 17(2):300–311. https://doi.org/10.1016/j.jsxm.2019.11.259

Lou CH, Zhao Q, Gao ES, Shah IH (2006) Can the internet be used effectively to provide sex education to young people in China? J Adolesc Health 39:720–728

Malandain L, Blanc JV, Ferreri F, Thibaut F (2020) Pharmacotherapy of sexual addiction. Curr Psychiatry Rep 22(6):30. https://doi.org/10.1007/s11920-020-01153-4 PMID: 32377953

Marshall EA, Miller HA (2019) Consistently inconsistent: a systematic review of the measurement of pornography use. Aggress Violent Beh. https://doi.org/10.1016/j.avb.2019.08.019

McKelland MJ (2002) Virtual ethnography using the internet to study gay culture in Japan. Sexualities 5:387–406

Mechelmans DJ, Irvine M, Banca P, Porter L, Mitchell S, Mole TB, Lapa TR, Harrison NA, Potenza MN, Voon V (2014) Enhanced attentional bias towards sexually explicit cues in individuals with and without compulsive sexual behaviours. PLoS ONE 9(8):e105476. https://doi.org/10.1371/journal.pone.0105476

Miller W, Rollnick S (1991) Motivational interviewing: preparing people to change addictive behavior. Guilford Press, New York, NY

Moskowitz DA, Roloff ME (2007) The ultimate high: sexual addiction and the bug chasing phenomenon. Sex Addict Compuls 14:21–40

Myers WA (1995) Addictive sexual behavior. Am J Psychother 49:473–483

Nuttall B (2000) Emotion, affect and feeling. Retrieved 21 Jan 2013, from http//:www.affectivetherapy.co.uk/news.htm

Odlaug BL, Lust K, Schreiber LRN, Christenson G, Derbyshire K, Harvanko A, Golden D, Grant JE (2013) Compulsive sexual behavior in young adults. Ann Clin Psychiatry: Official J Am Acad Clin Psychiatrists 25(3):193–200

Orford J (1978) Hypersexuality: implications for a theory of dependence. Br J Addict 73:299–310

Qiu J, Shen B, Zhao M, Wang Z, Xie B, Xu Y (2020) A nationwide survey of psychological distress among Chinese people in the COVID-19 epidemic: implications and policy recommendations. Gen Psychiatry 33(2):e100213. https://doi.org/10.1136/gpsych-2020-100213

Quadland MC (1985) Compulsive sexual behavior: definition of a problem and an approach to treatment. J Sex Marital Ther 11:121–132

Ragan PW, Martin PR (2000) The psychobiology of sexual addiction. Sex Addict Compuls 7:161–175

Raymond NC, Coleman E, Miner MH (2003) Psychiatric comorbidity and compulsive/impulsive traits in compulsive sexual behavior. Compr Psychiatry 44:370–380
Raymond NC, Grant JE, Coleman E (2010) Augmentation with naltrexone to treat compulsive sexual behavior: a case series. Ann Clin Psychiatry 22:56–62
Reid RC (2007) Assessing readiness to change among clients seeking help for hypersexual behaviour. J Sexual Addict Compulsivity 14:167–186
Reid RC (2013) Personal perspectives on hypersexual disorder. Sex Addict Compuls 20:4–18
Reid RC, Carpenter BN (2009) Exploring relationships of psychopathology in hypersexual patients using the MMPI-2. J Sex Marital Ther 35:294–310
Reid RC, Meyer MD (2016) Substance use disorders in hypersexual adults. Curr Addict Rep 3:400–405. https://doi.org/10.1007/s40429-016-0124-9
Reid RC, Harper JM, Anderson EH (2009) Coping strategies used by hypersexual patients to defend against the painful effects of shame. Clin Psychol Psychother 16:125–138
Reid RC, Carpenter BN, Gilliland R, Karim R (2011) Problems of self-concept in a patient sample of hypersexual men with attention- deficit disorder. J Addict Med 5:134–140
Reid RC, Dhuffar MK, Parhami I, Fong TW (2012) Exploring facets of personality in a patient sample of hypersexual women compared with hypersexual men. J Psychiatr Pract 18:262–268
Reid RC, Bramen JE, Anderson A, Cohen MS (2014) Mindfulness, emotional dysregulation, impulsivity, and stress proneness among hypersexual patients. J Clin Psychol 70(4):313–321. https://doi.org/10.1002/jclp.22027
Reid RC, Grubbs JB, Kraus SW (2020) Prevention and treatment of sex addiction. In: Sussman SY (ed) The Cambridge handbook of substance and behavioral addictions
Rickards S, Laaser M (1999) Sexual acting-out in borderline women: impulsive self-destructiveness or sexual addiction/compulsivity? Sex Addict Compuls 6:31–45
Rinehart NJ, McCabe MP (1997) Hypersexuality: Psychopathology or normal variant of sexuality? Sex Marital Ther 12:45–60
Rinehart NJ, McCabe MP (1998) An empirical investigation of hypersexuality. Sex Marital Ther 13:369–384
Rissel C, Richters J, de Visser RO, McKee A, Yeung A, Caruana T (2017) A profile of pornography users in Australia: findings from the second Australian study of health and relationships. J Sex Res 54(2):227–240
Sadiza J, Varma R, Jena SPK, Singh TB (2011) Group cognitive behaviour therapy in the management of compulsive sex behaviour. Int J Crim Justice 6:309–325
Salzman L (1972) The highly sexed man. Med Aspects Hum Sex 6:36–49
Sansone R, Muennich E (2008) Borderline personality symptomatology and sexual impulsivity. Int J Psychiatry Med 38:53–60
Schneider JP (1991) How to recognise the signs of sexual addiction—asking the right questions may undercover serious problems. Postgrad Med 90:171–174
Schneider JP (2000) A qualitative study of cybersex participants: gender differences, recovery issues, and implications for the therapist. Sex Addict Compuls 7:249–278
Schneider JP, Irons R (2001) Assessment and treatment of addictive sexual disorders: relevance for chemical dependency relapse. Subst Use Misuse 36:1795–1820
Schwartz MF (2004) Review of sexual addiction: an integrated approach. Arch Sex Behav 33:520
Schwartz MF, Brasted WS (1985) Sexual addiction: self-hatred, guilt, and passive rage contribute to this deviant behavior. Med Aspects Hum Sex 19:103–107
Schwartz MF, Southern S (2000) Compulsive cybersex: the new tea room. Sex Addict Compuls 7(1–2):127–144. https://doi.org/10.1080/10720160008400211
Seegers J (2003) The prevalence of sexual addiction symptoms on the college campus. Sex Addict Compuls 10:247–258
Skegg K, Nada-Raja S, Dickson N, Paul C (2010) Perceived "out of control" sexual behavior in a cohort of young adults from the Dunedin multidisciplinary heath and development study. Arch Sex Behav 39:968–978

Stulhofer A, Jelovica V, Ruzic J (2008) Is early exposure to pornography a risk factor for sexual compulsivity? Findings from an online survey among young heterosexual adults. Int J Sex Health 20:270–280

Sussman S, Lisha N, Griffiths MD (2011) Prevalence of the addictions: a problem of the majority or minority? Eval Health Prof 34:3–56

Travin S (1995) Compulsive sexual behaviors. Psychiatr Clin North Am 18:155–169

Turkle S (1995) Life on the screen. Simon & Schuster, New York

van der Kolk BA, Fisler RE (1994) Childhood abuse and neglect and loss self-regulation. Bulletin Menninger Clinic 58:145–168

Voon V, Mole TB, Banca P, Porter L, Morris L, Mitchell S, Lapa TR, Karr J, Harrison NA, Potenza MN, Irvine M (2014) Neural correlates of sexual cue reactivity in individuals with and without compulsive sexual behaviours. PLoS ONE 9(7)

Wan M, Finlayson R, Rowles A (2000) Sexual dependency treatment outcome study. Sex Addict Compuls 7:177–196

Weintraub D, Koester J, Potenza MN, Siderowf AD, Stacy M, Voon V, Whetteckey J, Wunderlich GR, Lang AE (2010) Impulse control disorders in Parkinson disease: a cross-sectional study of 3090 patients. Arch Neurol 67(5):589–595. https://doi.org/10.1001/archneurol.2010.65

Weiss R (Nov 2013a) An introduction to the assessment, diagnosis and treatment of sex and porn addiction. CPD Workshop, Day 1, London, UK

Weiss R (Nov 2013b) Advanced clinical work with porn and sex addiction. CPD Workshop, Day 2, London, UK

Weiss R, Schneider J (2006) Untangling the web. Sex porn, and fantasy obsession in the internet age. New York, NY: Alyson

Weiss R (Sep 2010) Sex addiction and the new media frontier. Presentation at society for the advancement of sexual health conference, Boston, SA

Weiss R (2011) Can women be sex addicts? Retrieved 29 Sep 2020, from http://blogs.psychcentral.com/sex/2020/09/can-women-be-sex-addicts www.sash.net. The society for the advancement of sexual health

Weissberg JH, Levay AN (1986) Compulsive sexual behavior. Med Aspects Hum Sex 20:129–132

Wind TR, Rijkeboer M, Andersson G, Riper H (2020) The COVID-19 pandemic: the 'black swan' for mental health care and a turning point for e-health. Internet Interv 20:100317. https://doi.org/10.1016/j.invent.2020.100317

Winters J, Christoff K, Gorzalka BB (2010) Dysregulated sexuality and high sexual desire: distinct constructs? Arch Sex Behav 39:1029–1043

Womack SD, Hook JN, Ramos M, Davis DE, Penberthy JK (2013) Measuring hypersexual behavior. Sex Addict Compuls 20:65–78. https://doi.org/10.1080/10720162.2013.768126

World Health Organization (2019) International statistical classification of diseases and related health problems, 11th edn. WHO, Geneva

Wright PJ (2012) A longitudinal analysis of US adults' pornography exposure. Sexual socialization, selective exposure, and moderating role of unhappiness. J Media Psychol 24:67–76

Young KS (1998) Internet addiction: the emergence of a new clinical disorder. Cyberpsychol Behav 1:237–244

Young KS (2007) Cognitive behavioural therapy with internet addicts: treatment outcomes and implications. CyberPsychol Behav 10:671–679

Young KS (2011) CBT-IA: the first treatment model for internet addiction. J Cogn Psychother 4:304–331

Chapter 8
Exercise Addiction

Attila Szabo and Krisztina Ábel

Abstract This chapter examines compulsive and obsessive exercise behavior from the perspective of behavioral addictions. First, the theoretical interpretation of the dysfunction is presented considering psychological and biological models for the etiology of the morbidity. Next, the research on biological aspects of exercise addiction is presented briefly, which is followed by the discussion of the most common comorbidities, including a distinction between primary and secondary exercise addiction, personality characteristics and other addictions. The overview then continues with the presentation of the key assessment tools used in exercise addiction research. Subsequently, the prevalence rates of exercise addiction are presented. The distinction between the estimated risk of exercise addiction and a full-blown disorder is highlighted to aid the conceptual clarification of research in the field. Finally, in the last section, possible treatment options are also presented. The chapter ends with a conclusion section highlighting the need to use an idiographic method in studying exercise addiction.

Keywords Exercise addiction · Commitment · Compulsion · Dependence · Obsessive passion · Withdrawal symptoms

A. Szabo (✉) · K. Ábel
Faculty of Education and Psychology, Institute of Psychology and Institute of Health Promotion and Sport Sciences, ELTE Eötvös Loránd University, Prielle Kornélia 47, Suite 3, 1117 Budapest, Hungary
e-mail: szabo.attila@ppk.elte.hu

K. Ábel
e-mail: abel.krisztina@ppk.elte.hu

H. M. Pontes (ed.), *Behavioral Addictions*, Studies in Neuroscience, Psychology and Behavioral Economics, https://doi.org/10.1007/978-3-031-04772-5_8

8.1 Introduction

8.1.1 Physical Activity and Exercise in the Contemporary Lifestyle

The industrial revolution and the progressive advancement of information technology contributed to the reduction, and even elimination of the need for physical exertion in the contemporary human lifestyle. Therefore, the biological laws of evolution, preparing humans for physically active life such as hunting, fishing, fighting, escaping predators, building shelter, and manual work for gathering food are suddenly challenged by major lifestyle changes, rendering modern life increasingly sedentary (Freese et al. 2017). To buffer the increasing health risks and ill effects of sedentary behavior, regular physical activity must be incorporated into people's daily activities (Bushman and American College of Sports Medicine 2017). Most scientists agree that in addition to physical work and active commuting (e.g., walking, cycling), only regular sports and exercise might compensate for the lost physical activity that was part of the earlier survival activities (Péronnet and Szabo 1993). The scholastic standpoint penetrates the mass media as well (Berry et al. 2020), and, consequently, most people realize and agree with the need to incorporate regular exercise, sports, and/or other forms of physical activity in their lifestyle, but still a significant proportion of the world's population is not sufficiently physically active (Guthold et al. 2018).

Physical activity is conceptualized as *all forms of movements* which require energy expenditure to sustain work performed by the skeletal muscles; sports and exercise are two of its main forms (Caspersen et al. 1985). While all movements qualify for physical activity, based on this definition, they may be sub-classified into planned and unplanned forms which serve different goals. Unplanned but necessary movements such as buying groceries, washing dishes, climbing the stairs, shoveling snow, or merely walking to a destination serve survival activities (Szabo et al. 2019b). Exercise is usually a planned physical activity aimed at benefiting one's health and mastering a physical skill. Regular exercise has a relatively stable pattern, as well as volume, which can be described in terms of frequency, duration, and intensity. Obviously, sports are also specific forms of exercises, but in addition to purposeful and planned movement, they also involve 'rules' and 'contest', making them more mastery-oriented in contrast to the freely planned or self-organized forms of exercise, which are most often enjoyment, fun, and/or health-oriented (Szabo et al. 2019a).

Research reveals that regular physical activity has numerous health benefits (Malm et al. 2019; Reiner et al. 2013). There is also convincing empirical evidence which links regular exercise and sport participation to positive mental wellbeing (Biddle et al. 2019; Kekäläinen et al. 2020). For example, exercise relieves anxiety and aids in dealing with stress (Lucibello et al. 2019; Kim et al. 2019), which could be one of the several motives for engaging in regular exercise. Stress relief motivated exercise is therapeutic in nature (Szabo et al. 2019a, b). Exercising individuals experiencing chronic or re-occurrent stress may adopt exercise as a means of coping (Egorov and Szabo 2013). For these people exercise becomes 'obligatory', a must-do activity to function well in their daily lives (Szabo 2010). If exercise is not possible for any reason, like the temporary closure of the usual training facility, injury, work or family commitment, and the like, obligatory exercisers experience severe withdrawal symptoms, such as irritability, anxiety, negative affect, and similar unpleasant emotions (Szabo 2010). These symptoms reflect one of the six typical symptoms of addiction as outlined in the components model of addiction (Griffiths 2005).

8.1.2 Exercise Addiction: Definition and Terminology

Obligatory exercise is characterized by progressively increasing training volume due to tolerance. To experience the same stress-alleviating effect of exercise as before, obligatory exercisers need to increase their exercise volume continuously. At this stage, in accord with the 'components model of addictions' (Griffiths 2005), obligatory exercisers exhibit all the typical symptoms of addiction. Another group of hard training exercisers is competitive athletes who train for long hours every day within a balanced schedule while maintaining control over their sport behavior. Similar exercise patterns may also be observed in most dedicated recreational exercisers. However, a small proportion of them might lose control over the exercise behavior and experience a compulsive need to engage in the activity at unpredictable times, which is dictated by an inner urge. This 'inner urge' could be the attempt to achieve something behind the individual's ability (like running a marathon in a subjectively expected time that conflicts with the physical ability of the person), or a need to escape from some noxious emotions or feeling states such as work-related stress, relationship or health problems, and the like. Exercising to the point at which an individual loses control over the behavior that becomes obligatory and results in physical and/or mental damage can be defined as exercise addiction (Griffiths 1997; Thaxton 1982).

Exercise addiction is often termed as exercise dependence by some researchers (Cockerill and Riddington 1996; Hausenblas and Downs 2002). Further, some scholars have often referred to the condition as obligatory exercising (e.g., Pasman and Thompson 1988), by emphasizing the compulsive aspect of the behavior. Indeed, in the mass media, exercise addiction is often termed as compulsive exercise (Eberle 2004) or as exercise abuse (Davis 2000). It is important to stress that all these synonymous terms are theoretically intended to denote the same concept. However, there are many reasons why alternating the terms may be unproductive. While the term dependence is often used as a synonym for addiction, the latter incorporates the former and also includes compulsion (Goodman 1990). Accordingly, addiction may be described as the two-dimensional function of dependence and compulsion. Consequently, when using the term 'dependence', one misses an integral component of exercise addiction, the urge or compulsion, being the main propelling force behind the disorder. Consequently, in line with Goodman, not all dependencies and compulsions may be classified as addiction (see Fig. 8.1).

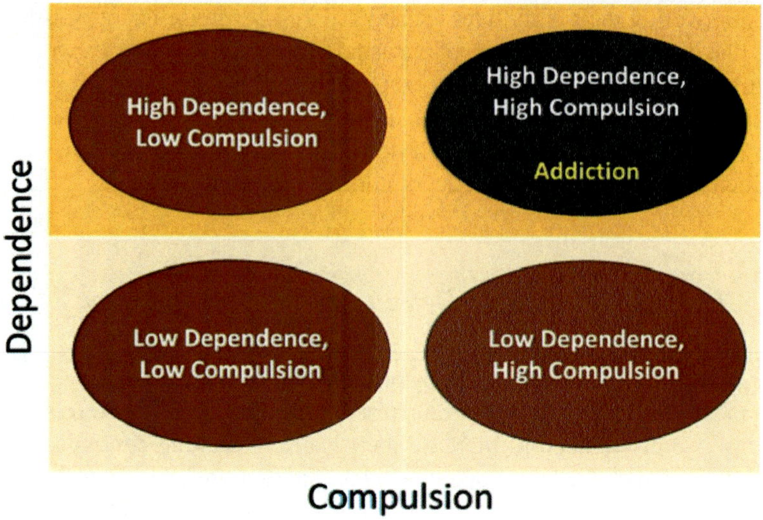

Fig. 8.1 Dependence and compulsion in addiction

8.1.3 Theoretical Models for Exercise Addiction

8.1.3.1 Psychological Models

The Cognitive Appraisal Hypothesis

An early psychological model proposed by Szabo (1995) purports that exercise addiction surfaces when exercise becomes a means of coping with mental stress. In this model, repeated intensive sessions of exercise provide an escape from chronic stress. Once the exercise is adopted as a means of coping, the individual depends on it to function normally in everyday life. When exercise is prevented for a certain reason(s), withdrawal symptoms are experienced because the loss of exercise triggers loss of control over the stressful situation. Therefore, the exerciser experiences greater vulnerability to stress through the exacerbated negative feelings associated with the inability to exercise. The increasing emotional hardship can only be resolved via the resumption of the previous exercise pattern often at the expense of other obligations in the individual's daily life. While this model perceives exercise addiction as a means of coping or escape, it only accounts for the persistence of addiction, but not for its onset. The model is illustrated in Fig. 8.2.

The Four Phase Model for Exercise Addiction

The 'four phase' model for exercise addiction (Freimuth et al. 2011) depicts the progressive change from healthy into addictive exercise. The pleasurable activity

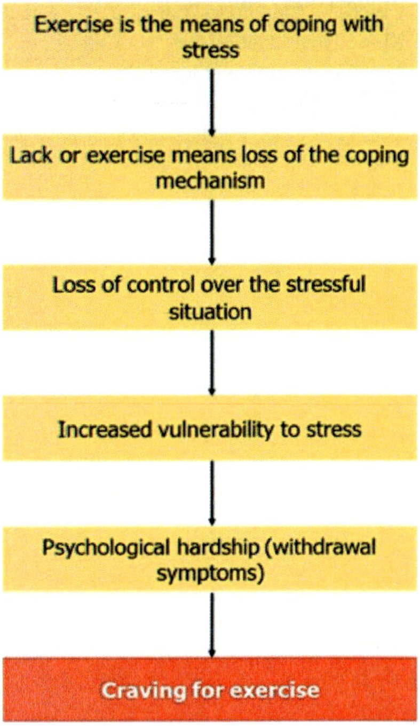

Fig. 8.2 The cognitive arousal hypothesis (Szabo 1995)

characterizes the first phase while the behavior is under control. In this phase, there are no negative consequences in general, but rarely muscles soreness or minor strains may occur. In phase two, the psychological benefits of exercise are realized, and the mood-modifying effects may be used for emotion regulation. Addiction is most likely to occur when exercise becomes the primary and/or the sole means of coping with stress. This part of the model attempts to identify the onset of exercise addiction, but it does not specify two important issues: (1) some distress must exist, whether progressively mounting or suddenly appearing, and (2) under what specific conditions or influences will exercise be adopted for emotion regulation? The third phase is characterized by the rigid organization of daily obligations around exercise, negative consequences due to exaggerated exercise, and several forms of exercise either for replacing or complementing the habitual mode of exercise. Further, exercise is performed individually, rather than with friends, in a team, or during scheduled fitness classes. The fourth and last stage comprises the typical symptoms of fully manifested addiction like salience, tolerance, conflict, need for mood modification, withdrawal symptoms, and relapse. Although the model is logical, it does not account for the onset and motive beyond emotion regulation, and it does not explain who and why will become addicted to exercise (Fig. 8.3).

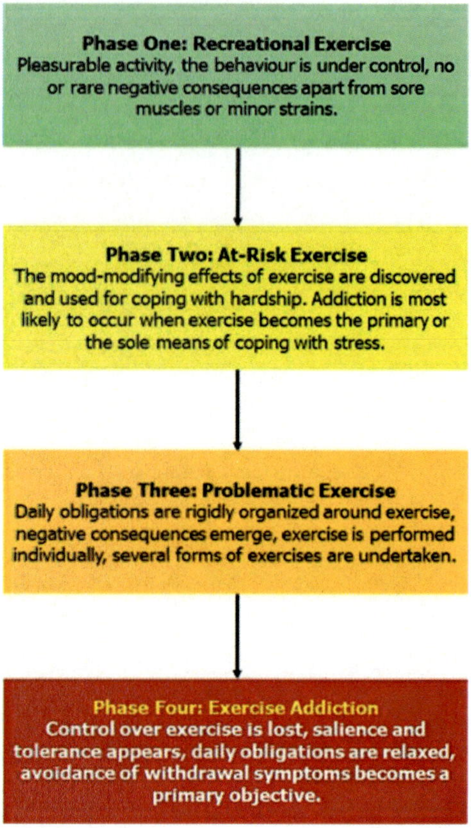

Fig. 8.3 The "Four-Phase Model" explanation for exercise addiction (Freimuth et al. 2011)

The Interactional Model for Exercise Addiction

An idiographic model forwarded for exercise addiction is the interactional model (Egorov and Szabo 2013). The model purports that interaction between personal values, social image, past exercise experience, and life situation jointly determine whether one will use exercise for coping or resort to other means of dealing with stress. The possible number of interactions between personal and situational factors is large and, hence, each case reflects an idiographic mental schema resembling a unique 'black-box'. This box could only be opened after a clinical diagnosis performed by a mental health professional. Indeed, unlike other chemical and/or behavioral addictions (except work addiction if the work is physical), exercise addiction has a unique characteristic not present in other addictions, which is the physical challenge or energy-expenditure. Based on preliminary laboratory evidence, it was proposed that exercise acts as a cathartic-buffer for stress (Szabo and Tsang 2003). Habitual exercisers, when experiencing stress—knowing the mood-improving effects of exercise from experience (Freimuth et al. 2011)—might resort to exercise to cope with the hardship. However, not all regular exercisers will resort to exercise to reduce stress but instead may adopt passive forms of escape behaviors or chemical addiction(s). Therefore, the interactional model considers the personal circumstances that interact with social-environmental factors and, thus, provides a comprehensive view on the genesis of exercise addiction in the affected individuals. Indeed, a positive link was disclosed between exercise addiction risk-scores and trait anxiety (Bircher et al. 2017), perfectionism (Bircher et al. 2017; Cook 1996; Lichtenstein et al. 2017), and obsessive-compulsiveness (Lichtenstein et al. 2017; Spano 2001). Further, it was reported that neuroticism, extraversion could predict symptoms of exercise addiction (Cook et al. 2020; Hausenblas and Giacobbi 2004). Finally, gender (Cook et al., 2013) and sex-role orientation (Rejeski et al. 1987) may also have mediating roles. A large number of possible combinations of subjective psychological factors interacting with situational variables may render it difficult—if not impossible—to scrutinize exercise addiction using a nomothetic approach. Thus, the gist of the interactional model is that each case of exercise addiction is unique and, therefore, can only be understood via an idiographic perspective. Figure 8.4 illustrates the interactional model of exercise addiction.

The Relevance of the Interactional Model to the Pragmatics, Attraction, Communication, Expectation (PACE) Model

The interactional model for exercise addiction (Egorov and Szabo 2013) is largely consistent with the earlier proposed 'Pragmatics, Attraction, Communication, Expectation' (PACE) model of addictions in general (Sussman et al. 2011b). In the interactional model (see Fig. 8.4), a complex set of personal factors interact with several environmental or situational factors to determine the main reason (i.e., key motivation) for exercise behavior. All exercise motives can be classified either into therapeutic or mastery categories (Robbins and Joseph 1985; Szabo et al. 2019a, b). A health (i.e., mental or physical) motivated person, for example, may run to improve health (e.g., gain health) and/or to prevent symptoms of morbidity like gaining weight, being lethargic, developing high blood pressure, having elevated blood glucose level, and the like. Both incentives are therapeutic. However, health motives could also have a mastery-orientation, like becoming stronger and/or lifting more weight (i.e., performance orientations) or concentrating better and/or being more productive at work. If a better concentration would be the aim, a therapeutic orientation would be established, but if the expected consequence of the better concentration (i.e., productivity) is the objective, then the mastery orientation is the driving force.

The most important component of the interactional model is the consideration of a suddenly emerging reaction, determined by a set of idiographic (i.e., personal and situational) interactions in the black-box to an ongoing and no longer bearable—or suddenly appearing—life stressor that causes psychological pain over which the individual has no control. This component accounts for the surmise that exercise addiction is not evolutionary or slowly progressing, but rather revolutionary or suddenly surfacing (Szabo 2010). When the situation gets out of control, a person might 'gravitate' towards a means of available coping in accord with the 'Pragmatics' phase of

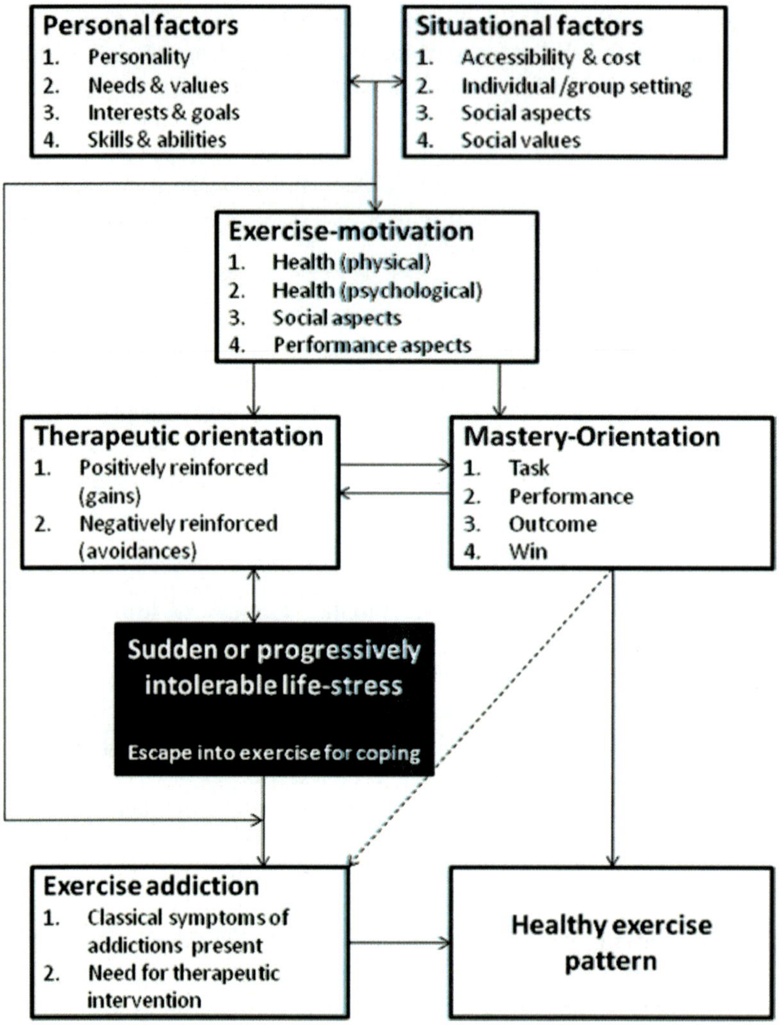

Fig. 8.4 The interactional model of exercise addiction (Egorov and Szabo 2013)

the PACE model (Sussman et al. 2011a)—see Fig. 8.5). The choice is determined by conscious and subconscious decisions (in the black-box) between individual characteristics, situational factors, life event antecedents, and current exercise behavior which is aligned with the 'Attraction' component of the PACE model, in a similar way as the motivation for exercise is initially determined. Accordingly, even mastery-oriented exercisers may shift focus to the therapeutic aspects of exercise and get more involved in it to get rid of distress. This attentional cognition is also in line with the 'Communication' part of the PACE model in that experience, inter- and intra-personal thought; beliefs, and convictions influence the escape path or the means of coping of the person. For example, lack of experience with alcohol, tobacco, or leisure drugs

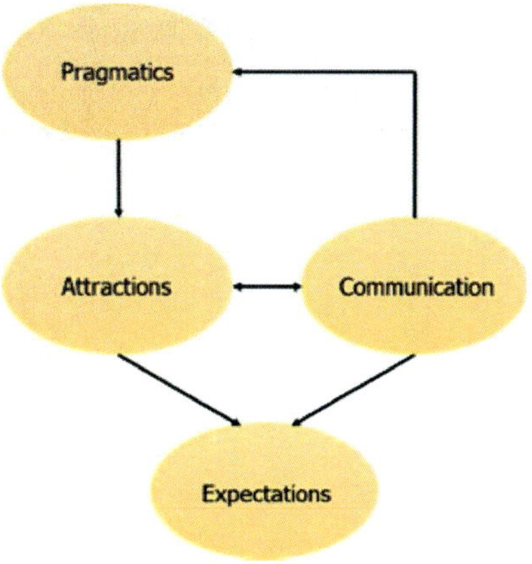

Fig. 8.5 Pragmatics, Attraction, Communication, Expectation (PACE) model (Sussman et al., 2011b)

in conjunction with long exercise history and positive beliefs about exercise (from media, friends, and health values) all interact with unique personal factors during the effort of coping. An already 'therapeutic' exerciser in the model is more likely to choose their exercise as a means of coping. Then, also in agreement with the PACE model, the greater the expectation linked to exercise, the more unlikely the exerciser will turn to other forms of addiction. Perceived as a positive addiction, it is much easier to hide behind exercise while maintaining one's reputation in the social environment than other forms of addictions bearing a strong negative social stigma. The PACE model is illustrated in Fig. 8.5.

The Passion Model

The most recent psychological model for exercise addiction is the passion model (Lichtenstein et al. 2020a). Recent research has revealed a significant overlap between the risk of exercise addiction and passion (Kovacsik et al. 2018, 2019; Kovácsik et al. 2020; Lichtenstein et al. 2020b; Lichtenstein et al. 2020b; Parastatidou et al. 2014; Sicilia et al. 2017; Szabo 2018). In fact, Szabo (2018) proposed that a high risk of exercise addiction often only reflects a high level of commitment and passion for sport and exercise. Lichtenstein et al. (2020b) produced evidence for the three dimensions of passion and suggested that there is a progressive transition from commitment into the three stages of passion that in some cases may culminate in addiction. The model is logical and appealing to explain the passion-addiction relationship, but it is still tentative because it is based on a single cross-sectional study of 1225 Danish fitness participants. The passion model is illustrated in Fig. 8.6.

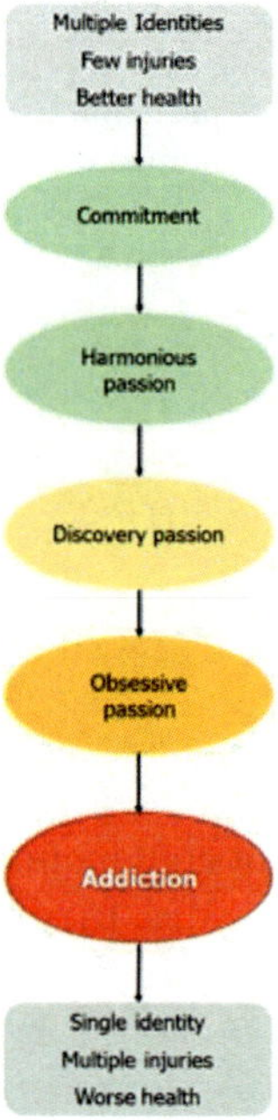

Fig. 8.6 The passion model of exercise addiction (Lichtenstein et al. 2020b)

8.1.3.2 Biological Models

The Sympathetic Arousal Hypothesis

The sympathetic arousal hypothesis is a physiological model based on adaptation to exercise. The sympathetic adaptation to regular exercise lowers the overall level of arousal (Thompson and Blanton 1987). According to this model, lower levels of arousal may be associated with lethargy, sluggishness, or low-energy states. This uncomfortable feeling prompts the person to increase the level of arousal. For the regular exercisers, their exercise activity is the most tangible way to increase arousal. However, since the arousal mediating effects of exercise are relatively short-lasting, increased bouts of exercise are necessary to generate an optimal state of arousal, leading to tolerance and the onset of an addictive pattern of exercising. An issue with this model is that sympathetic adaptation to exercise occurs in everyone, while surely not all exercisers become addicted. The model is illustrated in Fig. 8.7.

The Biopsychosocial Model

The biopsychosocial model is different from several conceptualizations of exercise addiction. The model suggests that certain biological factors, like Body Mass Index (BMI), mediate the development and maintenance of exercise addiction (McNamara and McCabe 2012). In this model, the biological factors are closely connected to

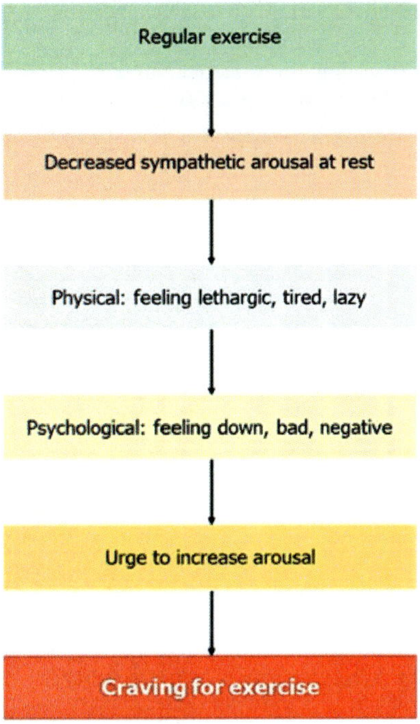

Fig. 8.7 The sympathetic arousal hypothesis (Thompson and Blanton 1987)

social and psychological factors, contributing to the etiology of exercise addiction. Social factors in the model may include the coach-athlete relationship, teammate pressure, socio-cultural pressure, and social support. Self-esteem and beliefs about exercise generally characterize psychological factors. In the biopsychosocial model of exercise addiction, social forces are thought to share a reciprocal relationship with psychological factors (Fig. 8.8). It is posited that social and psychological factors mediate the onset and maintenance of exercise addiction. The model has been employed in the explanation of exercise addiction among elite athletes (McNamara and McCabe 2012). At its core, BMI as a biological factor, might be the route of the origin of exercise addiction in elite athletes. The interaction between social and psychological processes then determines whether exercise (sport) addiction occurs or not. This interaction is linked to socially and psychologically demanding athletic training at high levels of competition. However, Freimuth et al. (2011) clarified that hard training for long hours and ambitious strivings for a noble goal to become the best – which characterizes successful elite athletes – should not be confused with the symptoms of addiction. The model is illustrated in Fig. 8.8.

The Interleukin Six (IL-6) Model

Another theoretical model has highlighted the possible role of interleukin six (IL-6) in exercise addiction (Hamer and Karageorghis 2007). The IL-6 is a pro-inflammatory

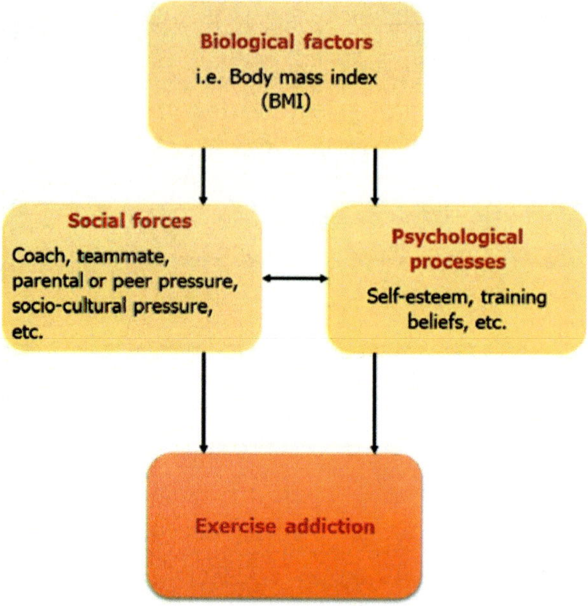

Fig. 8.8 The biopsychosocial model (based on McNamara and McCabe 2012, pg. 756)

and anti-inflammatory cytokine secreted by T cells and macrophages to increase the immune response to trauma, such as burning or other types of tissue damage leading to inflammation. The blood concentration level of IL-6 increases during exercise (Aguiló et al. 2014), and higher levels of IL-6 are associated with increased cardiovascular mortality, depression, and negative affect (Puterman et al. 2014). Hamer and Karageorghis (2007) suggested that IL-6 links the periphery to the brain. This link may mediate the components of exercise addiction. In people prone to morbidity, exercise results in a momentary reduction in negative affect. However, at the same time, it raises the synthesis of IL-6 and activates the neuroendocrine pathways, which contribute to the negative feelings manifested through the experiencing of withdrawal symptoms. Therefore, exercise acts as a vicious circle by lowering and increasing negative affect. This is a psycho-neuroimmunological model that deserves further research attention. The model is presented in Fig. 8.9.

The Monoamine Model

The monoamine model is based on the observation that exercise triggers an increase in catecholamine levels in the peripheral blood circulation (Cousineau et al. 1977). For example, it was shown that a 30-min bout of medium to high intensity aerobic exercise

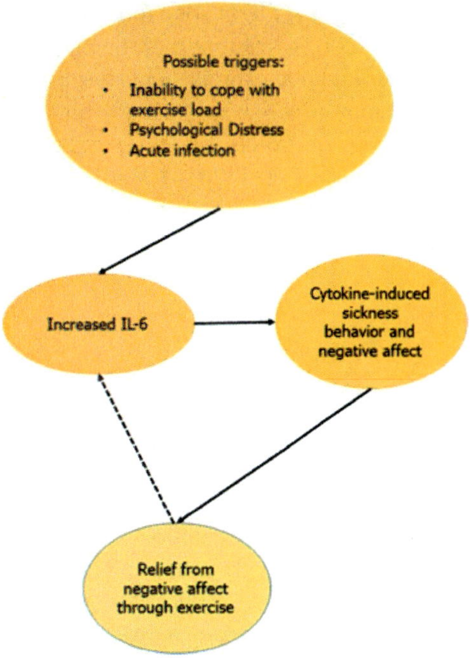

Fig. 8.9 The interleukin six (IL-6) model (Hamer and Karageorghis 2007)

increases the uric phenylacetic acid levels – reflecting the phenylethylamine concentration – in healthy males habituated to exercise (Szabo et al. 2001). While catecholamines, among other functions, are involved in the stress response, phenylethylamine is more closely linked to changes in mood. In light of the monoamine hypothesis, it is thought that in addition to an increase in monoamines in the peripheral circulation, the central aminergic activity may also rise in response to exercise. Since brain monoamines are involved in the regulation of mood and affect, their alteration by exercise seems to be an attractive explanation for the role of exercise in stress response. This is a biological model that is probably more closely linked to the positive mood-enhancing effects of exercise than exercise addiction per se. However, in light of this model exercise may act as a buffer of stress in the addiction process, in that the negative emotional experiences resulting from life stress are soaked up by the positive effects of exercise.

The Endorphin Model

The endorphin model is perhaps the most popular in the literature. It posits that exercise leads to increased levels of beta-endorphins in the brain, which acts as internal psychoactive agents by generating feelings of euphoria. This hypothesis may be analogous to a substance or recreational drug addiction (e.g., heroin, morphine, etc.) except that the psychoactive agent (i.e., beta-endorphin) is generated internally during exercise instead of being administered externally. Endogenous opioids are involved in modulating several sensory, motivational, emotional, and mental functions (McNally and Akil 2002). In support of the model, a study with positron emission tomography (PET) found that exercise – performed between the aerobic and anaerobic threshold for 60 min – increased the availability of μ-opioid receptors in the anterior cingulate cortex, prefrontal, and temporal cortex of young, healthy recreational exercising men (Saanijoki et al. 2014). While further research in this area is needed, the opioid response to exercise is likely to be workload- or dose-dependent in addition to individual variability. It could be one of the several explanations for exercise addiction in connection with stress management.

8.2 Biological Bases of Exercise Addiction

The biological concomitants of exercise addiction were rarely investigated to date. The few studies linking biological factors to exercise addiction are cross-sectional, and cause and effect relationships cannot be established on their basis. One study examined the link between electroencephalography (EEG) activity and exercise addiction (Beh et al. 1996). The authors found that the power distribution within the alpha band differed between exercise addicted and non-addicted individuals recording higher frequencies for the addicted group. This result was interpreted as a difference in cortical arousal. Later in a study with 28 women, Gapin et al. (2009) found that a greater relative left frontal activity correlated with greater exercise addiction scores. The authors concluded that there could be a relationship between frontal asymmetry, which is an index of negative emotions, and exercise addiction.

8.3 Specific Co-morbidities

8.3.1 Primary and Secondary Exercise Addiction (Eating Disorders)

There is an unfortunate and conceptually incorrect distinction between two similarly compulsive and exaggerated patterns of exercise behavior in the literature. The terms primary and secondary exercise addiction, or dependence as used in the literature (e.g., De Coverley Veale 1987) attempts to differentiate between exercise addiction without (primary) and with (secondary) accompanied eating disorders (e.g., anorexia nervosa, bulimia, binge eating, and so on). Since exercise is a means of caloric expenditure and, hence, weight control, many scholars believe that a high volume of exercise in eating disorders is a manifestation of exercise addiction. Diet and exercise are means of losing weight in the various eating disorders. Therefore, exercise's role is weight-loss. While weight may be important in exercise addiction, the main goal is to regulate one's emotions (Gapin et al. 2009). In exercise addiction, one 'runs away' from stress and/or other psychological hardship, and exercise represents the solution. In eating disorders, exercise is a means for achieving weight loss. Therefore, the distinguishing feature between the two concepts is that in primary exercise addiction, the exercise is the objective, whereas in secondary exercise addiction, weight loss is the objective, while exaggerated exercise is the means of achieving both objectives. We, the authors, strongly oppose the use of the term 'secondary exercise addiction' because there is no evident addiction in eating disorders. Perhaps one can measure a compulsion to exercise or otherwise diet alone is not sufficient to lose weight. Such compulsion is motivated by the need to lose weight and to feel comfortable. Therefore, a high volume of exercise simply acts as a complimentary behavior in eating disorders.

8.3.2 Personality Characteristics

8.3.2.1 The Big Five Personality Traits and Exercise Addiction

The Big Five traits are extraversion, neuroticism, conscientiousness, openness to experience and agreeableness (Matsumoto 2009). In most of the studies reviewed by Bircher et al. (2017), exercise addiction was positively associated with extraversion. Hausenblas and Giacobbi (2004) reported that conscientiousness had no association with exercise addiction. In contrast, Costa and Oliva (2012) found that it was negatively related to continuance, loss of control, reductions in engaging in other activities, and the total dependence score. Hausenblas and Giacobbi (2004) also found that openness was independent of exercise addiction, while Costa and Oliva (2012) found an association between exercise addiction and intention effects. Further, Kern (2009) reported that the strongest predictor traits of exercise addiction were openness to experience with emotional stability or with agreeableness. Kern also suggested the existence of an addictive personality type.

8.3.2.2 Perfectionism and Exercise Addiction

The most frequently explored trait in exercise addiction research is perfectionism (Bircher et al. 2017). It implies a trait-based tendency to have extremely high expectations of the self or others (Matsumoto 2009). Hagan and Hausenblas (2003) found that individuals at risk of exercise addiction were more perfectionists than those at lower risk levels. The high exercise addiction group also showed more exercise addiction symptoms than low scoring subjects. The link between perfectionism and exercise addiction was demonstrated in several other studies as well (see Bircher et al. 2017).

8.3.2.3 Narcissism and Exercise Addiction

Narcissism refers to the grandiosity of the self and preoccupation with dreams or fantasies about success and power. Narcissistic people have an exaggerated sense of entitlement, and they approach others exploitatively (Matsumoto 2009). Spano (2001) reported that narcissism resulted in greater physical activity but not a commitment to exercise. Bruno et al. (2014) found that participants showing a high risk for exercise addiction scored higher on both self-esteem and narcissism when compared to participants with a low risk of exercise addiction. Further, Miller and Mesagno (2014) also revealed that exercise addiction had a positive relationship with narcissism.

8.3.2.4 Self-Esteem and Exercise Addiction

Self-esteem risk is an index of how people evaluate their body, mental processes, history, behavior, and what their attitude towards them and opinion about them are (Matsumoto 2009). A work by Groves et al. (2007) found a positive relationship between self-esteem and the risk of exercise addiction. However, the nature of the association differed across the universities from where the participants were recruited. According to additional interviews performed by the authors, the symbolic interactionist identity theory (Stryker 2001) might explain the observed differences. For example, the origin of self-esteem is linked to identity reinforcement, and exercise addiction was more strongly related to self-esteem at universities where identity was closely associated with sport and exercise (Groves et al. 2007). The link between self-esteem and exercise addiction was further supported by a more recent study which also reported a positive relationship between body shame and exercise addiction (Ertl et al. 2017).

8.3.3 Exercise Addiction and Other Addictions

The co-occurrence between various addictions exists (Cook 1987), but they were rarely studied in the context of exercise addiction. Studies reported that exercise addiction may co-occur with other behavioral addictions, including compulsive shopping (Müller et al. 2015; Lejoyeux et al. 2008; Villella et al. 2011). In two studies (Müller et al. 2015; Villella et al. 2011) a positive relationship was found between these two behavioral addictions. In one study (Müller et al. 2015), the relationship was stronger in women than in men. In the study by Villella et al. (2011), which was based on 2853 young participants in addition to the connection between exercise and shopping addiction, a positive relationship emerged between the risk of exercise addiction, internet addiction, pathological gambling, and work addiction. However, the association was based on relatively weak correlations yielding only about 5% common variance in all instances. Supporting the association between exercise and shopping addiction, Lejoyeux et al. (2008) found that the frequency of individuals showing compulsive shopping tendencies was higher (63%) in exercisers who were at-risk of exercise addiction than in those who were not at-risk (38%). However, the reported prevalence of the risk of exercise addiction was extremely high in this investigation (42%), which sheds doubt on the means of assessment of this dependent measure. Indeed, in a later study, the authors failed to replicate their earlier results on the relationship between the risk for exercise addiction and compulsive shopping (Lejoyeux et al. 2012).

The risk for exercise addiction was also researched in relation to co-occurrence with chemical or substance addictions. A study with university students reported that the risk for exercise addiction was positively related to alcohol consumption and alcohol-related problems (Martin et al., 2008). However, the results were based on a statistically significant but weak correlation accounting for less than 5% shared variance between the risk for exercise addictions and alcohol drinking and associated problems. Further, the authors reported that only three out of the eight subscales assessing the risk for exercise addiction were consistently associated with alcohol use and related problems. Another correlational work could not find an association between the risk for exercise addiction and alcohol use disorder (Müller et al. 2015). Similarly, Lejoyeux et al. (2008) reported no difference in the prevalence of alcohol consumption between those at-risk and not at-risk for exercise addiction. This finding was supported in later research; however, in this study, the severity of the reliance on alcohol was greater in the former group than the latter (Lejoyeux et al. 2012). Concerning the co-occurrence of the risk of exercise addiction and nicotine use, Lejoyeux et al. (2008) found that nicotine dependence did not differ among those at-risk and at no risk for exercise addiction, but cigarette smokers in the former group smoked less than those in the latter group. These findings were confirmed by a cross-sectional work in which nicotine consumption among individuals at risk of exercise addiction was the lowest (Szabo et al. 2018). Further, this study found no difference in the frequency and severity of alcohol and illicit drug use between those at risk of exercise addiction and those who were not at risk. Consequently, based on the current empirical evidence, it appears that the risk of exercise addiction rarely co-occurs with other addictions, especially with substance or chemical addictions. However, a relatively weak relationship appears to exist between the risk of exercise addiction and some other behavioral addictions.

8.4 Prevalence Rates of Exercise Addiction

The reported prevalence of the risk of exercise addiction varies from 0.3% (Mónok et al. 2012) to 42% (Lejoyeux et al. 2008). The former value was based on robust population-wide research, while most other figures come from smaller convenience samples. An analysis of the literature has concluded that approximately 3% of the American population might be addicted to exercise (Sussman et al., 2011a). This figure is in line with the population-wide study reported by Mónok et al. (2012) in exercisers. Still, it is ten times higher than that reported for the Hungarian general population. However, these prevalence rates are unreliable because most exercisers classified as '*at risk of exercise addiction*' will never exhibit a mental dysfunction or problematic exercise behavior. Nevertheless, a recent study reporting 100 internet-based cases qualifying for exercise addiction (Juwono and Szabo 2020) suggested that there may be significantly more problematic cases than reported in the literature (Szabo 2018). Despite this conjecture, until there are no agreed diagnostic criteria, the prevalence of exercise addiction cannot be correctly established, with questionnaire-based 'risk' scores (inaccurately reflecting problematic exercise behavior) varying in accord with personal (i.e., exercise history, stress level, personality, etc.) and social-environmental (i.e., type of sport, training environment, etc.) factors giving rise to a wide and unreliable estimate range.

8.5 Assessment of Exercise Addiction

Current research assessment of exercise addiction is primarily based on two validated questionnaires. One is the Exercise Dependence Scale (EDS; Downs et al. 2004), which conceptualizes exercise addiction based on the DSM-IV's criteria for substance abuse. Empirical research reveals that the EDS can categorize exercisers into at-risk, dependent, and non-dependent categories. The EDS has seven subscales: (1) tolerance, (2) withdrawal, (3) intention effect, (4) lack of control, (5) time, (6) reduction of other activities, and (7) continuance. The other instrument, the Exercise Addiction Inventory (EAI), was developed to quickly and easily screen the risk of exercise addiction (Terry et al. 2004). The EAI is a concise 6-item instrument assessing six common symptoms of addictive behaviors: (1) salience, (2) mood modification, (3) tolerance, (4) withdrawal symptoms, (5) conflict, and (6) relapse. Both measures have been psychometrically investigated and proved psychometrically sound (Mónok et al. 2012). There are several other paper-and-pencil tools used to assess exercise addiction, but they have been used less often than the EDS and the EAI, especially in the past decade. It is important to stress that questionnaires are not diagnostic instruments and, hence, might provide an inaccurate interpretation of the risk of exercise addiction (Szabo 2018). Currently, a clinical diagnosis of exercise addiction is not possible because the dysfunction has no clinical diagnosis criteria. Addictive or problematic exercise behavior causing physical, psychological and/or social (Juwono and Szabo 2020) impairments to affected individuals can only be identified through consultation using an in-depth interview (Szabo 2018; Szabo et al. 2015).

8.6 Treatment Approaches

Treatment of exercise addiction has not been established as there are currently no officially recognized diagnosis criteria for the dysfunction. Severe symptoms of exercise addiction are treated in conjunction with co-morbidities (if any), or with various psychological interventions. An important treatment component is patient education (Berczik et al. 2012). Self-control is the gist of the treatment. Berczik and colleagues purported that while parents and/or teachers can help in the educational process, the first education sessions or consultations should be offered by a qualified mental health professional. A cognition-oriented intervention and behavioral management should be systematic. Adams et al. (2003) provided specific guidance for professionals working with the individual with problematic exercise behavior:

> 1) *Accept the role and responsibility of primary support for the person and participate in the management process; 2) Recognize that the addiction is likely to cause a breakdown in communication with significant others; 3) Recognize that the likely response is intense fear of losing control, helplessness, and that this may show itself through disorganized behavior through compulsions; and 4) Psychotherapeutic intervention utilized individualized approaches depending on the psychopathology noted in the patient. The common core of therapeutic intervention strategies includes the following:*
>
> *1) Identifying and interrupting the compulsive behavior through supportive individual psychotherapy.*
>
> *2) Engage the patients in understanding the health benefits and importance of moderation.*
>
> *3) Empower the patient to develop a self-management strategy.*
>
> *4) Understanding the organization of the person's defense structure and how the patient is coping with the addictive nature.*
>
> *5) Increasing the tolerance of the patient in adapting or accommodating to the compulsion through modification of their psychological defenses and acceptance and understanding of their response to the gaining of control and appropriate self-management skills.*
>
> *6) Unlink the compulsion and process specific triggers related to exercise dependence.*
>
> *7) Rebuilding the coping behaviors and enhancing the support system for the patient with respect of exercise.* (p. 103–104)

In addition, treatment approaches should focus on the transition from unhealthy to healthy levels of moderate exercise, balancing regular exercise with a healthy social life, coping with emotions resulting from decreased exercise to prevent relapse, and learning to enjoy other forms of exercise than the one(s) to which the person was addicted (Landolfi 2013).

8.7 Conclusions

Exercise addiction is characterized by excessive amounts of uncontrolled training that is salient in the person's life who also experiences psychological, physical and/or social impairments due to the maladaptive exercise behavior. At this time, clinical diagnosis of exercise addiction is not possible. Research evaluation criteria are based on several models, with the most popular being the components model of addiction (Griffiths 2005). In addition to the six symptoms (i.e., salience, conflict, mood modification, withdrawal, tolerance, and relapse) in the components model, addiction to exercise can be suspected if the exerciser loses control over the exercise behavior and there is some sort of damage to her or his wellbeing. Several psychological and physiological models try to account for the etiology of exercise addiction. The interactional model (Egorov and Szabo 2013) and its recently revised form (Dinardi et al. 2021) indicate that each case is unique because numerous personal and situational factors account for the onset of the dysfunction. Based on this model, exercise addiction cannot be studied with the nomothetic approach.

In accord with other authors (Müller et al. 2015; Szabo et al. 2015), each case of exercise addiction can be established via an idiographic approach resorting to in-depth interview. Nomothetic research only assesses the risk of exercise addiction which is an inaccurate predictor of problematic exercise behavior and, therefore, the actual prevalence of exercise addiction is equivocal. The biological bases of exercise addiction are unclear because the limited number of studies do not reflect causal relationships at this time. Empirical research suggests that exercise addiction may co-occur with other addictions, especially behavioral addictions, but more research is needed in a wider cross-cultural spectrum. Given that there is no official clinical diagnosis for exercise addiction at this point, treatment of the dysfunction follows those used in other addictions and behavioral and cognitive management techniques through the individual's education. Significantly more research is needed for the establishment of exercise addiction as a specific mental dysfunction. However, a symptom and damage criteria-based classification (Juwono and Szabo 2020) may be a starting point for clinical professionals to differentiate exercise addiction from other psychopathology and classify it as a distinct mental dysfunction.

References

Adams JM, Miller TW, Kraus RF (2003) Exercise dependence: diagnostic and therapeutic issues for patients in psychotherapy. J Contemp Psychother 33:93–107. https://doi.org/10.1023/A:1022883104269

Aguiló A, Monjo M, Moreno C et al (2014) Vitamin C supplementation does not influence plasma and blood mononuclear cell IL-6 and IL-10 levels after exercise. J Sports Sci 32:1659–1669. https://doi.org/10.1080/02640414.2014.912759

Beh HC, Mathers S, Holden J (1996) EEG correlates of exercise dependency. Int J Psychophysiol 23:121–128. https://doi.org/10.1016/0167-8760(96)00039-6

Berczik K, Szabó A, Griffiths MD et al (2012) Exercise addiction: symptoms, diagnosis, epidemiology, and etiology. Subst Use Misuse 47:403–417. https://doi.org/10.3109/10826084.2011.639120

Berry T, Yun L, Faulkner G et al (2020) Implicit and explicit evaluations of a mass media physical activity campaign: dDoes everything get better? Psychol Sport Exerc 49. https://doi.org/10.1016/j.psychsport.2020.101684

Biddle S, Ciaccioni S, Thomas G, Vergeer I (2019) Physical activity and mental health in children and adolescents: an updated review of reviews and an analysis of causality. Psychol Sport Exerc 42:146–155. https://doi.org/10.1016/j.psychsport.2018.08.011

Bircher J, Griffiths MD, Kasos K et al (2017) Exercise addiction and personality: a two-decade systematic review of the empirical literature (1995–2015). Balt J Sports Health Sci 3:19–33

Bruno A, Quattrone D, Scimeca G et al (2014) Unraveling exercise addiction: the role of narcissism and self-esteem. J Addict. https://www.hindawi.com/journals/jad/2014/987841/

Bushman B, American College of Sports Medicine (2017) ACSM's complete guide to fitness & health exercise nutrition children elderly youth. Hum-Kinet. http://www.humankinetics.com/press/press-archive/acsms-complete-guide-to-fitness-and-health

Caspersen CJ, Powell KE, Christenson GM (1985) Physical activity, exercise, and physical fitness: definitions and distinctions for health-related research. Public Health Rep 100:126–131

Cockerill IM, Riddington ME (1996) Exercise dependence and associated disorders: a review. Couns Psychol Q 9:119–129. https://doi.org/10.1080/09515079608256358

Cook DR (1987) Self-identified addictions and emotional disturbances in a sample of college students. Psychol Addict Behav 1:55–61. https://doi.org/10.1037/h0080429

Cook RH, Griffiths MD, Pontes HM (2020) Personality factors in exercise addiction: a pilot study exploring the role of narcissism, extraversion, and agreeableness. Int J Ment Health Addict 18:89–102. https://doi.org/10.1007/s11469-018-9939-z

Cook B, Hausenblas H, Rossi J (2013) The moderating effect of gender on ideal-weight goals and exercise dependence symptoms. J Behav Addict https://pubmed.ncbi.nlm.nih.gov/26165771/.

Cook CA (1996) The psychological correlates of exercise dependence in aerobics instructors. Unpublished master's thesis

Costa S, Oliva P (2012) Examining relationship between personality characteristics and exercise dependence. Rev Psychol 19:7

Cousineau D, Ferguson RJ, de Champlain J et al (1977) Catecholamines in coronary sinus during exercise in man before and after training. J Appl Physiol. https://doi.org/10.1152/jappl.1977.43.5.801

Davis C (2000) Exercise abuse. Int J Sport Psychol 31:278–289

De Coverley Veale DM (1987) Exercise dependence. Br J Addict 82:735–740. https://doi.org/10.1111/j.1360-0443.1987.tb01539.x

Dinardi JS, Egorov AY, Szabo A (2021) The expanded interactional model of exercise addiction. J Behav Addict (online First). https://doi.org/10.1556/2006.2021.00061

Downs DS, Hausenblas HA, Nigg CR (2004) Factorial validity and psychometric examination of the exercise dependence scale-revised. Meas Phys Educ Exerc Sci. https://doi.org/10.1207/s15327841mpee0804_1

Eberle SG (2004) Compulsive exercise: too much of a good thing? 1

Egorov AY, Szabo A (2013) The exercise paradox: an interactional model for a clearer conceptualization of exercise addiction. J Behav Addict 2:199. https://doi.org/10.1556/JBA.2.2013.4.2

Ertl MM, Longo LM, Groth GH et al (2017) Running on empty: high self-esteem as a risk factor for exercise addiction. Addict Res Theor 1347257. https://doi.org/10.1080/16066359.2017

Freese J, Klement RJ, Ruiz-Núñez B et al (2017) The sedentary (r)evolution: have we lost our metabolic flexibility? F1000Research 6. https://doi.org/10.12688/f1000research.12724.2

Freimuth M, Moniz S, Kim SR (2011) Clarifying exercise addiction: differential diagnosis, co-occurring disorders, and phases of addiction. Int J Environ Res Public Health 8:4069. https://doi.org/10.3390/ijerph8104069

Gapin J, Etnier JL, Tucker D (2009) The relationship between frontal brain asymmetry and exercise addiction. J Psychophysiol 23:135–142. https://doi.org/10.1027/0269-8803.23.3.135

Goodman A (1990) Addiction: definition and implications. Br J Addict 85:1403–1408. https://doi.org/10.1111/j.1360-0443.1990.tb01620.x

Griffiths M (2005) A "components" model of addiction within a biopsychosocial framework. J Subst Use 10:191–197. https://doi.org/10.1080/14659890500114359

Griffiths MD (1997) Exercise addiction: a case study | Request PDF. In: ResearchGate. https://www.researchgate.net/publication/232050272_Exercise_Addiction_A_Case_Study

Groves M, Biscomb K, Nevill A, Matheson H (2007) Exercise dependence, self-esteem and identity reinforcement: a comparison of three universities in the United Kingdom. Sport Soc. https://doi.org/10.1080/17430430701717772

Guthold R, Stevens GA, Riley LM, Bull FC (2018) Worldwide trends in insufficient physical activity from 2001 to 2016: a pooled analysis of 358 population-based surveys with 1·9 million participants. Lancet Glob Health 6:e1077–e1086. https://doi.org/10.1016/S2214-109X(18)30357-7

Hagan AL, Hausenblas HA (2003) The relationship between exercise dependence. Am J Health Stud 18(2/3):133–137. In: Symptoms and perfectionism. University of Northampton. https://northampton.rl.talis.com/items/6BA49B89-78A4-AB62-3750-5B3E106A163B.html

Hamer M, Karageorghis C (2007) Psychobiological mechanisms of exercise dependence. Sports Med Auckland NZ. https://pubmed.ncbi.nlm.nih.gov/17503874/

Hausenblas HA, Downs DS (2002) How much is too much? The development and validation of the exercise dependence scale. Psychol Health 17:387–404. https://doi.org/10.1080/0887044022000004894

Hausenblas HA, Giacobbi PR Jr (2004) Relationship between exercise dependence symptoms and personality. Personal Individ Differ 36:1265–1273. https://doi.org/10.1016/S0191-8869(03)00214-9

Juwono ID, Szabo A (2020) 100 Cases of exercise addiction: more evidence for a widely researched but rarely identified dysfunction. Int J Ment Health Addict, pp 1–13. https://doi.org/10.1007/s11469-020-00264-6

Kekäläinen T, Freund AM, Sipilä S, Kokko K (2020) Cross-sectional and longitudinal associations between leisure time physical activity, mental well-being and subjective health in middle adulthood. Appl Res Qual Life 15:1099–1116. https://doi.org/10.1007/s11482-019-09721-4

Kern L (2009) Relationship between exercise dependence and big five personality. L'encephale 36:212–218. https://doi.org/10.1016/j.encep.2009.06.007

Kim H, Oh S, Lee D et al (2019) The effects of intense physical activity on stress in adolescents: findings from Korea youth risk behavior web-based survey (2015–2017). Int J Environ Res Public Health. https://pubmed.ncbi.nlm.nih.gov/31137869/

Kovacsik R, Griffiths MD, Pontes HM et al (2019) The role of passion in exercise addiction, exercise volume, and exercise intensity in long-term exercisers. Int J Ment Health Addict 17:1389–1400. https://doi.org/10.1007/s11469-018-9880-1

Kovácsik R, Tóth-Király I, Egorov A, Szabo A (2020) A longitudinal study of exercise addiction and passion in new sport activities: the impact of motivational factors. Int J Ment Health Addict. https://doi.org/10.1007/s11469-020-00241-z

Kovacsik R, Soós I, Vega R de la et al (2018) Passion and exercise addiction: Healthier profiles in team than in individual sports. Int J Sport Exerc Psychol

Landolfi E (2013) Exercise addiction. Sports Med Auckland NZ. https://pubmed.ncbi.nlm.nih.gov/23329605/

Lejoyeux M, Guillot C, Chalvin F et al (2012) Exercise dependence among customers from a Parisian sport shop. J Behav Addict 1:28–34. https://doi.org/10.1556/jba.1.2012.1.3

Lejoyeux M, Avril M, Richoux C et al (2008) Prevalence of exercise dependence and other behavioral addictions among clients of a Parisian fitness room. Compr Psychiatry. https://pubmed.ncbi.nlm.nih.gov/18555055/.

Lichtenstein MB, Jensen ES, Larsen PV et al (2020a) Passion for exercise has three dimensions: psychometric evaluation of the passion scale in a Danish fitness sample. Transl Sports Med. https://doi.org/10.1002/tsm2.173

Lichtenstein MB, Jensen ES, Szabo A (2020b) Exercise addiction, obsessive passion, and the use of nutritional supplements in fitness center attendees. Transl Sports Med 3:188–195. https://doi.org/10.1002/tsm2.131

Lichtenstein M, Hinze C, Emborg B, et al (2017) Compulsive exercise: links, risks and challenges faced. Psychol Res Behav Manag. https://pubmed.ncbi.nlm.nih.gov/28435339/

Lucibello KM, Parker J, Heisz JJ (2019) Examining a training effect on the state anxiety response to an acute bout of exercise in low and high anxious individuals. J Affect Disord 247:29–35. https://doi.org/10.1016/j.jad.2018.12.063

Malm C, Jakobsson J, Isaksson A (2019) Physical activity and sports-real health benefits: a review with insight into the public health of Sweden. Sports Basel Switz. https://pubmed.ncbi.nlm.nih.gov/31126126/

Martin JL, Martens MP, Serrao HF, Rocha TL (2008) Alcohol use and exercise dependence: co-occurring behaviors among college students? J Clin Sport Psychol 2:381–392. https://doi.org/10.1123/jcsp.2.4.381

Matsumoto D (2009) The Cambridge dictionary of psychology. Cambridge University Press, New York, NY, US

McNally GP, Akil H (2002) Role of corticotropin-releasing hormone in the amygdala and bed nucleus of the stria terminalis in the behavioral, pain modulatory, and endocrine consequences of opiate withdrawal. Neuroscience 112:605–617. https://doi.org/10.1016/S0306-4522(02)00105-7

McNamara J, McCabe MP (2012) Striving for success or addiction? Exercise dependence among elite Australian athletes. J Sports Sci 30:755–766. https://doi.org/10.1080/02640414.2012.667879

Miller KJ, Mesagno C (2014) Personality traits and exercise dependence: Exploring the role of narcissism and perfectionism. Int J Sport Exerc Psychol

Mónok K, Berczik K, Urbán R et al (2012) Psychometric properties and concurrent validity of two exercise addiction measures: a population wide study. Psychol Sport Exerc 13:739–746. https://doi.org/10.1016/j.psychsport.2012.06.003

Müller A, Loeber S, Söchtig J et al (2015) Risk for exercise dependence, eating disorder pathology, alcohol use disorder and addictive behaviors among clients of fitness centers. J Behav Addict 4:273. https://doi.org/10.1556/2006.4.2015.044

Parastatidou IS, Doganis G, Theodorakis Y, Vlachopoulos SP (2014) The mediating role of passion in the relationship of exercise motivational regulations with exercise dependence symptoms. Int J Ment Health Addict 12:406–419. https://doi.org/10.1007/s11469-013-9466-x

Pasman L, Thompson JK (1988) Body image and eating disturbance in obligatory runners, obligatory weightlifters, and sedentary individuals. Int J Eat Disord 7:759–769. https://doi.org/10.1002/1098-108X(198811)7:6%3c759::AID-EAT2260070605%3e3.0.CO;2-G

Péronnet F, Szabo A (1993) Sympathetic response to acute psychosocial stressors in humans: linkage to physical exercise and training. Exercise psychology: exercise psychology: the influence of physical exercise on psychological processes. Wiley, New York, NY, pp 172–217

Puterman E, Epel ES, O'Donovan A et al (2014) Anger is associated with increased IL-6 stress reactivity in women, but only among those low in social support. Int J Behav Med 21:936–945. https://doi.org/10.1007/s12529-013-9368-0

Reiner M, Niermann C, Jekauc D, Woll A (2013) Long-term health benefits of physical activity—a systematic review of longitudinal studies. BMC Public Health. https://pubmed.ncbi.nlm.nih.gov/24010994/

Rejeski W, Best D, Griffith P, Kenney E (1987) Sex-role orientation and the responses of men to exercise stress. /paper/Sex-Role-Orientation-and-the-Responses-of-Men-to-Rejeski-Best/ebf80e89c17fac32701c52e2f1cfdea5bdc7d2ee

Robbins JM, Joseph P (1985) Experiencing exercise withdrawal: possible consequences of therapeutic and mastery running. J Sport Exerc Psychol 7:23–39. https://doi.org/10.1123/jsp.7.1.23

Saanijoki T, Tuominen L, Nummenmaa L et al (2014) Physical exercise activates the μ-opioid system in human brain. J Nucl Med 55:1909–1909

Sicilia Á, Alcaraz-Ibáñez M, Lirola M-J, Burgueño R (2017) Influence of goal contents on exercise addiction: analysing the mediating effect of passion for exercise. J Hum Kinet 59:143. https://doi.org/10.1515/hukin-2017-0154

Spano L (2001) The relationship between exercise and anxiety, obsessive-compulsiveness, and narcissism. Personal Individ Differ 30:87–93. https://doi.org/10.1016/S0191-8869(00)00012-X

Stryker S (2001) Traditional symbolic interactionism, role theory, and structural symbolic interactionism: the road to identity theory. In: Turner JH (ed) Handbook of sociological theory. Springer, US, Boston, MA, pp 211–231

Sussman S, Leventhal A, Bluthenthal RN et al (2011a) A framework for the specificity of addictions. Int J Environ Res Public Health 8:3399. https://doi.org/10.3390/ijerph8083399

Sussman S, Lisha N, Griffiths M (2011b) Prevalence of the addictions: a problem of the majority or the minority? Eval Health Prof 34:3. https://doi.org/10.1177/0163278710380124

Szabo A (2010) Addiction to exercise: a symptom or a disorder? Nova Science Publishers, New York

Szabo A (2018) Addiction, passion, or confusion? new theoretical insights on exercise addiction research from the case study of a female body builder. Eur J Psychol 14:296. https://doi.org/10.5964/ejop.v14i2.1545

Szabo A, Tsang E (2003) Motivation for increased self-selected exercise intensity following psychological distress: laboratory-based evidence for catharsis? J Psychosom Res 55:133. https://doi.org/10.1016/S0022-3999(03)00308-8

Szabo A, Billett E, Turner J (2001) Phenylethylamine, a possible link to the antidepressant effects of exercise? Br J Sports Med 35:342. https://doi.org/10.1136/bjsm.35.5.342

Szabo A, Griffiths MD, de Marcos R, LV, et al (2015) Methodological and conceptual limitations in exercise addiction research. Yale J Biol Med 88:303

Szabo A, Griffiths MD, Høglid RA, Demetrovics Z (2018) Drug, nicotine, and alcohol use among exercisers: does substance addiction co-occur with exercise addiction? Addict Behav Rep 7:26. https://doi.org/10.1016/j.abrep.2017.12.001

Szabo A, Boros S, Bősze JP (2019a) Are there differences in life-satisfaction, optimism, pessimism and perceived stress between therapeutic and mastery exercisers? A preliminary investigation. Balt J Sport Health Sci 3. https://doi.org/10.33607/bjshs.v3i114.807

Szabo A, Griffiths MD, Demetrovics Z (2019b) Psychology and exercise. Nutr Enhanc Sports Perform Muscle Build Endur Strength, pp 63–72. https://doi.org/10.1016/B978-0-12-813922-6.00005-9

Szabo A (1995) The impact of exercise deprivation on wellbeing of habitual exercises. Aust J Sci Med Sport. https://pubmed.ncbi.nlm.nih.gov/8599747/

Terry A, Szabo A, Griffiths M (2004) The exercise addiction inventory: a new brief screening tool. Addict Res Theory 12:489–499. https://doi.org/10.1080/16066350310001637363

Thaxton L (1982) Physiological and psychological effects of short-term exercise addiction on habitual runners. J Sport Exerc Psychol 4:73–80. https://doi.org/10.1123/jsp.4.1.73

Thompson J, Blanton P (1987) Energy conservation and exercise dependence: a sympathetic arousal hypothesis. Med Sci Sports Exerc. https://pubmed.ncbi.nlm.nih.gov/3574055/. Accessed 20 Sep 2020

Villella C, Martinotti G, Di Nicola M et al (2011) Behavioural addictions in adolescents and young adults: results from a prevalence study. J Gambl Stud 27:203–214. https://doi.org/10.1007/s10899-010-9206-0

Chapter 9
Work Addiction

Paweł Andrzej Atroszko

Abstract The available data leaves no doubt that compulsive overworking is a genuine problem related to significant harm. Thus far, most cases were recognized in relation to the official diagnosis of anankastia/obsessive–compulsive personality disorder (OCPD). However, while this personality domain may be a risk factor, the available evidence suggests that work addiction is, to some extent, an independent clinical entity and addictive disorder with its own etiology, symptomatology, epidemiology, and course. Work addiction has substantial epidemiological significance due to its high prevalence rates (ranging from 6.6 to 20%) and impairments that it causes. Currently, no well-established theoretical models explaining the biological underpinnings of work addiction exist, and there are no related neuroimaging, physiological or genetic studies to date. It is comorbid with numerous mental disorders and potentially associated with stress-related health problems, particularly cardiovascular disease (CVD). There are appropriate psychometric measures, such as the Bergen Work Addiction Scale, grounded in the addiction framework. There is paucity of high-quality data on the effectiveness of treatments, with most empirical evidence supporting a mindfulness-based approach. There is some indirect support for self-help groups such as Workaholics Anonymous and case reports on individual and family therapies. Treatments for other behavioral addictions and substance use disorders, as well as for anankastia/OCPD, may be adapted for work addiction because of shared similarities. These may include, notably, cognitive behavioral therapy and motivational interviewing. Prevention initiatives directed at young populations, are indispensable to decrease the high prevalence of this disorder in industrialized countries. Perhaps, the greatest challenge currently facing research, prevention and treatment of this disorder is the social and institutional resistance to acknowledge it.

Keywords Anankastia · Obsessive–compulsive personality disorder · Perfectionism · Study addiction · Work addiction · Workaholism

P. A. Atroszko (✉)
Institute of Psychology, University of Gdańsk, Street address Bażyńskiego 4, 80-309 Gdańsk, Poland
e-mail: p.atroszko@ug.edu.pl

H. M. Pontes (ed.), *Behavioral Addictions*, Studies in Neuroscience, Psychology and Behavioral Economics, https://doi.org/10.1007/978-3-031-04772-5_9

9.1 Introduction

Work is one of the most valued, if not the most important, activity in industrialized societies. Occupation forms the basis for social roles and personal identities. Labor is considered to be fundamental to the organization of social structures and is arguably the most important civilization-building factor (Kranzberg and Gies 1975). Work is indispensable for most people, and hard work often is externally imposed. For these reasons, the concept of compulsive overworking as a mental health disorder can be difficult to understand for many people. It is reflected in the resistance among some researchers, the general population, and institutions to acknowledge that uncontrollable overworking might lead to an addictive disorder that can elicit substantial impairments. This reluctance to accept is both explicitly stated (see Griffiths et al. 2018), as well as implicit but evident. The surprisingly low number of journal articles on compulsive overworking published in high profile scientific journals can be considered a good indicator reflecting the multiplicity of factors affecting low social, institutional, and scientific recognition of this problem (see Atroszko et al. 2020a, b).

At the same time, there is no doubt about two facts. Firstly, high workload and occupational stress are significant contributors to the diseases and disorders constituting major components of the global burden of disease (Atroszko et al. 2020b). Overworking is a recognized causal factor in sudden deaths from cardiac events (*karoshi*) and suicide (*karojisatsu*) (Kondo and Oh 2010; see Atroszko et al. 2019). Secondly, compulsive overworking is already officially acknowledged as a facet of one of the mental health disorders since the mid-twentieth century (American Psychiatric Association [APA] 1952). Symptoms related to an undue preoccupation with productivity to the exclusion of pleasure and interpersonal relationships are recognized as criteria for obsessive–compulsive personality disorder/anankastic personality disorder (OCPD/APD) or, more recently, anankastia domain in personality disorder, in official classifications of diseases and disorders, *International Classification of Diseases* (ICD 11, World Health Organization 2019), and *Fifth Edition of the Diagnostic and Statistical Manual for Mental Disorders* (DSM-5; APA 2013). Recently, it was suggested that, in many cases, loss of control over engagement in work shows addictive qualities and should be better conceptualized and clinically recognized as work addiction. In such cases, anankastia/OCPD may be a risk factor. However, the available studies suggest that work addiction is, to some extent, an independent clinical entity and addictive disorder with its own etiology, symptomatology, epidemiology, and course (Atroszko 2019a, b; Atroszko et al. 2019).

9.1.1 *History*

Compulsive overworking is likely to be present in human societies since ancient times; however, to the author's knowledge, no systematic studies on this issue are available. The excessive preoccupation with work and productivity, often linked to greed, excessive worry, and anxiety can be arguably traced back at least around 2500 years ago. Advice on proper attitude to activity and inactivity is present in the early cultural documents from that period. Tao Te Ching, attributed to Lao Tsu, states, “The Tao never does anything, yet through it all things are done. If powerful men and women could center themselves in it, the whole world would be transformed by itself, in its natural rhythms.” (Mitchell 1988, p. 37). The Bhagavad Gita, from the same period, teaches, “One who sees inaction in action, and action in inaction, is intelligent among men.” (Prabhupada 1997, p. 208).

DSM-I (APA 1952) included compulsive personality with features such as “an inordinate capacity for work,” and a “lack [of] a normal capacity for relaxation” (APA 1952, p. 37). Excessive work devotion was one of the few criteria of OCPD that have persisted from DSM-III through to the DSM 5 (APA 2013), and is often termed as “workaholism” in the OCPD literature (Grilo et al. 2004). The features similar to OCPD, particularly perfectionistic concerns, were first described by Pierre Janet in 1903 as the “psychasthenia”, and later adopted by Freud (Mancebo et al. 2005). Psychasthenics were identified to experience physical problems such as headaches, backaches, or insomnia. Similar and other physical symptoms were observed among some individuals on days where they were unable to work in a paper by Ferenczi from (1919). The notion of work addiction was introduced in the psychological literature about 50 years ago by Oates (1968, 1971). As early as in the 1970s it has been acknowledged in high-profile medical literature (Eliot and Forker 1976). It was followed with a period of a few decades of conceptual ambiguities and several, often ad hoc created, theoretical models with poor empirical support. Simultaneously, it has been gradually clinically recognized and treated (Berglas 2004; Fassel 1990; Robinson 2014; Schaef and Fassel 1988).

Only in the last decade, work addiction was more explicitly and systematically grounded in behavioral addiction theory and research (Atroszko 2012; Griffiths 2011), with a measurement tool developed based on the common addiction criteria (Andreassen et al. 2012; Griffiths 2005). It resulted in the gradual accumulation of higher quality data. The coinciding rapid development of the behavioral addiction field led to a recent debate among scholars researching work addiction (Atroszko et al. 2019; Griffiths et al. 2018). Some general agreements were established stating that: (i) compulsive overworking is a genuine problem, (ii) that there is enough empirical data to support its relationship with impaired psychosocial functioning of clinical relevance, (iii) compulsive over-working is not a transient behavioral pattern and that there is evidence for the persistence of work addiction among a minority of individuals, (iv) that the factors contributing to work addiction go far beyond personality alone and that more research on the contribution of meso-level and macro-level factors to work addiction is urgently needed (see Atroszko et al. 2019).

9.1.2 Definition

A tendency for compulsive overworking is approached from two broad frameworks. On the one hand, there is the clinical psychology and behavioral addiction approach, and on the other hand, there is the organizational and work psychology framework. In the latter case, the term "workaholism" is often preferred to denote a dysfunctional and pathological form of heavy work investment constituting a significant problem for the functioning of employees and organizations. Most research in the organizational literature refrains from explicit clinical frameworks in describing this phenomenon. Moreover, similar constructs exist within the psychological literature, such as obsessive passion concerning work-activities in the well-recognized Vallerand's (2015) dualistic model.

Throughout the decades, various definitions were suggested by authors with different backgrounds and theoretical approaches, including conceptualizations as an attitude, a trait, an obsessive–compulsive behavior, and as an addiction (Andreassen 2014; Griffiths et al. 2018; Sussman 2012). Several typologies of workaholics have been proposed based on clinical observations (Fassel 1990; Robinson 1999), statistical analyses (Spence and Robbins 1992), and theoretical considerations (see Andreassen 2014; Sussman 2012). Spence and Robbins' (1992) distinction between enthusiastic and non-enthusiastic workaholics, among a few other types, gained much attention, especially in the organizational literature. It influenced subsequent theoretical models and empirical studies, and currently still inspires researchers. However, the original classification was based on a scale that was not grounded in strong theory, was not investigated in terms of factorial structure, and the types were identified based on cluster analysis in a relatively small sample of social workers. This does not fulfill basic requirements for robust scientific inquiry (Wagenmakers et al. 2012). Nevertheless, the scale is the most widely used workaholism measure (Andreassen 2014; Burke 1999), despite clear problems with replicability of its structure and, consequently, with types based on subscales (Huang et al., 2010). Arguably, differentiation between enthusiastic and non-enthusiastic workaholics, to a large extent, reflects differences between healthy work engagement and work addiction (Di Steffano and Gaudiino 2019; Griffiths et al. 2018). Though more studies are necessary to delineate differences between these constructs precisely, currently, it is a widely accepted distinction among work addiction researchers (Atroszko et al. 2019).

More clinically oriented conceptualizations of work addiction as an addictive disorder begun with Oates (1971), and comprise works of Robinson (2014), Fassel (1990), Porter (1996), Griffiths (2011), Atroszko (2012), and Andreassen and colleagues (2012). They include the elements of preoccupation with work/compulsion or addiction to work and the negative consequences of excessive work. These are often also found in definitions grounded in organizational literature (Schaufeli et al. 2009; Spence and Robins 1992; for an overview see Andreassen 2014; Sussman 2012). In recent years substantial developments in the conceptualization of behavioral addictions have been made (Brand et al. 2020), which allowed suggesting a general definition of work addiction as a behavioral addiction:

> Work addiction is characterized by a compulsion to work and preoccupation with work activities leading to a significant harm and distress of a functionally impairing nature to the individual and/or other significantly relevant relationships (friends, family). The behavior is characterized by the loss of control over the working activity and persists over a significant period of time. This problematic work-related behavior can have varying intensity from mild to severe. (Atroszko et al. 2019, p. 9)

Propositions concerning the development of specific diagnostic symptoms were also delineated:

> Loss of control over the working activity involves working more than planned despite the negative consequence and/or unsuccessful attempts to reduce the activity and/or progressive increase in time spent on working. Withdrawal symptoms (including irritability, negative feelings, sleep problems, etc.) are frequent if the planned/desired amount of work is hindered or appear when attempts at reduction of the amount of work are undertaken. The work activity often serves to reduce negative feelings and/or avoid interpersonal and/or intrapersonal conflicts. (Atroszko et al. 2019, p. 9)

This definition takes into account that the core element of addiction is compulsivity and refers to the common elements to most of the existing definitions of addictions. It is envisaged that this definition will facilitate studies on the valid diagnostic criteria of work addiction that could be used in future official classifications of disorders.

9.1.3 Theoretical Model

Throughout decades, different, clinically oriented conceptualizations used alcohol and substance use disorders as models to define work addiction criteria. The most recent and impactful model referred to common addiction symptoms (Griffiths 2005) in describing work addiction (Andreassen et al. 2012; Griffiths 2011). The core components include: (i) salience, (ii) mood modification, (iii) tolerance, (iv) withdrawal, (v) conflict, (vi) relapse, (vii) health and/or other problems. Nevertheless, specific symptoms related to work addiction may go beyond the common components framework as it is typical for other behavioral addictions (Atroszko et al. 2019). These may include factors such as the enjoyment/satisfaction from work, involvement in work, self-efficacy in work, socioeconomic status, and financial situation (i.e., significant socioeconomic repercussions as a consequence of reducing the number of hours worked), and self-awareness of the problem.

In reference to the general addiction model, it was suggested that work addiction could not be conceptualized as anankastia/OCPD (Atroszko 2019a, b; Atroszko et al. 2019). Firstly, not all cases of anankastia/OCPD are related to addictive work, and not all cases of compulsive overworking are related to rigid perfectionism. Secondly, there are characteristic addiction symptoms present, which are not definitional for a disordered personality. Mainly, craving and engaging in work activities despite negative consequences, loss of control, and withdrawal symptoms in relation to work-related activities are typical for addiction and not a definitional for a trait domain grounded in core perfectionistic concerns (WHO 2019). Disordered personality may be an underlying problem and a risk factor for work addiction, as it is for substance use disorders and behavioral addictions, with some estimates suggesting that the majority of addicts have some personality disorder (DeJong et al. 1993). The elimination of the explicit symptom related to workaholism in the description of anankastia in the latest revision of the ICD-11 (WHO 2019) seems to reflect, to some extent, the growing recognition of the distinction between disordered personality and compulsive work-related behavior.

9.1.4 Behavioral Addiction Criteria

Recently three meta-level criteria for considering potential addictive behaviors as fulfilling the category of "other specified disorders due to addictive behaviors" were suggested (Brand et al. 2020). It can be argued that currently, the first criterion requiring empirical evidence of clinically relevant consequences and functional impairments in daily life due to addictive behavior is to a noteworthy extent fulfilled in relation to work addiction (Atroszko et al. 2019). The other two require further investigation. They posit congruence of the candidate phenomenon with current theoretical models of addictive process and require data suggesting its similarity to other addictive behaviors in terms of psychological (and neurobiological) mechanisms. The specificity of work addiction, and other addictions, should be taken into account when investigating these similarities.

9.1.5 Relevance: Burnout and the Global Burden of Disease

Two notable factors point to the considerable relevance of work addiction as an epidemiological concern (Atroszko et al. 2020b). First is that compulsive overworking is a highly prevalent behavior in a professionally active population. The second factor concerns the fact that high workload and occupational stress are closely related to burnout and are recognized risk factors for a wide range of diseases and disorders constituting major components of the global burden of disease (e.g., CVD, diabetes and autoimmune disorders, cancer). Based on these premises, it can be expected that work addiction is involved in stress-related diseases, in many cases via burnout. A general model of compulsive overworking in relation to the global burden of disease was suggested (Atroszko et al. 2020b). Moreover, indirect harm caused by compulsive overworking on other people may exceed the direct effects on individuals manifesting the behavior. It is because work addiction is unequivocally related to problematic functioning in families and intimate relations, causing distress for partners and children of work addict (Clark et al. 2016). Moreover, work addicted managers may have a substantial negative effect on their co-workers, organization, and the recipients of the organization's work (Atroszko and Atroszko 2020).

9.1.6 *The Need for Early Prevention and the Construct of Study Addiction*

Because of the potential high epidemiological significance of compulsive overworking, there is a need for early prevention. In order to facilitate this, study addiction has been conceptualized as an early form of work addiction, and the data gathered so far, though limited (see Atroszko et al. 2019), is highly consistent with this hypothesis (see Atroszko, 2015, 2019b). This has vital importance from the perspective of investigating and understanding the development of compulsive overworking. Addictions tend to appear in youth; therefore, effective prevention and intervention programs should be directed as early as possible, arguably in high schools since study addiction was identified as a potential issue in this population (Wróbel 2020).

9.2 Biological Bases

The biological bases for work addiction have not been systematically investigated. Currently, there are no available neuroimaging or electrophysiological studies on similarities with the psychological processes and underlying neural correlates established for substance-use disorders and gambling/gaming disorders (Brand et al. 2020). Furthermore, while there are some studies showing the relationship between parent and child work addiction (Kravina et al. 2014), no data on genetic risk factors was published, neither from twin studies nor from molecular genetic studies on candidate genes.

One biological hypothesis postulates "addiction to adrenaline" underlying work addiction (Robinson 2014). It reflects insights from clinical observations of patients seeking treatment for work addiction and suggests that work may act as a form of stimulant. While the notion of adrenaline as an addictive agent may be controversial, there is evidence for the involvement of the adrenergic system in stimulant-induced behaviors (Schmidt and Weinshenker 2014). It may have crucial significance for future studies on work addiction because addictions to stimulants have different clinical features, such as withdrawal, than addictions to opioids, benzodiazepines, or alcohol (APA 2013). Most importantly, withdrawal in the case of stimulants such as cocaine typically does not present a threat to life and is less specific than in cases of other substances. It may manifest as a dysphoric mood and physiological changes, such as fatigue, insomnia or hypersomnia, increased appetite, psychomotor agitation, or retardation. They may linger for weeks or months.

Data on withdrawal in work addiction mostly comes from case studies (Robinson 2014), epidemiological surveys (Atroszko et al. 2017; Andreassen et al. 2014), and indirect research such as that on leisure sickness (Van Heck and Vingerhoets 2007). These are consistent with withdrawal from stimulants, comprise most of the previously mentioned symptoms, and in the case of leisure sickness, include viral infections causing colds, fever, and flu-like symptoms. One possible avenue for investigation is related to the potential complex and paradoxical analgesic effects of endogenous catecholamines (Yaksh 1985) and corticosteroids (McEwan and Kalia 2010). Since work generates physiological activation and stress, it may serve as a pain management strategy such that increased levels of adrenaline and cortisol are involved in pain reduction. This may be, to some extent, similar to the mechanisms involved in the well-documented analgesic effects of exercise (Naugle et al. 2012). Despite these potential temporary effects, in the long run chronic stress typically leads to failure of physiological systems and disease such as CVD (see Atroszko et al. 2020b).

Moreover, the fact that it is a goal-driven behavior links it clearly to dopaminergic system activity as dopamine is involved in motivated behavior (Olivetti et al. 2020). Nevertheless, currently, there are no comprehensive theoretical models based on addiction research explaining the biological underpinnings of work addiction. Specifically, there are no hypotheses on how this behavior could bypass the adaptive mechanisms of habituation in the reward system. It seems that there might be an overlooked mechanism of hooking up on the anticipation of reward in the absence of real reinforcements. In other words, since dopamine is involved in anticipation of reward, being in a constant process towards some goal (e.g., working or studying) may create a specific neural mechanism in vulnerable individuals, that is, those with high perseverance and ability to delay gratification.

Nevertheless, some researchers argued that neither withdrawal nor dopamine system dysregulation are sufficient criteria for addiction (Heinz et al. 2020). Drug craving, seeking and consumption despite harmful consequences is required for the diagnosis of substance dependence. There is evidence for such phenomenological presentation in work addiction related to work-activities (Atroszko et al. 2019). Biological models need to be integrated with behavioral and clinical effects of compulsive overworking, taking into account its coping role. From this perspective, work addiction poses a fascinating challenge to addiction theory and research, and may significantly broaden our understanding of compulsivity.

9.3 Specific Comorbidities

Most data on the co-occurrence of work addiction with other disorders comes from epidemiological surveys, some of which included large national samples. Few clinical samples and case studies corroborated reported findings from large scale epidemiological investigations. Most of these results can be considered highly consistent. Because most data come from survey studies, very little is known about the comorbidity of work addiction with particular clinically diagnosed disorders. In some cases, the co-occurring disorder may play a role of an underlying cause or risk factor (e.g., anankastia/OCPD or attention-deficit hyperactivity disorder [ADHD]), and in other instances of potential consequence (e.g., sleep disorders or CVD). In some cases, it might be both (e.g., generalized anxiety or depression), as well as it may mostly share common etiology with work addiction (e.g., eating disorders or other addictions).

9.3.1 Obsessive-Compulsiveness

OCPD has been suggested as an underlying explanatory mechanism of work addiction (see McMillan et al. 2001). The core feature of anankastia/OCPD is rigid perfectionism, which is increasingly identified as a transdiagnostic process. It is associated with a strong need for control (APA 2013; WHO 2019). A meta-analysis found a considerable association between work addiction and perfectionism ($\rho = 0.55$; Clark et al. 2016). Studies indicate that the so-called "workaholism" symptom in OCPD shows good diagnostic efficiency and is present in a significant portion of cases (Grilo et al. 2004). A large survey showed a relationship between work addiction and obsessive–compulsive disorder (Andreassen et al., 2016). Study addiction was related to psychasthenia (Atroszko 2015). While similarities between work addiction and anankastia/OCPD are profound, thus far, there are no clinical studies directly investigating relationships between these two conditions. Only recently, a more systematic effort towards theoretical clarification of their relationship was undertaken (Atroszko 2019a, b; Atroszko et al. 2019, 2020b).

9.3.2 Attention-Deficit Hyperactivity Disorder

Work addiction was suggested to be a result of a compensating mechanism for ADHD (Andreassen et al. 2016). All studies on the relationship between work addiction and ADHD used the Bergen Work Addiction Scale (BWAS) (Andreassen et al. 2012) and the Adult ADHD Self-Report Scale (ASRS-v1.1) developed by the WHO. The results showed consistent co-occurrence in Poland, Norway, and among Swiss men with around two to three times higher risk for ADHD among work addicts (Atroszko et al. 2017; Andreassen et al. 2016; Marmet et al. 2019a, b). In the case of ADHD, the usage of a screening scale for adults instead of official diagnosis presents some advantages as ADHD tends to be underdiagnosed in the adult population. ADHD was not related to study addiction in a large sample of Polish high school students, which points to potential differences in risk factors between study and work addiction (Wróbel 2020).

9.3.3 Anxiety or Fear-Related Disorders

The relationship between work addiction and negative emotions is well-established, with meta-analysis reporting moderate effect size for trait ($\rho = 0.31$) and state ($\rho = 0.32$) negative affect (Clark et al. 2016). All studies on the relationship between work addiction and anxiety included surveys on non-clinical samples (Andreassen et al. 2016; Bartczak and Ogińska-Bulik 2012; Robinson 1999), using common anxiety scales, and finding a significant relationship. One large survey found a significant co-occurrence of work addiction with social anxiety disorder (SAD; Marmet et al. 2019a, b). Several studies found a consistent relationship between study addiction and SAD among undergraduate students (Atroszko 2015), students of music academies and high school students (Wróbel 2020). The fear of evaluations is congruent with rigid perfectionism and anankastia/OCPD.

9.3.4 Mood Disorders

In recent years, large scale, good quality epidemiological surveys using BWAS and different measures of depressive symptoms, congruent with major depression, showed a consistent relationship between work addiction and depression in Norway, Switzerland, and China (Andreassen et al. 2016; Marmet et al. 2019a, b; Yang et al. 2020). In a Polish study, work addiction based on the BWAS cut-off was related to the previous diagnosis of depression (Atroszko et al. 2017), with 10.4% of work addicts who reported having ever been diagnosed with depression compared to 4.2% for non-work addicts. Studies of varying quality using other measures of work addiction corroborated these findings in diverse samples in Iran, Poland, Japan, and China (e.g., Bartczak and Ogińska-Bulik 2012). Furthermore, a meta-analysis showed moderate effect sizes for the relationships between work addiction and burnout ($\rho = 0.40$) and its components: emotional exhaustion ($\rho = 0.42$), cynicism ($\rho = 0.29$), and depersonalization ($\rho = 0.35$) (Clark et al. 2016). Burnout and depression overlap with a strong indication that they represent the same core construct (Bianchi et al. 2019). Furthermore, independent studies with different measures of both study addiction and depression showed their substantial relationships, with estimates of about three times higher risk for clinical depression among individuals addicted to studying (after recalculation of results in Atroszko 2015).

Based on previous observations that workaholics' behavior is often intense, highly energetic, driven, and they are in constant haste, it was suggested that work addiction might be related to hypomania (McMillan et al. 2001). A study on a sample of bipolar disorder (BD) outpatients showed high scores on a work addiction scale (Di Nicola et al., 2010). This finding was recently corroborated in a large survey among Swiss men in which the BWAS and a screener for BD were used (Marmet et al. 2019a, b). It suggested that at least in some cases, work addiction can be related to impulsivity typical for bipolar patients. It is noteworthy because work addiction is commonly associated with a high need for control (anankastia/OCPD) and conscientiousness ($\rho = 0.16$, 95% CI −0.007. to 331; Clark et al. 2016; Griffiths et al. 2018). This ambiguity might be related to clinically observed; however, scarcely empirically studied heterogeneity of work addiction types grounded in the clinical framework (Fassel 1990; Robinson 2014). Furthermore, due to the specificity of BD, it may be the case that in particular circumstances it should be considered an exclusion diagnosis for work addiction (Durand-Moreau et al. 2018), in a similar manner that it is in compulsive shopping. Nevertheless, stress is a triggering factor in BD, and overworking may be a cause for a manic episode (Umeoka et al. 2020).

9.3.5 Sleep–Wake Disorders

Work addiction has been consistently linked to poor sleep quality (Andreassen et al. 2017; Griffiths et al. 2018; Kubota et al. 2014; Salanova et al. 2016); however, little is known about its relationship to particular sleep disorders. Recently, there has been an increasing number of studies on the mechanisms by which work addiction affects poor sleep (see Atroszko and Atroszko 2020). It may have profound clinical significance since sleep is strictly related to health and mortality. The latest research found that 80% of professionally active women with work addiction comorbid with eating disorders reported experiencing sleepwalking or night terrors (Atroszko et al. 2020a). In this study, work addicted women showed higher rates of restless leg syndrome symptoms than the non-work addicted group. Work addiction tends to be related to eveningness (Andreassen et al. 2017; Atroszko et al. 2020a). Chronotypes have been linked to health; particularly, evening types are frequently found to show worse indicators of health than morning types. Eveningness may often be incompatible with work schedules and behaviors, poor sleep quality, sleep debt, and work-related chronic fatigue (see Andreassen et al. 2017).

9.3.6 Disorders Due to Substance Use or Addictive Behaviors

Based on the assumption that addiction is one underlying process with different manifestations (Shaffer et al. 2004), it can be expected that work addiction is comorbid with substance use disorders and other addictive behaviors. In the late 1970s, Canadian medical literature described addictive problems among physicians wherein workaholism and alcoholism were reported as frequently adopted coping mechanisms among doctors (see Atroszko and Atroszko 2020). Progress from compulsive overworking based in obsessive–compulsive personality to alcohol and drug addiction was explicitly suggested as a typical dynamic of ineffective self-regulation (Cohen 1980). Today, 40 years later, both substance use disorders and professional burnout have been shown to be prevalent among medical professionals (Hey 1994; Imo 2017; Warner et al. 2013). Thus far, only one epidemiological study, among Swiss men investigated the co-occurrence of work addiction and alcohol dependence showing higher rates of alcohol dependence among work addicts (Marmet et al. 2019a, b). A study on a clinical sample showed a relationship between compulsive working and problematic alcohol use (Christo et al. 2003). Moreover, case studies of French patients from the Occupational Disease Centre showed significant patterns of alcohol and tobacco use, suggesting dependence to these substances in work addicts (Durand-Moreau et al. 2018).

Most of the studies on work addiction comorbid with multiple other addictive behaviors and substances have frequently employed single-item questions concerning problematic work, as well as a person centered approach in the analyses of the data (Thege et al. 2016; Tsai et al. 2016). These studies reported clusters of work addiction co-occurring with substance use disorders and other behavioral addictions. A study using the BWAS corroborated these results (Marmet et al. 2019b), indicating that work addiction is prevalent among individuals with a general proclivity towards addictive behaviors. Moreover, it particularly tends to co-occur with other socially sanctioned behaviors, including exercise addiction, eating addiction, and shopping addiction.

9.3.7 Eating and Food Disorders

Eating disorders and work addiction, to a significant extent, share etiological factors and clinical features, notably rigid perfectionism. In a recent study, five out of six females (83%) previously diagnosed with an eating disorder reached the cut-off score on the BWAS for work addiction (Atroszko et al. 2020a). Of all women who reached this cut-off, 5.5% reported eating disorders, which significantly exceeds prevalence in the general population. This preliminary study suggests that work addiction may be highly prevalent among professionally active women diagnosed with eating disorders.

9.3.8 Other Personality Disorders

Some studies showed a relationship between work addiction and narcissism, potentially linking it to narcissistic personality disorder. However, the role of grandiosity and interpersonal hostility in work addiction is likely to be complex and moderated by other factors (Falco et al. 2020). A relationship between work addiction and borderline personality disorder (BPD) was also found among young Swiss men (Marmet et al. 2019b).

9.3.9 CVD and Other Stress-Related Health Problems

Work addiction was theoretically linked to CVD in the 1970s (Eliot and Forker 1976) and recently empirically linked to CVD (Salanova et al. 2016). However, the data is very scarce and preliminary. A meta-analysis showed that work addiction is related to Type A Personality ($\rho = 0.43$; Clark et al. 2016), in which the competitiveness component is associated with CVD risk (Lohse et al. 2017). Since high workload and occupational stress are recognized risk factors for a wide range of diseases and disorders constituting major components of the global burden of disease, it can be expected that work addiction is involved in stress-related diseases (Atroszko et al. 2020b). Nevertheless, currently, there is almost no high-quality data supporting this assumption apart from its relationship with mental health disorders (e.g., depression, anxiety), generally worse health ($\rho = 0.33$; Clark et al. 2016; Griffiths et al. 2018), and isolated studies on physiological correlates, for example showing markers of an inflammatory response in work addicts (Girardi et al. 2019). In general, work addiction is not recognized as a causal risk factor in any disease or disorder despite very strong indirect data suggesting such a link.

9.4 Prevalence Rates and Course

Currently, work addiction is not recognized in the existing official classifications of disorders (APA 2013; WHO 2019), as such, the available data on prevalence is based on psychometric testing. In these circumstances, many, if not most, of the prevalence studies suffer from typical disadvantages related to ad hoc cut-off points and the use of non-representative data. A review of studies which analyzed data based on various cut-offs suggested a prevalence rate of work addiction of about 10% (Sussman et al. 2011). Moreover, the so-called "workaholism" symptom is present in a significant portion of cases of OCPD (Grilo et al. 2004), which is the most prevalent personality disorder in the general population (3%–8%) and outpatient groups (Diedrich and Voderholzer 2015).

Presently, the most theoretically sound and empirically valid cut-off score uses a polythetic approach applied to seven common addiction symptoms defined for work addiction-related behaviors (see Sect. 5.3; Andreassen et al. 2012). Prevalence estimates based on this method range from 6.6% in Denmark, 8.1% in young Swiss men, 8.3% in a nationally representative sample in Norway, 16.8% in a sample of the working population in Poland, and 20.6% in Hungary (Andreassen et al. 2014; Atroszko et al. 2017; Lichtenstein et al. 2019; Marmet et al. 2019a; Orosz et al. 2016). Interestingly, these estimates are very similar to the prevalence rates of study addiction based on the analogous tool and cut-off score for the Bergen Study Addiction Scale (Atroszko 2015; see Wróbel 2020). The available data suggests that work addiction is generally significantly more prevalent than most other addictive behaviors. Its occurrence substantially varies between countries, which may reflect cultural factors and their interaction with socioeconomic conditions (see Atroszko et al. 2020b). In terms of the course of work addiction, a study among high school youth suggests that it is a more stable and persevering problem than most of the addictive behaviors (Sussman et al. 2015).

9.5 Assessment Approaches

9.5.1 Overview of Measurement of Work Addiction

Almost all psychometric measures of work addiction either lack grounding in established addiction theory or adequate empirical validation (often no psychometric evaluation), and most tools presented both problems simultaneously. A more exhaustive description and history of work addiction measures can be found in the works of Andreassen (2014), or Sussman (2012), and even some less popular tools are referenced in Burke (1999). One of the first and probably most widely used scales is the Workaholism Battery (WorkBAT) by Spence and Robbins (1992), measuring three components: Drive, Work Involvement, and Work Enjoyment. Interestingly original scale was not investigated for factorial structure, which probably could account for typical problems with replicating three components (e.g., Huang et al. 2010). Also, in the 1990s another popular measure, the Work Addiction Risk Test (WART), more grounded in addiction framework, was developed by Robinson (1999). The structure of the scale was investigated with exploratory factor analysis, which again probably accounts for typical problems with replicating initially identified five factors. Based on these two widely used scales Schaufeli and colleagues (2009) later on developed the Dutch Work Addiction Scale (DUWAS).

Arguably, the most important breakthrough so far was introduced with BWAS (Andreassen et al. 2012), which is explicitly grounded in addiction theory and has acceptable psychometric properties. Despite its evident advantages, it is still relatively rarely used since most of the research on work addiction is carried in the organization and work psychology settings, which tend to prefer DUWAS over the more clinically oriented BWAS. There is very limited data on the relationship between these scales, and the available study in a Romanian sample showed their partial convergence (Shkoler et al. 2017). These two scales are described in more detail below.

Today, other distinctive measures have been developed (Shkoler et al. 2017), often in national languages other than English. While frequently empirically validated to some extent, they gain little recognition and empirical investigation, probably due to the language barrier and conceptual remoteness from the dominating current research. In general, thus far, work addiction research suffered from an inadequate psychometric approach to developing assessment tools. Currently, the BWAS, though still not perfect, is the most recommended scale and can serve as a benchmark for further developments in the measurement of work addiction. The scale is anchored in the addiction theory, is brief and convenient, and has the best psychometric properties from the available work addiction measures. It also allows for calculating more meaningful prevalence estimates. Even though it is clinically grounded, this does not limit its use in work and organizational settings.

9.5.2 *The Dutch Work Addiction Scale (DUWAS): Organizational Approach*

DUWAS is a 10-item questionnaire that measures work addiction (Schaufeli et al. 2009). The scale measures two components: Working Excessively and Working Compulsively assessed by five items each. It was based on six items taken from WART (Robinson 1999) and four items from WorkBAT (Spence and Robbins 1992). The scale is not based on explicit addiction theory. The answers are provided on a four-point scale that ranges from *(almost) never* to *(almost) always*. The DUWAS measures workaholism as an excessive and obsessive–compulsion toward work. Even though the scale is used widely and the two-factor solution, with correlated factors, has been validated cross-culturally in several countries, the hierarchical structure has not been tested (Balducci et al. 2015; see Rantanen et al. 2015). It poses significant concerns about whether the scale measures a higher-order factor of work addiction. Meaningfully testing this hypothesis is further limited by only two first-order factors that pose technical limitations on the confirmatory factor analytical approach. A study showed a single second-order factor of a different 20-item version of the scale with three first-order factors (Sharma and Sharma 2013). However, only three primary factors still put severe limitations on the relevance of the analysis (Lee and Cadogan 2013). The structure of the scale was tested in a more complex model, including the engagement scale showing some evidence for measurement invariance (Hu et al. 2014). Otherwise, the subscales show good concurrent and predictive validity (Kubota et al. 2014).

9.5.3 *The Bergen Work Addiction Scale (BWAS): Clinical Approach*

The BWAS is a seven-item tool developed by Andreassen et al. (2012) that is grounded in general addiction theory (Griffiths 2005). The items are worded to reflect seven common diagnostic addiction criteria. The answers are provided on a five-point scale that range from *never* to *always*. It asks how often *during the last year* the symptoms have occurred. The cut-off score is based on a polythetic approach, that is, scoring 4 (*often*) or 5 (*always*) on at least four of the seven items. It is congruent with modern nosological systems (APA 2013). The scale showed adequate factorial structure in different countries and good concurrent, divergent, and predictive validity (Andreassen et al. 2014; Atroszko et al. 2017; Lichtenstein et al. 2019; Marmet et al. 2019a, b; Orosz et al. 2016; Yang et al. 2020). More in-depth statistical analyses using Item Response Theory suggest that some refinements to the scale could increase its psychometric soundness (Bereznowski and Konarski 2020). Particularly rephrasing some items is recommended to better reflect clinically relevant symptoms with improved face and diagnostic validity.

9.6 Prevention and Treatment Approaches

9.6.1 Denial and Low Social and Institutional Recognition of the Problem

One of the most important impediments to the treatment of work addiction is the denial and resistance to search for treatment among those who might need it. In general, most of the addicted individuals do not recognize their problem or seek help (Goldstein et al. 2009); however, it has been argued by clinicians that this problem is even more pronounced among work addicts as their behavior is supported by the high value put on hard work in industrialized societies (Berglas 2004; Killinger 1992; Robinson 2014). Even some of the workaholism researchers did not initially perceive it as a problem, and today still such voices exist (see Andreassen 2014; Atroszko and Atroszko 2020; Sussman 2012). It may be at least partially responsible for the scarcity of high-quality data on treatment in work addiction, and a related curious paradox of mismatch between high need for help and the paucity of clinical research. Some clinicians who have rich experience in working with workaholic clients throughout decades, including psychiatrists and clinical psychologists are best-selling authors of self-help books on work addiction and occupational health (Berglas 2004; Fassel 1990; Killinger 1992; Schaef and Fassel 1990). Some of these guides reappear in new editions (e.g., Robinson 2014), and prestigious institutions recognized some of the authors for their work in the addiction field. However, only Robinson (2014) published more systematic research in work addiction, focusing on the family of work addicts. The high demand for self-help books shows a considerable need for support for compulsive overworking in the general population. It is juxtaposed with the low formal institutional recognition of the problem and scarcity of systematic research on the treatment options, despite work addiction being recognized in patients of occupational disease centers with compelling case studies (Durand-Moreau et al. 2018). Moreover, one study showed that 50% of men and 28% of women with first-episode depression among employed individuals recruited from occupational health care units were diagnosed with OCPD (Raiskila et al. 2013).

9.6.2 Treatment Approaches in Other Addictions and Anankastia/OCPD

Overall, work addiction is not a formal diagnosis, and so far, there are no randomized clinical trials (RCTs) on the efficacy of any treatment for this condition. However, there are case reports and isolated studies on some approaches showing promising results. Also, the effective interventions for other addictions may be adapted to the treatment of work addiction due to considerable similarities in addictive disorders (Myers 2019; Zajac et al. 2017). These may include motivational interviewing (MI), cognitive behavioral therapy (CBT), dialectical behavior therapy (DBT) tailored to addiction, group therapy, and/or a combination of different therapeutic strategies. Similarly, treatments for anankastia/OCPD (Grant et al. 2019), which may underline many of the cases of work addiction, might prove effective for managing work addiction and prevention of relapse. Diedrich and Voderholzer (2015) provided a review of OCPD treatment research. There is some evidence for carbamazepine, fluvoxamine, and citalopram. Data supports cognitive therapy, CBT, group CBT, interpersonal therapy, and schema therapy. There is at least some evidence for metacognitive interpersonal therapy, an adapted version of DBT, as well as supportive–expressive dynamic psychotherapy. Even in relation to anankastia/OCPD more RCTs on therapeutic interventions are highly warranted.

9.6.3 Macro-, Meso- and Micro-level Prevention and Treatment Approaches

Prevention and treatment of work addiction can be approached from the macro-, meso- and micro-level. The macro-level perspective is likely most important and practically neglected, and it involves the concept of universal prevention. The meso-level interventions center around work-place and organizational solutions. The micro-level approach concerns individual prevention and treatment solutions, which may be further categorized into self-help and professional treatments. Some of the solutions are not exclusive to just one level.

For example, the introduction of mindfulness classes in schools as part of obligatory activity from the early stages of formal education may result in an increase in self-regulatory skills on a population level. Subsequently, it may lead to a reduction not only of work addiction, but also other addictive behaviors and, generally, psychopathology. As such, it would be a macro-level educational policy. It is already realized to some extent, for example, in the United Kingdom (UK Government 2019). Nevertheless, organizations may include mindfulness training for their employees regardless of higher-level policies. Finally, each individual may decide to practice mindfulness to improve emotional self-regulation and decrease symptoms of work addiction. Mindfulness is only one example of a feasible intervention, but many other cross-level solutions could be implemented.

9.6.3.1 Macro-level Solutions: Universal Prevention

The current approach to psychopathology, in general, and addiction in particular, often focuses on the treatment of most severe cases. On the other hand, it can be argued that preventive measures show a considerable cost/benefit ratio (Miller and Hendrie 2009). Strictly speaking, prevention is focused on avoiding future illness for all the population, not just individuals at immediate risk (Haggerty and Mrazek

1994) since most cases of disease or disorder appear in low-risk groups. This effect is called "prevention paradox" (Rossow and Romelsjö 2006). Rose (2001) compellingly outlined the benefits of universal prevention showing that the prevalence of disease or disorder is in relation to the average level of particular behavior related to this disease or disorder in the whole population. The number of work addicts in a particular population can be expected to be in relation to the prevailing behaviors characteristic for work addiction (e.g., a culture of presenteeism, meaning working while ill) (see Atroszko and Atroszko 2020).

Universal prevention has been successfully applied in the field of addiction (Danielsson et al. 2012). In the work-related domain, this approach would benefit from integrated efforts of international health institutions and governments and may require specific work-related regulations and programs (Atroszko et al. 2020b). Changes in the macroeconomic environment and employment situation tend to affect workaholism (Atsuko and Mitsuru 2004). Proper job security seems essential for preventing work addiction, including death from overworking (Kondo and Oh 2010). Prevalence rates based on the same cut-off are about two- to three-fold higher in central European countries (e.g., Poland and Hungary) in comparison to Scandinavian countries (e.g., Norway and Denmark) or Switzerland, which differ considerably in terms of employment stability and social care. Specific regulations, for example, mitigating presenteeism, could be beneficial given it predicts work addiction (Karanika-Murray et al. 2015). Some steps towards enabling the early application of this approach were done with the conceptualization of study addiction as a precursor of work addiction.

9.6.3.2 Meso-level Solutions: Organizational and Work-Place Interventions

Presently, the majority of studies on work addiction have been conducted within the work psychology and organizational framework. Typically they concentrate on differentiating between healthy work engagement and unhealthy workaholism (Andreassen, 2014; Atroszko et al. 2019; Di Stefano and Gaudiino 2019). From this perspective, employers and managers should focus on promoting healthy engagement and reducing work addiction. Based on the existing findings, several solutions grounded in organizational literature were proposed. These focus on *norms* and *values* that encourage work engagement rather than workaholism. They should include interventions providing (i) fulfillment of employees' basic needs at work, (ii) proper management behavior such as transformational leadership style wherein leaders influence by example, use inspirational motivation, show individual consideration, and stimulate followers intellectually, (iii) career counseling, and (iv) work-life balance programs focusing on training employees in skills related to time management, stress reduction, and relaxation techniques (Andreassen 2014). Typically, recommendations include the usage of Employee Assistance Programs, enforced vacations, development opportunities for better engagement, or flexible roles (see Sussman 2012).

Despite a rich and still growing literature on solutions for a healthy workplace, in 2019, one of the crucial points on WHO's agenda for the next years was tackling burnout, which is a growing rather than diminishing phenomenon across countries and professions (Atroszko et al. 2020b). Occupational stress is the most significant source of stress for the adult working population and has been growing in the last decades. It points to a disproportion between expanding scientific literature on a healthy workplace and increasing work-related ill-health. One of the likely causes for such state of affairs is that individuals who should be responsible for introducing and managing healthy work-place behaviors, that is, leaders and managers, are the ones who show high risk for compulsive overworking (Clark et al. 2016). A recent review emphasized a few crucial points related to the role they play in developing and sustaining work addiction in organizations (Atroszko and Atroszko 2020).

To begin with, holding a managerial position is one of the vital socio-demographic risk factors for work addiction. Leaders and managers have a direct and indirect effect on employee well-being. They are influencing the organizational culture and climate, which are recognized factors contributing to work addiction. Therefore, managers are likely a major source of work addiction and its consequences in organizations. The available studies suggest a considerable contribution of work-addicted managers to the harm of employees in terms of chronic stress, mental and physical illness, and loss of productivity due to ineffective supervision. Taking into account the role of the environment in addiction, creating a climate of overworking may have substantial consequences, especially if the stressful settings last for a prolonged time. However, even if relatively transient (a few months or years), acute stress and unfavorable environment may have profound consequences.

Based on studies in substance use disorders, and some of the research into work addiction, several effects of severe stress should be taken into account (see Atroszko and Atroszko 2020). Firstly, it can trigger a long-lasting addictive pattern in vulnerable individuals, which may persist even after a change of environment. Secondly, it may, directly and indirectly, lead to consequences of overworking and compulsive overworking: (i) life-threatening phenomena such as death from overwork or from health complications, as well as suicide, (ii) other such as health problems and their complications, family problems, and loss of productivity. These consequences may linger even after a change of environment and resolution of addiction symptoms. It suggests a great need to systematically integrate the existing data on leadership behavior and its role in employees' well-being in the context of work addiction. The critical question is, how can we develop frameworks for effectively implementing healthy work environment interventions taking into account that leaders are at high risk of work addiction and addiction is, in most cases, related to denial? It is not only a question of reducing human suffering, but a crucial economic problem since occupational stress generates enormous costs for society and industry (European Agency for Safety and Health at Work 2014). For this reason, governments, social institutions, and organizations should be vitally interested in systematically studying this conundrum in order to overcome it. In this context, it needs to be acknowledged that the recently developed guidelines on a healthy work environment by the WHO do not account for compulsive overworking and its specificity (Atroszko et al. 2020b). Also, they do not take into consideration the large population of self-employed individuals.

9.6.3.3 Micro-level Interventions

Micro-level interventions focus on the individual and include self-help solutions and professional treatments. Self-help options include Workaholics Anonymous groups, self-help books (e.g., Robinson 2014), and options such as mindfulness, work-life balance programs, and a positive psychology approach (Burwell and Chen 2008). Practical self-help skills and strategies may focus on not working at home, organizing proper breaks and scheduling recovery time, setting limits, planning realistic goals, developing healthy attitudes towards the self, others, work, and modifying thought patterns. Also, they include cultivating social relationships and committing to leisure activities, as well as taking care of a healthy diet, exercise, and sleep habits.

Treatment options include family therapy, CBT with such variations as Rational Emotive Behavioural Therapy (REBT), MI, positive psychotherapy, and quality of life therapy (Burwell and Chen 2008). Nevertheless, these are mostly feasible theoretical propositions repeated in the literature without any systematic data on their application to work addiction (Andreassen 2014; Sussman 2012). Most indirect empirical support for some usefulness of self-help and individual and family therapy approaches comes from Workaholics Anonymous activity and works of professional counsellors, notably Robinson (2014), Fassel (1990), and Berglas (2004) describing case reports and specific approaches and problems in therapy.

Self-Help Options

Workaholics Anonymous groups are based on principles of sharing experiences and supporting each other in solving common problems and recovering from workaholism. The organization offers meetings worldwide online and offline, literature, and conferences. It is self-supporting through its own contributions. It is based on the 12-step program and support groups, and has provided its book designed to help with recovery (Workaholics Anonymous 2020). It reflects a spirituality based approach to addiction treatment (Miller and Bogenschutz 2007), and recovery from long working hours (Fry et al. 2006). Thus far, the effectiveness of this intervention has not been empirically evaluated. Nevertheless, the group exists and actively operates in numerous countries for almost 30 years, which provides some indirect support for its potential benefits.

Mindfulness-based interventions have well-documented efficacy in a wide range of mental health disorders, and are arguably suitable to improve work-related well-being (Bartlett et al. 2019). A case study (Shonin et al. 2014) followed by a controlled trial (not randomized) provided initial evidence of the effectiveness of meditation awareness training (MAT) in reducing work addiction symptoms and psychological distress (Van Gordon et al. 2017). Moreover, MAT participants demonstrated better job satisfaction, work engagement, and fewer hours spent working but without a decline in job performance. These interventions may also be used at a population level for the prevention of work addiction. Mindfulness training can be effectively used in schools from early education (Dunning et al. 2019), as well as in the workplace (Bartlett et al. 2019).

Therapeutic Interventions

CBT has been successfully used in addiction treatment, including behavioral addictions (Myers 2019; Zajac et al. 2017). Berglas (2004) and Chen (2006) described the principles of REBT applied to work addiction. Burwell and Chen (2002) provided a case example that includes cognitive reframing, emotive intervention, accepting self, shame-attacking exercise, behavioral modification, delegating tasks to others, setting boundaries, desensitizing anxiety, practicing substitution, and exercising leisure.

MI is a brief and highly cost-effective approach to modifying addictive behaviors with substantial empirical evidence (Lundahl et al. 2010). It includes a wide range of techniques used to help people explore and resolve ambivalence about behavioral change. The interventions can be as brief as one meeting.

Family therapy of work addicts was practiced and described by Robinson (1998). He analyzed the structural and dynamic characteristics of the workaholic family and provided treatment recommendations. These included acknowledging and working with denial, unrealistic expectations put on children, identifying the structure of a workaholic family, negotiating boundaries around work talk, developing effective problem solving, better communication, more clearly established family roles, greater affective responses, more affective involvement, and higher general family functioning, tracing intergenerational nature of addiction, and solving intimacy problems. Different approaches to family therapy are available.

9.7 Final Conclusions

The existing data leaves no doubt that compulsive overworking is a genuine problem related to significant harm. Thus far, most cases were recognized in relation to the official diagnosis of anankastia/OCPD. However, while this personality domain may be a risk factor, the available studies suggest that work addiction is, to some extent, an independent clinical entity and addictive disorder with its own etiology, symptomatology, epidemiology, and course. This disorder seems to have substantial epidemiological significance due to the high prevalence rates and impairment that it causes. Currently, no well-established theoretical models are explaining the biological underpinnings of work addiction and there are no related neuroimaging, physiological or genetic studies. Work activities typically create physiological activation. Based on clinical observations and epidemiological surveys, it is suggested that work addiction may serve a similar function to a stimulant and may share some clinical features with stimulant addiction such as symptomatology and course of withdrawal. Studies suggest that work addiction is comorbid with anankastia/OCPD, OCD, ADHD, depression and BD, anxiety and SAD, eating disorders, alcohol use disorder, and, generally, other addictive behaviors, and potentially narcissistic personality disorder and BPD. However, because most data come from surveys, very little is known about the comorbidity of work addiction with clinically diagnosed disorders. Since high workload and occupational stress are health-related risk factors, work addiction may be potentially associated with stress-related diseases, particularly CVD. There are good psychometric measures, such as the BWAS, grounded in the addiction framework. These may facilitate further screening and epidemiological studies.

There are no standard treatment approaches to work addiction, with scarcely any high-quality data on the effectiveness of interventions. Most empirical evidence is for the mindfulness-based approach. There is some indirect support for self-help groups such as Workaholics Anonymous, and case studies on individual and family therapies. Treatments for other behavioral addictions and substance-use disorders, as well as for anankastia/OCPD may be adapted for work addiction because of shared similarities. These may include, notably, CBT and MI. Furthermore, it is suggested that universal prevention programs, including directed at young populations, are indispensable to decreasing the high prevalence of this disorder in industrialized countries.

Work addiction has a relatively long standing history, in comparison to other behavioral addictions; however, there is remarkably little high-quality research on its biological underpinnings, clinical assessment, and treatment. At the same time, clinicians with rich experience in treating work addicts are best-selling authors of self-help guides for work addiction, showing substantial demand in the general population for systematic help with compulsive overworking. Perhaps, the greatest challenge currently facing research, prevention and treatment of this disorder is the social and institutional resistance to acknowledge that the most valued human activity and the basis for personal identity may possess addictive qualities. Alcoholism became a recognized medical problem in the eighteenth century with the advent of the industrial revolution. To some extent it was because the effects of excessive alcohol consumption begun to conflict with persons' availability for work and work performance, and addicts were viewed as unproductive social outcasts (Hanson 2013). Moreover, industrialization and urbanization increased crime, poverty, and infant mortality rates which contributed to substance abuse.

Work addiction may become recognized as a disorder in the future when suffering caused by it at the personal and family level, and its individual, social and economic costs in modern society, which seem to exceed any benefits from such type of high work investment, are acknowledged. For this established conceptualization within recognized addiction models and more good quality data is still needed. Particularly, greater interest among clinically oriented researchers and integrated efforts to incorporate the notion of harmful and unprofitable hard work into frameworks of free-market economy could be of benefit.

Funding Sources The author received no financial support for the research, authorship and/or publication of this article.

Authors' Contribution The manuscript was prepared by PA.

Conflict of Interest The author declared no potential conflicts of interest with respect to the research, authorship and/or publication of this article.

References

American Psychiatric Association (1952) Diagnostic and statistical manual of mental disorders, 1st edn. Author, Washington, DC

American Psychiatric Association (2013) Diagnostic and statistical manual of mental disorders, 5th edn. Author, Arlington, VA

Andreassen CS (2014) Workaholism: an overview and current status of the research. J Behav Addict 3(1):1–11. https://doi.org/10.1556/jba.2.2013.017

Andreassen CS, Griffiths MD, Hetland J et al (2012) Development of a work addiction scale. Scand J Psychol 53(3):265–272. https://doi.org/10.1111/j.1467-9450.2012.00947.x

Andreassen CS, Bakker AB, Bjorvatn B et al (2017) Working conditions and individual differences are weakly associated with workaholism: a 2–3-year prospective study of shift-working nurses. Front Psychol 8:2045. https://doi.org/10.3389/fpsyg.2017.02045

Andreassen CS, Griffiths MD, Hetland J et al (2014) The prevalence of workaholism: a survey study in a nationally representative sample of Norwegian employees. PLoS ONE 9(8). https://doi.org/10.1371/journal.pone.0102446

Andreassen CS, Griffiths, MD, Sinha R et al (2016) The relationships between workaholism and symptoms of psychiatric disorders: a large-scale cross-sectional study. PLoS ONE 11(5). https://doi.org/10.1371/journal.pone.0152978

Atroszko PA (2019a) Work addiction as a behavioural addiction: towards a valid identification of problematic behavior. Aust NZ J Psychiat 53(4):284–285. https://doi.org/10.1177/0004867419828496

Atroszko PA (2019b) Response to: Loscalzo and Giannini (2018). A boon of incoherence: insights on the relationship between study/work addiction and obsessive-compulsive personality disorder. Psychiatria i Psychologia Kliniczna 19(2):237–243. https://doi.org/10.15557/PiPK.2019.0025

Atroszko PA, Pallesen S, Griffiths MD et al (2017) Work addiction in Poland: adaptation of the bergen work addiction scale and relationship with psychopathology. Health Psychol Rep 5(4):345–355. https://doi.org/10.5114/hpr.2017.68759

Atroszko PA, Demetrovics Z, Griffiths MD (2020b) Work addiction, obsessive-compulsive personality disorder, burn-out, and global burden of disease: implications from the ICD-11. Int J Environ Res Public Health 17(2):660. https://doi.org/10.3390/ijerph17020660

Atroszko PA, Atroszko B (2020) The costs of work-addicted managers in organizations: Towards integrating clinical and organizational frameworks. Amfiteatru Econ 22 (in press)

Atroszko PA, Demetrovics Z, Griffiths MD (2019) Beyond the myths about work addiction: Toward a consensus on definition and trajectories for future studies on problematic overworking. A response to the commentaries on: Ten myths about work addiction (Griffiths et al, 2018). J Behav Addict 8(1):7–15. https://doi.org/10.1556/2006.8.2019.11

Atroszko PA, Mytlewska WM, Atroszko B (2020a) Majority of professionally active women diagnosed with eating disorders may be at risk of work addiction: An overlooked comorbidity. Health Psychol Rep (in press)

Atroszko PA (2012) Research on behavioural addictions: work addiction. In: Baranowska-Szczepańska M, Gołaszewski M (eds) Modern research trends of young scientists: current status, problems and prospects. Wydawnictwo Naukowe Wyższej Szkoły Handlu i Usług, Poznań, pp 11–24

Atroszko PA (2015) The structure of study addiction: selected risk factors and the relationship with stress, stress coping and psychosocial functioning. Dissertation, University of Gdańsk

Atsuko K, Mitsuru W (2004) Effects of economic environmental changes on job demands and workaholism in Japan. J Organ Change Manag 17(5):537–548. https://doi.org/10.1108/09534810410554533

Balducci C, Avanzi L, Consiglio C et al (2015) A cross-national study on the psychometric quality of the Italian version of the Dutch Work Addiction Scale (DUWAS). Eur J Psychol Assess 33:422–428. https://doi.org/10.1027/1015-5759/a000300

Bartczak M, Ogińska-Bulik N (2012) Workaholism and mental health among Polish academic workers. Int J Occup Saf Ergon 18(1):3–13. https://doi.org/10.1080/10803548.2012.11076910

Bartlett L, Martin A, Neil AL et al (2019) A systematic review and meta-analysis of workplace mindfulness training randomized controlled trials. J Occup Health Psychol 24(1):108–126. https://doi.org/10.1037/ocp0000146

Bereznowski P, Konarski R (2020) Is the polythetic approach efficient in identifying potential work addicts? Comparison of the polythetic approach with the item response theory framework. Pol Psychol Bull (in press)

Berglas S (2004) Treating workaholism. In: Coombs RH (ed) Handbook of addictive disorders: a practical guide to diagnosis and treatment. John Wiley & Sons, Hoboken, pp 383–407

Bianchi R, Schonfeld IS, Laurent E (2019) The trouble with burnout: an update on burnout-depression overlap. Am J Psychiatry 176(1):79. https://doi.org/10.1176/appi.ajp.2018.18091026

Brand M, Rumpf HJ, Demetrovics Z et al (2020) Which conditions should be considered as disorders in the International Classification of Diseases (ICD-11) designation of "other specified disorders due to addictive behaviors"? J Behav Addict. https://doi.org/10.1556/2006.2020.00035

Burke RJ (1999) Workaholism in organizations: Measurement validation and replication. Int J Stress Manag 6(1):45–55. https://doi.org/10.1023/A:1021910303807

Burwell R, Chen CP (2002) Applying REBT to workaholic clients. Couns Psychol Q 15(3):219–228. https://doi.org/10.1080/09515070210143507

Burwell R, Chen CP (2008) Positive psychology for work-life balance: A new approach in treating workaholism. In: Burke RJ, Cooper CL (eds) The long work hours culture. Causes, consequences and choices. Emerald, Bingley, p 295–313

Chen CP (2006) Improving work-life balance: REBT for workaholic treatment. In: Burke RJ (ed) Research companion to workaholism: current status working time and work addiction. Edward Elgar, Cheltenham, pp 310–329

Christo G, Jones SL, Haylett S et al (2003) The shorter PROMIS questionnaire: further validation of a tool for simultaneous assessment of multiple addictive behaviours. Addict Behav 28(2):225–248. https://doi.org/10.1016/S0306-4603(01)00231-3

Clark MA, Michel JS, Zhdanova L et al (2016) All work and no play? A meta-analytic examination of the correlates and outcomes of workaholism. J Manage 42:1836–1873. https://doi.org/10.1177/0149206314522301

Cohen S (1980) The conspiracy of silence. Can Fam Physician 26:847–849

Danielsson AK, Wennberg P, Hibell B et al (2012) Alcohol use, heavy episodic drinking and subsequent problems among adolescents in 23 European countries: does the prevention paradox apply? Addiction 107(1):71–80. https://doi.org/10.1111/j.1360-0443.2011.03537.x

DeJong CA, Van den Brink W, Harteveld FM et al (1993) Personality disorders in alcoholics and drug addicts. Compr Psychiatry 34(2):87–94. https://doi.org/10.1016/0010-440X(93)90052-6

Di Stefano G, Gaudiino M (2019) Workaholism and work engagement: how are they similar? How are they different? A systematic review and meta-analysis. Eur J Work Organ Psychol 28(3):329–347. https://doi.org/10.1080/1359432X.2019.1590337

Di Nicola M, Tedeschi D, Mazza M et al (2010) Behavioural addictions in bipolar disorder patients: role of impulsivity and personality dimensions. J Affect Disord 125(1–3):82–88. https://doi.org/10.1016/j.jad.2009.12.016

Diedrich A, Voderholzer U (2015) Obsessive–compulsive personality disorder: a current review. Curr Psychiatry Rep 17(2):1–10. https://doi.org/10.1007/s11920-014-0547-8

Dunning DL, Griffiths K, Kuyken W et al (2019) Research review: the effects of mindfulness-based interventions on cognition and mental health in children and adolescents—a meta-analysis of randomized controlled trials. J Child Psychol Psychiatry 60(3):244–258. https://doi.org/10.1111/jcpp.12980

Durand-Moreau Q, Le Deun C, Lodde B et al (2018) The framework of clinical occupational medicine to provide new insight for workaholism. Ind Health 56:441–451. https://doi.org/10.2486/indhealth.2018-0021

Eliot RS, Forker AD (1976) Emotional stress and cardiac disease. JAMA 236(20):2325–2326. https://doi.org/10.1001/jama.1976.03270210051030

European Agency for Safety and Health at Work (2014) Calculating the cost of work-related stress and psychosocial risks. European Risk Observatory Literature Review. https://osha.europa.eu/pl/tools-and-publications/publications/literature_reviews/calculating-the-cost-of-work-related-stress-and-psychosocial-risks. Accessed 2 June 2019

Falco A, Girardi D, Di Sipio A et al (2020) Is narcissism associated with heavy work investment? The moderating role of workload in the relationship between narcissism, workaholism, and work engagement. Int J Environ Res Public Health 17(13):4750. https://doi.org/10.3390/ijerph17134750

Fassel D (1990) Working ourselves to death: the high cost of workaholism, the rewards of recovery. Harper, San Francisco

Ferenczi S (1919) II Sonntagsneurosen. Internationale Zeitschriftfür Psychoanalyse 5:46–48

Fry LW, Matherly LL, Vitucci S (2006) Spiritual leadership theory as a source for future theory, research, and recovery from workaholism. In: Burke RJ (ed) Research companion to working time and work addiction. Edward Elgar, Cheltenham, pp 330–352

Girardi D, De Carlo A, Dal Corso L et al (2019) Is workaholism associated with inflammatory response? The moderating role of work engagement. TPM Test Psychom Methodol Appl Psychol 26(2):305–322. https://doi.org/10.4473/TPM26.2.9

Goldstein RZ, Bechara A, Garavan H et al (2009) The neurocircuitry of impaired insight in drug addiction. Trends Cogn Sci 13:372–380. https://doi.org/10.1016/j.tics.2009.06.004

Grant JE, Pinto A, Chamberlain SR (eds) (2019) Obsessive-compulsive personality disorder. American Psychiatric Association Publishing, Washington, DC

Griffiths M (2005) A 'components' model of addiction within a biopsychosocial framework. J Subst Use 10(4):191–197. https://doi.org/10.1080/14659890500114359

Griffiths MD (2011) Workaholism: a 21st century addiction. Psychologist 24:740–744

Griffiths MD, Demetrovics Z, Atroszko PA (2018) Ten myths about work addiction. J Behav Addict 7:845–857. https://doi.org/10.1556/2006.7.2018.05

Grilo CM, Skodol AE, Gunderson JG et al (2004) Longitudinal diagnostic efficiency of DSM-IV criteria for obsessive–compulsive personality disorder: a 2-year prospective study. Acta Psychiatr Scand 110(1):64–68. https://doi.org/10.1111/j.1600-0447.2004.00311.x

Haggerty RJ, Mrazek PJ (eds) (1994) Reducing risks for mental disorders: frontiers for preventive intervention research. National Academy Press, Washington, DC

Hanson D (2013) Historical evolution of alcohol consumption in society. Oxford University Press, Oxford

Heinz A, Daedelow LS, Wackerhagen C et al (2020) Addiction theory matters—why there is no dependence on caffeine or antidepressant medication. Addict Biol 25(2). https://doi.org/10.1111/adb.12735

Hey S (1994) Sick doctors. Workaholics harm families also. BMJ 309(6963):1235

Hu Q, Schaufeli W, Taris T et al (2014) East is East and West is West and never the twain shall meet: work engagement and workaholism across Eastern and Western cultures. J Behav Soc Sci 1(1):6–24

Huang JC, Hu C, Wu TC (2010) Psychometric properties of the Chinese version of the workaholism battery. J Psychol 144(2):163–183. https://doi.org/10.1080/00223980903472219

Imo UO (2017) Burn-out and psychiatric morbidity among doctors in the UK: a systematic literature review of prevalence and associated factors. Bjpsych Bull 41(4):197–204. https://doi.org/10.1192/pb.bp.116.054247

Karanika-Murray M, Pontes HM, Griffiths MD (2015) Sickness presenteeism determines job satisfaction via affective-motivational states. Soc Sci Med 139:100–106. https://doi.org/10.1016/j.socscimed.2015.06.035

Killinger B (1992) Workaholics: the respectable addicts. Fireside, New York

Kondo N, Oh J (2010) Suicide and karoshi (death from overwork) during the recent economic crises in Japan: the impacts, mechanisms and political responses. J Epidemiol Community Health 64(8):649–650. https://doi.org/10.1136/jech.2009.090787

Kranzberg M, Gies J (1975) By the sweat of thy brow: work in the Western World. G. P. Putnam's Sons

Kravina L, Falco A, De Carlo NA et al (2014) Workaholism and work engagement in the family: the relationship between parents and children as a risk factor. Eur J Work Organ Psychol 23(6):875–883. https://doi.org/10.1080/1359432X.2013.832208

Kubota K, Shimazu A, Kawakami N et al (2014) Workaholism and sleep quality among Japanese employees: a prospective cohort study. Int J Behav Med 21:66–76. https://doi.org/10.1007/s12529-012-9286-6

Lee N, Cadogan JW (2013) Problems with formative and higher-order reflective variables. J Bus Res 66(2):242–247. https://doi.org/10.1016/j.jbusres.2012.08.004

Lichtenstein MB, Malkenes M, Sibbersen C et al (2019) Work addiction is associated with increased stress and reduced quality of life: validation of the Bergen work addiction scale in Danish. Scand J Psychol 60(2):145–151. https://doi.org/10.1111/sjop.12506

Lohse T, Rohrmann S, Richard A, Bopp M, Faeh D, Swiss National Cohort Study Group (2017) Type A personality and mortality: Competitiveness but not speed is associated with increased risk. Atherosclerosis 262:19–24. https://doi.org/10.1016/j.atherosclerosis.2017.04.016

Lundahl BW, Kunz C, Brownell C et al (2010) A meta-analysis of motivational interviewing: twenty-five years of empirical studies. Res Soc Work Pract 20: 137–160. https://doi.org/10.1177/1049731509347850

Mancebo MC, Eisen JL, Grant JE et al (2005) Obsessive compulsive personality disorder and obsessive compulsive disorder: clinical characteristics, diagnostic difficulties, and treatment. Ann Clin Psychiatry 17(4):197–204

Marmet S, Studer J, Wicki M et al (2019b) Unique versus shared associations between self-reported behavioral addictions and substance use disorders and mental health problems: a commonality analysis in a large sample of young Swiss men. J Behav Addict 8(4):664–677. https://doi.org/10.1556/2006.8.2019.70

Marmet S, Studer J, Lemoine M, Grazioli VS, Bertholet N, Gmel G (2019a) Reconsidering the associations between self-reported alcohol use disorder and mental health problems in the light of co-occurring addictions in young Swiss men. PloS ONE 14(9). https://doi.org/10.1371/journal.pone.0222806

McEwen BS, Kalia M (2010) The role of corticosteroids and stress in chronic pain conditions. Metabolism 59:S9–S15. https://doi.org/10.1016/j.metabol.2010.07.012

McMillan LH, O'Driscoll MP, Marsh NV et al (2001) Understanding workaholism: data synthesis, theoretical critique, and future design strategies. Int J Stress Manag 8(2):69–91. https://doi.org/10.1023/A:1009573129142

Miller TR, Hendrie D (2009) Substance abuse prevention dollars and cents: a cost-benefit analysis. US Department of Health and Human Services, Substance Abuse and Mental Health Services Administration, Center for Substance Abuse Prevention, Rockville, MD

Miller WR, Bogenschutz MP (2007) Spirituality and addiction. South Med J 100(4):433–437
Mitchell S (1988) Tao Te Ching (Lao Tzu). Harper Perennial, New York
Myers B (2019) Psychotherapy for substance use disorders. In: Stein DJ, Bass JK, Hofmann SG (eds) Global mental health and psychotherapy: Adapting psychotherapy for low- and middle-income countries. Elsevier Academic Press, pp 241–256. https://doi.org/10.1016/B978-0-12-814932-4.00011-2
Naugle KM, Fillingim RB, Riley JL III (2012) A meta-analytic review of the hypoalgesic effects of exercise. J Pain 13(12):1139–1150. https://doi.org/10.1016/j.jpain.2012.09.006
Oates WE (1968) On being a "workaholic." Pastoral Psychol 19(8):16–20. https://doi.org/10.1007/BF01785472
Oates WE (1971) Confessions of a workaholic. World Books, New York
Olivetti PR, Balsam PD, Simpson EH et al (2020) Emerging roles of striatal dopamine D2 receptors in motivated behaviour: implications for psychiatric disorders. Basic Clin Pharmacol 126:47–55. https://doi.org/10.1111/bcpt.13271
Orosz G, Dombi E, Andreassen CS et al (2016) Analyzing models of work addiction: single factor and bi-factor models of the Bergen work addiction scale. Int J Ment Health Ad 14(5):662–671. https://doi.org/10.1007/s11469-015-9613-7
Porter G (1996) Organizational impact of workaholism: suggestions for researching the negative outcomes of excessive work. J Occup Health Psychol 1(1):70–84
Prabhupada ABS (1997) Bhagavad-Gita as it is. Bhaktivedanta Book Trust, Los Angeles
Raiskila T, Sequeiros SB, Kiuttu J et al (2013) Obsessive-compulsive personality disorder is common among occupational health care clients with depression. J Occup Environ Med 55(2):168–171. https://doi.org/10.1097/JOM.0b013e3182717e6d
Rantanen J, Feldt T, Hakanen JJ et al (2015) Cross-national and longitudinal investigation of a short measure of workaholism. Ind Health 53:113–123. https://doi.org/10.2486/indhealth.2014-0129
Robinson BE (1998) The workaholic family: a clinical perspective. Am J Fam Ther 26:65–75. https://doi.org/10.1080/01926189808251087
Robinson BE (1999) The work addiction risk test: development of a tentative measure of workaholism. Percept Mot Skills 88(1):199–210. https://doi.org/10.2466/pms.1999.88.1.199
Robinson BE (2014) Chained to the desk: a guidebook for workaholics, their partners and children, and the clinicians who treat them. NYU Press, New York
Rose G (2001) Sick individuals and sick populations. Int J Epidemiol 30:427–432. https://doi.org/10.1093/ije/30.3.427
Rossow I, Romelsjö A (2006) The extent of the 'prevention paradox' in alcohol problems as a function of population drinking patterns. Addiction 101(1):84–90. https://doi.org/10.1111/j.1360-0443.2005.01294.x
Salanova M, Lopez-Gonzalez AA, Llorens S et al (2016) Your work may be killing you! Workaholism, sleep problems and cardiovascular risk. Work Stress 30:228–242. https://doi.org/10.1080/02678373.2016.1203373
Schaef AW, Fassel D (1988) The addictive organization. Harper & Row Publishers, New York
Schaufeli WB, Shimazu A, Taris TW (2009) Being driven to work excessively hard: the evaluation of a two-factor measure of workaholism in the Netherlands and Japan. Cross-Cult Res 43(4):320–348. https://doi.org/10.1177/1069397109337239
Schmidt KT, Weinshenker D (2014) Adrenaline rush: the role of adrenergic receptors in stimulant-induced behaviors. Mol Pharmacol 85(4):640–650. https://doi.org/10.1124/mol.113.090118
Shaffer HJ, LaPlante DA, LaBrie RA et al (2004) Toward a syndrome model of addiction: multiple expressions, common etiology. Harv Rev Psychiatry 12(6):367–374. https://doi.org/10.1080/10673220490905705
Sharma P, Sharma J (2013) A confirmatory factor analysis of Dutch work addiction scale (DUWAS). Glob Bus Rev 14(2):211–223. https://doi.org/10.1177/0972150913477449
Shkoler O, Rabenu E, Vasiliu C et al (2017) Organizing the confusion surrounding workaholism: new structure, measure, and validation. Front Psychol 8:1803. https://doi.org/10.3389/fpsyg.2017.01803

Shonin E, Van Gordon W, Griffiths MD (2014) The treatment of workaholism with meditation awareness training: a case study. Explore (NY) 3(10):193–195

Spence JT, Robbins AS (1992) Workaholism: definition, measurement, and preliminary results. J Pers Assess 58:160–178. https://doi.org/10.1207/s15327752jpa5801_15

Sussman S (2012) Workaholism: a review. J Addict Res Ther Suppl 6(1):4120. https://doi.org/10.4172/2155-6105.S6-001

Sussman S, Lisha N, Griffiths M (2011) Prevalence of the addictions: a problem of the majority or the minority? Eval Health Prof 34:3–56. https://doi.org/10.1177/0163278710380124

Sussman S, Pokhrel P, Sun P et al (2015) Prevalence and co-occurrence of addictive behaviors among former alternative high school youth: a longitudinal follow-up study. J Behav Addict 4(3):189–194. https://doi.org/10.1556/2006.4.2015.027

Thege BK, Hodgins DC, Wild TC (2016) Co-occurring substance-related and behavioral addiction problems: a person-centered, lay epidemiology approach. J Behav Addict 5(4):614–622. https://doi.org/10.1556/2006.5.2016.079

Tsai J, Huh J, Idrisov B et al (2016) Prevalence and co-occurrence of addictive behaviors among Russian and Spanish youth: a replication study. J Drug Educ 46(1–2):32–46. https://doi.org/10.1177/0047237917704635

UK Government (2019). https://www.gov.uk/government/news/one-of-the-largest-mental-health-trials-launches-in-schools. Accessed 6 Sept 2020

Umeoka EH, van Leeuwen JM, Vinkers CH, Joëls M (2020) The role of stress in bipolar disorder. Curr Top Behav Neurosci. https://doi.org/10.1007/7854_2020_151

Van Gordon W, Shonin E, Dunn TJ et al (2017) Meditation awareness training for the treatment of workaholism: a controlled trial. J Behav Addict 6(2):212–220. https://doi.org/10.1556/2006.6.2017.021

Van Heck GL, Vingerhoets AJ (2007) Leisure sickness: A biopsychosocial perspective. Psihologijske Teme 16(2):178–200

Vallerand RJ (2015) The psychology of passion: a dualistic model. Oxford University Press, New York. https://doi.org/10.1093/acprof:oso/9780199777600.001.0001

Wagenmakers EJ, Wetzels R, Borsboom D et al (2012) An agenda for purely confirmatory research. Perspect Psychol Sci 7(6):632–638. https://doi.org/10.1177/1745691612463078

Warner DO, Berge K, Sun H et al (2013) Substance use disorder among anesthesiology residents, 1975–2009. JAMA 310(21):2289–2296. https://doi.org/10.1001/jama.2013.281954

Workaholics Anonymous (2020). https://www.workaholics-anonymous.org/. Accessed 6 Sept 2020

World Health Organisation (2019) International statistical classification of diseases and related health problems, 11th Revision, Beta Draft. https://icd.who.int./browse11/l-m/en. Accessed 2 June 2020

Wróbel WK (2020) Study addiction among high school students: Measurement and relationship with psychopathology, personality, quality of life, and school variables. Unpublished master's thesis, University of Gdańsk

Yaksh TL (1985) Pharmacology of spinal adrenergic systems which modulate spinal nociceptive processing. Pharmacol Biochem Behav 22:845–858. https://doi.org/10.1016/0091-3057(85)90537-4

Yang X, Qiu D, Lau MC et al (2020) The mediation role of work-life balance stress and chronic fatigue in the relationship between workaholism and depression among Chinese male workers in Hong Kong. J Behav Addict 9(2):483–490. https://doi.org/10.1556/2006.2020.00026

Zajac K, Ginley MK, Chang R et al (2017) Treatments for internet gaming disorder and Internet addiction: a systematic review. Psychol Addict Behav 31(8): 979–994. https://doi.org/10.1037/adb0000315

Chapter 10
Shopping Addiction

Aniko Maraz and Sebastiano Costa

Abstract Shopping is a common activity that is part of everyday life. Most people experience it as a rewarding activity, but there are some individuals, who lose control over their shopping habits, and develop shopping addiction (also termed as buying-shopping disorder, compulsive buying or buying disorder) whereby the shopping experience is characterized by uncontrollable behaviors, attitudes, and thoughts that interfere with daily activities and generate financial difficulties, problems at school or at work, disruption in relationships, among other issues. Although shopping addiction has a long history of research, there are still many aspects that remain unexplored or poorly understood within the literature. There are few studies, that explored the psycho-biological processes underpinning the behavior, and existing evidence indicates that shopping addiction shares several characteristics with other behavioral addictions. Furthermore, shopping addiction appears to be associated with certain and specific indicators of psychopathology and comorbidity. Shopping addiction is not currently recognized as a diagnosable disorder which makes the development of consensual definition and assessment tools challenging. This is still the need for further epidemiological studies, especially those assessing the effectiveness of existing treatment and intervention strategies.

Keywords Shopping addiction · Compulsive buying · Buying disorder

"Although the chapter is titled 'Shopping Addiction', other terminologies are now being adopted and considered more appropriate to describe this phenomenon. For instance, 'buying-shopping disorder' or "compulsive buying-shopping disorder" has been deemed as a more suitable term in light of the recent evidence (Kyrios et al. 2018; Muller et al. 2019; Muller et al. 2021). Notwithstanding this, due to the scope of the book, the term 'Shopping Addiction' will be adopted throughout the chapter for the sake of simplicity and to align with the scope of the book."

A. Maraz (✉)
Institute of Psychology, Humboldt-Universität Zu Berlin, Berlin, Germany
e-mail: aniko.maraz@gmail.com

S. Costa
Department of Psychology, Università Degli Studi Della Campania Luigi Vanvitelli, Caserta, Italy

H. M. Pontes (ed.), *Behavioral Addictions*, Studies in Neuroscience, Psychology and Behavioral Economics, https://doi.org/10.1007/978-3-031-04772-5_10

10.1 Introduction

Shopping is part of everyday life. Although the activity of acquiring is common and for the majority of people harmless, for some individuals consumption becomes an excessive activity, and for an even smaller number of these it may even fulfill the criteria of abuse or addiction (Maraz et al. 2016). The latter phenomenon is defined as shopping addiction (also termed as compulsive buying or buying disorder). Shopping addiction consists of a series of compulsive and uncontrollable behaviors, attitudes, and thoughts aimed at shopping to the point of interfering with daily activities generating financial difficulties, problems at school or at work, disruption in relationships, among other issues. Numerous scholars proposed that shopping addiction is strongly associated with indicators of psychopathology and both internalizing and externalizing problems (O'Guinn and Faber 1989; McElroy et al. 1994; Christenson et al. 1994).

The first description of this phenomena was reported by Kraepelin in (1915) who referred to it as 'oniomania', from the Greek *onios* that means for sale and *mania* that means insanity. After this first denomination, several different terms were provided, such as compulsive buying, buying mania, compulsive shopping, shopaholism, excessive shopping, compulsive spending, compulsive consumption, impulsive buying, impulsive purchasing, buying-shopping disorder, and shopping addiction. Common in these descriptions is the excessive, impulsive, and compulsive tendency to buy. Besides the various labels of shopping addiction, there are some phenomenological characteristics as well, that are considered typical of this phenomenon.

Generally, people with a shopping addiction devote increasing time and effort to buy products that are unnecessary, and often remain unused. However, they rarely return or give away products they bought even if they were never used or they are unusable (Schlosser et al. 1994; Black 2012). Rather than focusing on the product, people with a buying-shopping disorder are primarily focused on the process of buying (i.e., premeditation, selection, and purchasing) which they perceive as urgent, irresistible, and indispensable. Financial problems are very frequent among persons with shopping addiction because they tend to buy more than their budget allows, and to buy the most expensive and exclusive products that in addition to creating tensions and problems in everyday life can cause large debts, legal problems, and relationships crises.

People with a shopping addiction act based on their negative emotions unrelated to shopping, such as negative mood, anxiety, and distress that are appeased only after the purchase has been concluded (Miltenberger et al. 2003; Black 2007). Although it is usual for people with a buying-shopping disorder to feel regret and guilt after shopping (i.e., post-purchase guilt), they are not able to reduce or stop their shopping activities and these failures in the regulation of their behaviors lead to and increase the negative emotions (e.g., regret, guilt, later negative mood states). This process could constitute an addictive cycle, whereby shopping takes place in order to alleviate distress (e.g., psychological tension and frustration). After making a purchase, the sense of guilt and regret increases the need for more shopping to diminish negative emotions (Desarbo and Edwards 1996). Buying then could become an addictive process allowing the individual to escape from negative feelings that are characterized by an underestimation of long-term consequences and lack of self-regulation and impulse control (Desarbo and Edwards 1996).

With the advent of the internet, a form of online shopping addiction has become increasingly frequent and with specific characteristics compared to the forms of shopping in physical stores. Generally, those who show signs of online shopping addiction using new technologies such as smartphone, smartwatch, laptop or tablet share similar characteristics with people that buy on TV programs by phone, and/or mail catalogue. The similarity steams from the fact that with the lack of social interaction it has become possible to buy quickly, via limited sale and offers. Furthermore, online shopping enables one to keep the purchase private and hidden from others, further facilitating and promoting shopping addiction (Lejoyeux and Weinstein 2010). On the other hand, people with non-internet shopping addiction often enjoy the process of selecting and discussing with a sales assistant as a comparison between experts on the product (Müller et al. 2015).

However, the differences between offline and online shopping addiction are still poorly explored. For example, materialistic individuals who seek to enhance their emotions and identity reported the strongest tendencies toward compulsive buying on the internet (Dittmar et al. 2007). Those with low self-esteem are especially prone to excessive online shopping (Zheng et al. 2020). However, online shopping has features that are distinctly different from offline, brick-and-mortar shopping. Perhaps one of the most important differences is the lack of social contact, which diminishes shame usually associated with pathological buying mechanisms (Yi 2012). Online shopping enables the added value of positive anonymity, which may increase compulsive buying as individuals may feel less inhibited about their shopping when they feel that acting on impulse is appropriate (Rook and Fisher 1995). Finally, as opposed to non-internet shopping, which requires time and the opportunity to leave home, online shopping can be done on the spur of the moment at any context, including home, thus completely changing the nature of pathological buying, especially regarding impulsive purchases.

Although the first description of shopping addiction is over a hundred years old, an univocal and consensual definition still does not exist because sufficient research is lacking. The fifth revision of the Diagnostic and Statistical Manual of Mental Disorders (DSM-5) included shopping addiction as an example of Addictive Disorders: " *[...] groups of repetitive behaviors, which some term behavioral addictions, with such subcategories as "sex addiction," "exercise addiction," or "shopping addiction," are* ***not included because at this time there is insufficient peer-reviewed evidence to estab-*** *lish the diagnostic criteria and course descriptions needed to identify these behaviors as mental disorders.*" (American Psychiatric Association and American Psychiatric Association 2013, p. 481). Furthermore, several authors (Brand et al. 2020; Muller et al. 2019; Muller et al. 2021) have suggested the inclusion of the buying-shopping disorder as a candidate for the category "***Other specified disorders due to addictive*** *behaviours*" of the ICD-11.

Further research is therefore encouraged. There are several studies published in the field that have found a series of common aspects that characterize the individuals who buy compulsively. The first widely agreed-upon definition was published by McElroy et al. (1994) based on the study of O'Guinn and Faber (1989). The proposed preliminary, operational criteria was "*maladaptive preoccupation with buying or shopping, or maladaptive buying or shopping impulses or behavior, as indicated by at least one of the following: (1) frequent preoccupation with buying or impulses to buy that are experienced as irresistible, intrusive, and/or senseless. (2) frequent buying of more than can be afforded, frequent buying of items that are not needed, or shopping for longer periods of time than intended.*

The buying preoccupations, impulses, or behaviors cause marked distress, are time-consuming, significantly interfere with social or occupational functioning, or result in financial problems (e.g. indebtedness or bankruptcy). The excessive buying or shopping behavior does not occur exclusively during periods of hypomania or mania." (McElroy et al. (1994), p. 247).

Diagnostic criteria excludes buying episodes during manic phases (McElroy et al. 1995), which may serve as a compensatory mechanism to control mood states. Recently, several scholars (Hartston 2012; Davenport et al. 2012; Andreassen et al. 2015) started to conceptualize shopping addiction within the behavioral addiction spectrum, using the components model of addiction proposed by Griffiths (2005, 2019) and the six core dimensions: salience, tolerance, mood modification, relapse, withdrawal, and conflict. Other conceptualizations of behavioral addiction have used the reward-deficiency model (Comings et al. 1997) that posits that any behavioral addiction may develop because of an under-functioning dopaminergic system and that rewarding behaviors such as shopping addiction compensates the hypo-functioning of the reward pathway (Kotyuk et al. 2020). Finally, using a Delphi expert consensus study (Muller et al. 2021, p. 208), the following criteria were also proposed: "*(a) intrusive and/or irresistible urges and/or impulses and/or cravings and/or preoccupations for buying/shopping; (b) diminished control over buying/shopping; (c) excessive purchasing of items without utilizing them for their intended purposes; (d) use of buying-shopping to regulate internal states; (e) negative consequences and impairment in important areas of functioning due to buying/shopping; (f) emotional and cognitive symptoms upon cessation of excessive buying/shopping; and (g) maintenance or escalation of dysfunctional buying/shopping behaviors despite negative consequences.*"

10.2 Biological Bases

Shopping addiction shares common characteristics and causes with other behavioral addictions and excessive behaviors. Although there are few studies that examined the biological bases of shopping addiction, there are several results that suggest a specific underlying neurobiological basis (Faber 1992, Kyrios et al. 2018). The most specific study on the neurobiological processes of shopping addiction was conducted by Raab et al. (2011) using functional magnetic resonance imaging. They compared people with compulsive and non-compulsive buying while performing purchasing decisions. The results of this study showed that people with compulsive buying reported a lower activation of the insula when they viewed products with prices compared to people with non-compulsive buying. Considering that the insula generally activates as a result of pain and loss sensations, the inhibition of the insula in participants with compulsive shopping when exposed to the product price indicates that they are mostly at risk of underestimating the impact of money loss. On the other hand, people with compulsive buying reported a higher activation in the nucleus accumbens during the presentation of products, which is likely to play a role in the reward circuit and its increased activation tends to promote the desire for the object and overestimate the evaluation of the benefits of purchasing the product.

Furthermore, Raab et al. (2011) described that the mechanism of the nucleus accumbens was also typical of other behavioral and substance addictions. The interplay of high activation of the nucleus accumbens and the inhibition of insula could explain the loss of control generally reported in excessive buying episodes (Raab et al. 2011). This study confirms previous studies on the neurobiological factors of buying behaviors in healthy shoppers (Knutson et al. 2007; Solnais et al. 2013) suggesting that preference of a product stimulates the activation of the nucleus accumbens, while a highly priced product promotes the activation of the insula (Knutson et al. 2007). Moreover, the study of the physiological process showed that people with a shopping addiction tended to have higher skin conductance responses, higher arousal, greater urge to buy, and an increase in craving compared with healthy controls during the presentation of shopping stimuli in a cue-reactivity paradigm (Trotzke et al. 2014). Lawrence et al. (2014) also examined the neurophysiology of shopping addiction showing electroencephalogram (EEG) coherence in the left shifted lateralization for preferred shopping items during visual memory tasks. Additionally, shopping addiction severity was related to greater levels of sensitivity to reward during presentation of preferred shopping items.

Moreover, buying or going shopping as all other experiences that provide enjoyment are generally associated with increased dopamine release (Wise and Rompré 1989; Olsen 2011), that is a neurotransmitter related to pleasure activities and to the reward mechanism that has a salient role in several behavioral and substance addictions (Comings et al. 1997; Moore et al. 2014). Generally, a substance or a behavior that stimulates the dopamine release system if consumed or performed excessively hyperstimulates and alters the dopamine reward system by reducing the number of dopamine receptors to maintain physiological balance. As a consequence, in order to obtain the same level of pleasure, the amount of dopamine needs to further increase for adequate stimulation of the reward system. Hartston (2012) suggests that each behavior or substance hyperstimulates the dopamine system to a different degree, and that although substances have a stronger effect on the dopamine release system than pleasant daily activities (i.e., shopping, sex, exercise), addictive behaviors are more frequently engaged and easily available, allowing to reach a similar consequence as drugs through at a higher frequency of exposure. In this way, for the development of a shopping addiction the frequency of the behaviors are key, because buying is a common activity that may be carried out without limitation and without the negative stigma that generally accompanies the one-time use of substances or behaviors such as gambling (Hartston 2012).

Finally, Devor et al. (1999) conducted a study that compared two deoxyribonucleic acid (DNA) sequence polymorphisms in the serotonin transporter gene (5-HTT) in twenty-one people with a shopping addiction against a group of healthy controls. Results showed that no significant differences emerged between the two groups in the genetic variation in the 5-HTT gene. Although this study found no significant risk factor in this gene for shopping addiction, some preliminary findings (Black et al. 1998) suggested that compulsive buyers have a higher prevalence of psychiatric disorders in the family history than the comparison groups, and that about 9.5% of first-degree relatives of people with a shopping addiction suffer from shopping addiction as well, suggesting an intergenerational transmission of the risk. However, these results need to be interpreted with caution because genetic, and neurobiological studies on shopping addiction are still limited and preliminary, not allowing a full understanding of its biological bases.

10.3 Specific Comorbidities

Shopping addiction is associated with several comorbid conditions such as depression, obsessive compulsive disorder (OCD), eating disorders, anxiety disorders, and others (Weinstein et al. 2016). Persons with shopping addiction frequently meet criteria for Axis I disorders, particularly mood disorders (21–100%), anxiety disorders (41–80%), substance use disorders (21–46%), and eating disorders (8–35%) (Black 2007). Black et al. (1998) found that overall, among 33 people with a shopping addiction, 79% had any Axis I disorder compared to 55% in the control group.

In another study, based on a representative sample, people with a buying-shopping disorder reported experiencing more general distress, impulsivity, sensation-seeking, lower self-esteem and wellbeing than their non-compulsive counterparts. More specifically, people with a shopping addiction reported higher values on each subscale of the Brief Symptom Inventory (i.e., depression, OCD, anxiety, paranoia, interpersonal sensitivity, hostility, psychoticism, somatization, and phobia) (Weinstein et al. 2016). Given the pervasive nature of shopping addiction (i.e. conflicts within family over budget or shame due to loss of control), it is reasonable for people with a shopping addiction to meet more than one diagnosis. However, the high rates of comorbidity also raise the question whether shopping addiction can be regarded as a stand-alone disorder or rather a symptom of other mental health issues (Black 2007).

The first and most frequent psychiatric comorbidity to screen for in shopping addiction is depression. For example, McElroy et al. (1994) found that 19 patients out of 20 met the lifetime diagnosis of a major mood disorder, mostly bipolar disorder, which usually preceded the onset of the buying symptoms. Shopping addiction may therefore be triggered or directly induced by depression. For most patients, buying typically increases during a mild or moderately severe depressive episode and decreases during hypomanic, manic, and severe depressive phases. Patients reported that when depressed, only shopping made them feel good. Among others, Black et al. (2012) also reported that people with a shopping addiction had lifetime mood, anxiety, impulse control disorders, and elevated levels of self-reported depression, attention deficit hyperactivity disorder symptoms, trait impulsivity, and novelty seeking compared with healthy control participants.

McElroy et al. (1995) first described shopping addiction as a type of compulsion. Lejoyeux et al. (2005), reported that the prevalence of shopping addiction was as high as 23% among patients with OCD and 6% in those without OCD. However, the symptoms of shopping addiction are rather heterogeneous. Yi (2013) found for example three distinct clusters among shoppers: buyers with high buying compulsivity and impulsivity ("compulsive–impulsive buyers"), buyers with low buying compulsivity and high impulsivity ("impulsive excessive buyers"), and the cluster of ordinary buyers. It is also possible that the diversity of symptoms reflects the heterogeneity of severity, and the more severe the disorder, the more compulsive symptoms prevail over impulsive ones (see Brand et al. 2016).

Eating disorder is also common among those having the symptoms of shopping addiction. Black et al. (1998) reported that about 20% of people with a shopping addiction also present with eating disorders. More specifically, about 17.6% of binge eaters present with shopping addiction disorders. Fernandez-Aranda et al. (2008) assessed 59 participants with restrictive anorexia, 33 with purgative anorexia, 29 with binge eating disorders, and 252 with both purging and non-purging types. The authors found that the lifetime prevalence of all impulse control disorders in the entire sample was 16.6%, of which compulsive buying disorder was the most common diagnosis (11.8%). People with a shopping addiction were also found to be three times more likely to have comorbid borderline personality disorder than people without a shopping addiction.

Several studies have analyzed the role of anxiety in shopping addiction (Miltenberger et al. 2003; Billieux et al. 2008; Davenport et al. 2012; Williams and Grisham 2012; Otero-López and Villardefrancos 2013). Weinstein et al. (2015) assessed shopping addiction, state and trait anxiety, and general obsessive–compulsive measures among 120 habitual internet shoppers (i.e., shopping more than twice a week) including 70 men and 50 women. They found a positive association between trait anxiety, shopping addiction, and OCD. Finally, several studies found that shopping addiction co-occurs with other behavioral and substance addictions (Mestre-Bach et al. 2017). Specifically, Black et al., (1998) found that those with shopping addiction have a higher risk of abusing substances than healthy controls. This result was corroborated in a study conducted by Di Nicola et al. (2015) reporting that alcohol use disorder patients exhibited significantly higher scores of shopping addiction when compared to controls. Lejoyeux et al. (2005) reported in a sample of nicotine-dependent women that shopping addiction was presented in about 46% of them, and in a sample of shopping mall visitors, Maraz et al. (2015b) showed that people with a shopping addiction awere more likely to use licit (e.g., medication and alcohol) and illicit substances (e.g., cannabis, amphetamine, and cocaine) than people without a shopping addiction.

Regarding the comorbidity with other behavioral addictions, Granero et al. (2016) reported in a sample of people with a shopping addiction that 19% also presented with comorbid gambling disorder. This is in line with a previous preliminary finding by Grant and Kim (2001) that found higher rates of shopping addiction in patients with problems of gambling. Finally, Yeh et al. (2008) reported a clinical case of a man who suffered concurrent symptoms of sexual addiction and shopping addiction. Several other studies (Mueller et al. 2011; Weinstein et al. 2015; Trotzke et al. 2015; Jiang and Shi 2016) also found problematic, excessive internet and mobile phone use in persons with shopping addiction.

10.4 Prevalence Rates

Clear estimations of the prevalence for behavioral addictions are generally difficult to obtain, because the quality of epidemiological studies in this field is often considered below the standards applied to other problematic behaviors (Rumpf et al. 2019). However, based on a meta-analysis conducted in 2015, the pooled prevalence of shopping addiction was found to be about 4.9% (99% CI: 3.4–6.9%) (Maraz et al. 2016). This estimation was based on data from 10,000 participants living in Hungary (1.85%), Spain (7.1%), United States (5.8% in 2006 and 8.1% in 1992), and Germany (West Germany in 1991: 5.1%, East Germany in 1991: 1%; West Germany in 2001: 8%, East Germany in 2001: 6.5%). Mueller and colleagues (Mueller et al. 2008) reported a prevalence rate of 6.9% in Germany. There are currently no representative data available from Asia even though although Asian students or shoppers generally appear to have higher scores on shopping addiction than European students or shoppers (Maraz et al. 2016). On the other hand, based on 19 prevalence estimates, the prevalence of shopping addiction among university students has been reported to be about 8.3% (99% CI: 5.9–11.5%), and even higher, 16.2% (99% CI: 8.8–27.8%) in shopping-specific samples (i.e. among visitors of a shopping mall).

Out of thirteen samples (five of which were representative adult samples) where the effect of age was studied, seven reported that shopping addiction tendency decreased with age, five reported no relationship, and one reported that older students were more likely to engage in compulsive buying than younger students. However, whether these differences are related to age, the change in the level of income, or whether they may reflect generational differences remains unclear. The latter notion is supported by data from Germany, which reports a six-fold increase in shopping addiction tendencies in East Germany and 1.5-fold increase in West Germany over 10 years (in 1991 and 2001) in representative samples (Neuner et al. 2005).

In terms of gender-based prevalence rates, women appear to be more affected than men. Out of 17 studies (of which 4 in representative samples) reporting the relationship between gender and shopping addiction, 12 found that women had higher shopping addiction scores than men, 4 reported no gender differences, and one reported that shopping addiction was more prevalent among men. However, when tested in the meta-analysis, this effect was either non-significant or very weak, suggesting that despite the tendency, at the moment there is not enough evidence supporting the notion that women are significantly more prone to compulsive buying than men (Maraz et al. 2016).

10.5 Assessment Approaches

One of the existing challenges in estimating reliable prevalence rates of shopping addiction relates to the lack of robust diagnostic symptoms and a clear definition as to what constitutes shopping addiction. Typically, studies use different measures to assess shopping addiction, each having a different conceptual background. Most definitions include cognitive–affective indicators as well as maladaptive behavioral consequences when defining the disorder (e.g. debts). The screening instruments used across studies tend to differ in terms of the indicators of financial consequences (e.g. credit card use, debts, loan, etc.) and are subject to differences according to countries, subcultures, and/or age groups.

Given the challenges of establishing a reliable cut-off value for the existing shopping addiction assessment tools, about twice as many studies (n = 73) used rating scales as indicators of symptom severity (ignoring the cut-off values) as opposed to those that reported categorical data (n = 40) in 2016 (Maraz et al. 2016). A categorical approach has the advantage of identifying potentially self-harming individuals as well as the ability to provide an estimation of the extent of the problem in a given population. Furthermore, knowledge of the proportion of shopping addiction in the study sample would enable comparison across studies.

Moreover, many assessment tools contain items which mix lifetime, meaning if the individual has ever experienced problems with buying behavior (such as the Compulsive Buying Scale [CBS], (Faber and O'Guinn 1992), Minnesota Impulse Disorders Interview (Odlaug and Grant 2010), and the Shorter PROMIS Questionnaire [SPQ] (Pallanti et al. 2006)) with current prevalence, that is, problems with buying behavior at the time of assessment (such as the German Compulsive Buying Scale [G-CBS], (Mueller et al. 2010), Richmond Compulsive Buying Scale [RCBS], (Ridgway et al. 2008), Edwards Compulsive Buying Scale [ECBS] (Edwards 1993), [ECBS-R] (Maraz et al. 2015a), and the Passion Scale [PS] (Aiken et al. 2018)). This results in instruments measuring lifetime prevalence to report 1.6 times higher prevalence rates compared to current measures (Maraz et al. 2016). This difference is in line with other studies reporting lifetime (6.9%) and 12-month prevalence (3.4%) of any psychiatric disorder (Takeuchi et al. 1998).

Another issue around assessing shopping addiction is the cultural aspect of overshopping. For example, some countries have an extended tradition of bank check usage while others do not have such tradition. This difference is ignored when administering an instrument where using a bank check is an indicator of shopping addiction (e.g., CBS). Typically, studies apply a fixed cut-off value (i.e., 1.34 for the CBS) to categorize at-risk compulsive buyers, which sets the same standard of shopping addiction across countries without taking into consideration the instrument's local validity and reliability. Some scales, such as the CBS recommend using a cut-off score 2 standard deviations above the mean to adjust for cultural differences. However, this approach disregards distortions coming from sampling. As an attempt for a solution, scales, such as the RCBS suggest using items referring to psychosocial consequences rather than financial ones, which are relatively robust across cultures. Given the lack of official diagnostic criteria, clinically valid measures do not currently exist. However, there are at least three measures which were validated with a clinical population: the CBS, the GCBS and the RCBS.

In terms of online shopping addiction, at the moment only two instruments measure the behavior. The Compulsive Online Shopping Scale (COSS) (Manchiraju et al. 2017) was developed from the Bergen Shopping Addiction Scale by a simple change of wording, an approach which probably disregards the complexity of online shopping and its differences from the offline context beyond wording.

10.6 Treatment Approaches

The limited understanding of the phenomenon of shopping addiction clearly has repercussions on possible interventions and treatments. Although the gold standard for intervention of people with shopping addiction is still non-existing, some studies have shown promising results of certain pharmacological and psychological treatments. Naturally, the first drugs tested to treat patients with shopping addiction symptoms have been those successfully used in the treatment of patients with similar phenotypic characteristics (e.g., OCD, other impulse control disorders, and mood disorders) (Soares et al. 2016).

Given the common features of shopping addiction and OCD, some serotonin reuptake inhibitors (SSRI) have been tested, but the results showed minor or non-significant effects in reducing the symptoms of shopping addiction (Soares et al. 2016). Although fluvoxamine in a preliminary study appeared to be promising (McElroy et al. 1991; Black et al. 1997), double-blind placebo-controlled studies (Ninan et al. 2000; Black et al. 2000) showed that there were no significant differences in the symptomatology between fluvoxamine- and placebo-treated participants. Studies on two other SSRIs have achieved similar results, in randomized controlled trials, and placebo group members did not have differing clinical symptoms compared to shoppers who received citalopram (Koran et al. 2003) or escitalopram (Koran et al. 2007).

Furthermore, the efficacy of other antidepressants (e.g. bupropion, thymoleptics, nortriptyline) was tested only with single case reports (McElroy et al. 1991, 1994) that do not allow deriving concluding findings about the usefulness of SSRIs as a treatment for shopping addiction (Weinstein et al. 2016). Case reports and pilot trials showed preliminary evidence for the effectiveness of psychotropic drugs that influence neurotransmitter systems that have a role in the reward pathway. Specifically, Kim (1998) and Grant (2003) have shown that the use of opioid antagonists such as naltrexone reduced the craving effects and the urges to buy in single cases.

Two other single case reports (Guzman et al. 2007; Ye et al. 2014) concluded that the use of topiramate improved impulse control and reduced symptoms of shopping addiction. Furthermore, the manipulation of the glutamate system toward a memantine treatment showed less impulsive buying in nine participants (Grant et al. 2012). Although several treatments have been tested to date, there are currently no systematic and controlled studies that demonstrate the effectiveness of pharmacological interventions, and in conclusion the existing literature suggests that there is no evidence that pharmacological treatments are more effective than a placebo when treating shopping addiction at this point in time.

Compared to pharmacological treatments, psychological treatments are more promising in the treatment of shopping addiction (Lourenço Leite et al. 2014). Specifically, Mitchell and colleagues (Mitchell et al. 2006) conducted a pilot study with twenty-eight participants that received a cognitive-behavioral intervention and compared its efficacy to a waiting list control group. The twelve sessions during a period of ten weeks of cognitive-behavioral therapy (CBT) showed significant reductions in the symptoms of shopping addiction, number of compulsive buying episodes, and time spent buying compared to the control group, with the improvement being maintained at follow-up after six months.

The same intervention protocol (Mitchell et al. 2006) was applied in a randomized controlled trial (Mueller et al. 2008) to compare its efficacy to a waiting list control group for the treatment of shopping addiction. The intervention group was composed by thirty-one patients with shopping addiction and compared to a control group of 29 participants. Results confirmed that the intervention significantly reduced shopping addiction symptoms, which effects being maintained after six months. Another pilot study using CBT was conducted by Müller et al. (2013) comparing twenty-two patients in a CBT treatment group against a guided self-help group of twenty patients, and a group of fourteen patients on the waiting list. Results showed that the CBT treatment significantly reduced the severity of shopping addiction compared to both the guided self-help and the waiting list groups at the end of the intervention and at the six-month follow-up. Finally, Benson et al. (2014) conducted a preliminary randomized controlled trial of group therapy for shopping addiction using an integrated approach defined Stopping Overshopping model that showed that the eleven participants in the intervention group reported a reduction of symptoms, money, and time spent shopping compared to a waiting list control group after the intervention and after six months.

Overall, CBT shows promising results, and at this point in time, CBT is the only approach that has clinical trials to evaluate the significance of the treatment intervention, other studies on psychological treatments are based only on case studies. Several studies tested treatment approaches with single patients using psychoanalysis (Winestine 1985; Schwartz 1992), psychodynamic (Krueger 1988; Marcinko and Karlović 2005; Tavares et al. 2008), CBT approaches (Bernik et al. 1996; Hartston and Koran 2002; Marcinko and Karlović 2005; Kellett and Bolton 2009; Donahue et al. 2011), and family therapy approach (Tavares et al. 2008), and in some cases reported improvement and reduction of shopping addiction symptomatology, but overall the efficacy of these studies are often undermined by publication bias and vulnerable to many limitations that compromise the generalizability and the replication of the findings (Lourenço Leite et al. 2014; Hague et al. 2016).

Overall, findings on interventions for shopping addiction are still in progress and not conclusive. However, preliminary studies seem to suggest that the CBT approach may be more effective than other approaches, including pharmacological treatment in people with clinical-level shopping addiction. Thus, it is recommended to consider psychological treatment prior to starting pharmacotherapy (Black 2007; Lourenço Leite et al. 2014; Hague et al. 2016).

10.7 Final Conclusions

Shopping addiction has a long history of research, but there are still several aspects that remain unexplored or poorly understood within the existing literature. Shopping addiction is a complex condition that may be associated with severe financial, legal, social, emotional, and personal problems. The etiology of shopping addiction is not clear, but the few studies on the evidence of the psychobiological processes highlight the shared characteristics with other behavioral addictions.

Shopping addiction, in fact, is not a currently recognized behavioral addiction at this time, but the prevalence is growing routinely in modern society. Although epidemiological studies are still not conclusive, the risk of developing shopping addiction seems to be around 5%. The absence of conclusive findings stems from the lack of consensual assessment tools and large number of assessment tools that have been developed. The lack of robust assessment tools is partly due to the fact that there is a lack of well-validated diagnostic and clinical criteria for shopping addiction, making the study of this issue particularly difficult. Consequently, it is challenging to verify the effectiveness of the treatment studies. Although some treatments, especially the ones based on the CBT approach appear to be promising, studies with large samples and the methodological rigor necessary for clinical trials are still yet to be published.

In conclusion, it seems essential that future research on shopping addiction begins to acquire greater consensus and in-depth investigations to resolve the numerous issues that are still unresolved. Future investigations could develop the appropriate tools to clinicians and stakeholders to help people in need to diminish the impact of the negative consequences of compulsive, excessive, and unmanageable buying behaviors and to increase the quality of life among affected individuals.

References

Aiken KD, Bee C, Walker N (2018) From passion to obsession: development and validation of a scale to measure compulsive sport consumption. J Bus Res 87:69–79. https://doi.org/10.1016/j.jbusres.2018.02.019

American Psychiatric Association, American Psychiatric Association (eds) (2013) Diagnostic and statistical manual of mental disorders: DSM-5, 5th ed. Washington: American Psychiatric Association

Andreassen CS, Griffiths MD, Pallesen S, et al (2015) The Bergen Shopping Addiction Scale: reliability and validity of a brief screening test. Front Psychol 6. https://doi.org/10.3389/fpsyg.2015.01374

Benson AL, Eisenach D, Abrams L, van Stolk-Cooke K (2014) Stopping overshopping: a preliminary randomized controlled trial of group therapy for compulsive buying disorder. J Groups Addict Recovery 9:97–125. https://doi.org/10.1080/1556035X.2014.868725

Bernik MA, Akerman D, Amaral J, Braun R (1996) Cue exposure in compulsive buying. J Clin Psychiatry 57:90–90

Billieux J, Rochat L, Rebetez MML, Van der Linden M (2008) Are all facets of impulsivity related to self-reported compulsive buying behavior? Personal Individ Differ 44:1432–1442. https://doi.org/10.1016/j.paid.2007.12.011

Black DW (2012) Epidemiology and phenomenology of compulsive buying disorder. Oxford University Press

Black DW (2007) A review of compulsive buying disorder. World Psychiatry 6:14

Black DW, Gabel J, Hansen J, Schlosser S (2000) A Double-blind comparison of fluvoxamine versus placebo in the treatment of compulsive buying disorder. Ann Clin Psychiatry 12:205–211. https://doi.org/10.1023/A:1009030425631

Black DW, Monahan P, Gabel J (1997) Fluvoxamine in the treatment of compulsive buying. J Clin Psychiatry 58:159–163. https://doi.org/10.4088/JCP.v58n0404

Black DW, Repertinger S, Gaffney GR, Gabel J (1998) Family history and psychiatric comorbidity in persons with compulsive buying: preliminary findings. Am J Psychiatry 155:960–963. https://doi.org/10.1176/ajp.155.7.960

Black DW, Shaw M, McCormick B et al (2012) Neuropsychological performance, impulsivity, ADHD symptoms, and novelty seeking in compulsive buying disorder. Psychiatry Res 200:581–587. https://doi.org/10.1016/j.psychres.2012.06.003

Brand M, Young KS, Laier C et al (2016) Integrating psychological and neurobiological considerations regarding the development and maintenance of specific Internet-use disorders: an Interaction of person-affect-cognition-execution (I-PACE) model. Neurosci Biobehav Rev 71:252–266. https://doi.org/10.1016/j.neubiorev.2016.08.033

Brand M, Rumpf HJ, Demetrovics Z, MÜller A, Stark R, King DL, Goudriaan AE, Mann K, Trotzke P, Fineberg NA, Chamberlain SR, Kraus SW, Wegmann E, Billieux J, Potenza MN (2020) Which conditions should be considered as disorders in the International Classification of Diseases (ICD-11) designation of "other specified disorders due to addictive behaviors"? J Behav Addic. https://doi.org/10.1556/2006.2020.00035

Christenson GA, Faber RJ, de Zwaan M et al (1994) Compulsive buying: descriptive characteristics and psychiatric comorbidity. J Clin Psychiatry 55:5–11

Comings DE, Gade R, Wu S et al (1997) Studies of the potential role of the dopamine D1 receptor gene in addictive behaviors. Mol Psychiatry 2:44–56. https://doi.org/10.1038/sj.mp.4000207

Davenport K, Houston JE, Griffiths MD (2012) Excessive eating and compulsive buying behaviours in women: an empirical pilot study examining reward sensitivity, anxiety, impulsivity, self-esteem and social desirability. Int J Ment Health Addict 10:474–489. https://doi.org/10.1007/s11469-011-9332-7

Desarbo W, Edwards E (1996) Typologies of compulsive buying behavior: a constrained clusterwise regression approach. J Consum Psychol 5:231–262. https://doi.org/10.1207/s15327663jcp0503_02

Devor EJ, Magee HJ, Dill-Devor RM et al (1999) Serotonin transporter gene (5-HTT) polymorphisms and compulsive buying. Am J Med Genet 88:123–125. https://doi.org/10.1002/(sici)1096-8628(19990416)88:2%3c123::aid-ajmg5%3e3.0.co;2-s

Di Nicola M, Tedeschi D, De Risio L et al (2015) Co-occurrence of alcohol use disorder and behavioral addictions: relevance of impulsivity and craving. Drug Alcohol Depend 148:118–125. https://doi.org/10.1016/j.drugalcdep.2014.12.028

Dittmar H, Long K, Bond R (2007) When a better self is only a button click away: associations between materialistic values, emotional and identity-related buying motives, and compulsive buying tendency online. J Soc Clin Psychol 26:334–361. https://doi.org/10.1521/jscp.2007.26.3.334

Donahue CB, Odlaug BL, Grant JE (2011) Compulsive buying treated with motivational interviewing and imaginal desensitization. Ann Clin Psychiatry off J Am Acad Clin Psychiatr 23:226–227

Edwards EA (1993) Development of a new scale for measuring compulsive buying behavior. Financ Couns Plan 67–84

Faber RJ (1992) Money changes everything: compulsive buying from a biopsychosocial perspective. am Behav Sci 35:809–819. https://doi.org/10.1177/0002764292035006015

Faber RJ, O'Guinn TC (1992) A clinical screener for compulsive buying. J Consum Res 19:459. https://doi.org/10.1086/209315

Fernández-Aranda F, Pinheiro AP, Thornton LM et al (2008) Impulse control disorders in women with eating disorders. Psychiatry Res 157:147–157. https://doi.org/10.1016/j.psychres.2007.02.011

Granero R, Fernández-Aranda F, Steward T, et al (2016) Compulsive buying behavior: characteristics of comorbidity with gambling disorder. Front Psychol 7. https://doi.org/10.3389/fpsyg.2016.00625

Grant JE (2003) Three cases of compulsive buying treated with naltrexone. Int J Psychiatry Clin Pract 7:223–225. https://doi.org/10.1080/13651500310003219

Grant JE, Kim SW (2001) Demographic and Clinical features of 131 adult pathological gamblers. J Clin Psychiatry 62:957–962. https://doi.org/10.4088/JCP.v62n1207

Grant JE, Odlaug BL, Mooney M et al (2012) Open-label pilot study of memantine in the treatment of compulsive buying. Ann Clin Psychiatry off J Am Acad Clin Psychiatr 24:119–126

Griffiths M (2005) A 'components' model of addiction within a biopsychosocial framework. J Subst Use 10:191–197. https://doi.org/10.1080/14659890500114359

Griffiths MD (2019) The evolution of the 'components model of addiction' and the need for a confirmatory approach in conceptualizing behavioral addictions. Dusunen Adam J Psychiatry Neurol Sci 32:179. https://doi.org/10.14744/DAJPNS.2019.00027

Guzman CS, Filomensky T, Tavares H (2007) Compulsive buying treatment with topiramate, a case report. Rev Bras Psiquiatr 29:383–384. https://doi.org/10.1590/S1516-44462007000400020

Hague B, Kellett S, Sheeran P (2016) Testing the generalizability of impulse control problems in compulsive buying. J Soc Clin Psychol 35:269–288. https://doi.org/10.1521/jscp.2016.35.4.269

Hartston H (2012) The case for compulsive shopping as an addiction. J Psychoactive Drugs 44:64–67. https://doi.org/10.1080/02791072.2012.660110

Hartston HJ, Koran LM (2002) Impulsive behavior in a consumer culture. Int J Psychiatry Clin Pract 6:65–68. https://doi.org/10.1080/136515002753724045

Jiang Z, Shi M (2016) Prevalence and co-occurrence of compulsive buying, problematic Internet and mobile phone use in college students in Yantai, China: relevance of self-traits. BMC Public Health 16:1211. https://doi.org/10.1186/s12889-016-3884-1

Kellett S, Bolton JV (2009) Compulsive buying: a cognitive–behavioural model. Clin Psychol Psychother 16:83–99. https://doi.org/10.1002/cpp.585

Kim SW (1998) Opioid antagonists in the treatment of impulse-control disorders. J Clin Psychiatry 59:165–171. https://doi.org/10.4088/JCP.v59n0403

Knutson B, Rick S, Wimmer GE et al (2007) Neural predictors of purchases. Neuron 53:147–156. https://doi.org/10.1016/j.neuron.2006.11.010

Koran LM, Aboujaoude EN, Solvason B et al (2007) Escitalopram for compulsive buying disorder: a double-blind discontinuation study. J Clin Psychopharmacol 27:225–227. https://doi.org/10.1097/01.jcp.0000264975.79367.f4

Koran LM, Chuong HW, Bullock KD, Smith SC (2003) Citalopram for compulsive shopping disorder: an open-label study followed by double-blind discontinuation. J Clin Psychiatry 64:793–798. https://doi.org/10.4088/JCP.v64n0709

Kotyuk E, Magi A, Eisinger A et al (2020) Co-occurrences of substance use and other potentially addictive behaviors: epidemiological results from the psychological and genetic factors of the addictive behaviors (PGA) study. J Behav Addict 9:272–288. https://doi.org/10.1556/2006.2020.00033

Kraepelin E (1915) Psychiatrie; ein Lehrbuch für Studierende und Ärzte. Barth

Krueger DW (1988) On compulsive shopping and spending: a psychodynamic inquiry. Am J Psychother 42:574–584. https://doi.org/10.1176/appi.psychotherapy.1988.42.4.574

Kyrios M, Trotzke P, Lawrence L, Fassnacht DB, Ali K, Laskowski NM, Müller A (2018) Behavioral neuroscience of buying-shopping disorder: a review. Curr Behav Neurosci Rep 5(4):263–270. https://doi.org/10.1007/s40473-018-0165-6

Lawrence LM, Ciorciari J, Kyrios M (2014) Cognitive processes associated with compulsive buying behaviours and related EEG coherence. Psychiatry Res Neuroimaging 221:97–103. https://doi.org/10.1016/j.pscychresns.2013.10.005

Lejoyeux M, Bailly F, Moula H et al (2005) Study of compulsive buying in patients presenting obsessive-compulsive disorder. Compr Psychiatry 46:105–110. https://doi.org/10.1016/j.comppsych.2004.07.027

Lejoyeux M, Weinstein A (2010) Compulsive buying. Am J Drug Alcohol Abuse 36:248–253. https://doi.org/10.3109/00952990.2010.493590

Lourenço Leite P, Pereira VM, Nardi AE, Silva AC (2014) Psychotherapy for compulsive buying disorder: a systematic review. Psychiatry Res 219:411–419. https://doi.org/10.1016/j.psychres.2014.05.037

Manchiraju S, Sadachar A, Ridgway JL (2017) The Compulsive online shopping scale (COSS): development and validation using panel data. Int J Ment Health Addict 15:209–223. https://doi.org/10.1007/s11469-016-9662-6

Maraz A, Eisinger A, Hende B et al (2015a) Measuring compulsive buying behaviour: psychometric validity of three different scales and prevalence in the general population and in shopping centres. Psychiatry Res 225:326–334. https://doi.org/10.1016/j.psychres.2014.11.080

Maraz A, Griffiths MD, Demetrovics Z (2016) The prevalence of compulsive buying: a meta-analysis: prevalence of compulsive buying. Addiction 111:408–419. https://doi.org/10.1111/add.13223

Maraz A, van den Brink W, Demetrovics Z (2015b) Prevalence and construct validity of compulsive buying disorder in shopping mall visitors. Psychiatry Res 228:918–924. https://doi.org/10.1016/j.psychres.2015.04.012

Marcinko D, Karlović D (2005) Oniomania–successful treatment with fluvoxamine and cognitive-behavioral psychotherapy. Psychiatr Danub 17:97–100

McElroy SL, Keck P, Pope HG, et al (1994) Compulsive buying: a report of 20 cases. J Clin Psychiatry

McElroy SL, Keck PE, Phillips KA (1995) Kleptomania, compulsive buying, and binge-eating disorder. J Clin Psychiatry 56 Suppl 4:14–26; discussion 27

McElroy SL, Satlin A, Pope H et al (1991) Treatment of compulsive shopping with antidepressants: a report of three cases. Ann Clin Psychiatry 3:199–204. https://doi.org/10.3109/10401239109147991

Mestre-Bach G, Steward T, Jiménez-Murcia S, Fernández-Aranda F (2017) Differences and similarities between compulsive buying and other addictive behaviors. Curr Addict Rep 4:228–236. https://doi.org/10.1007/s40429-017-0153-z

Miltenberger RG, Redlin J, Crosby R et al (2003) Direct and retrospective assessment of factors contributing to compulsive buying. J Behav Ther Exp Psychiatry 34:1–9. https://doi.org/10.1016/S0005-7916(03)00002-8

Mitchell JE, Burgard M, Faber R et al (2006) Cognitive behavioral therapy for compulsive buying disorder. Behav Res Ther 44:1859–1865. https://doi.org/10.1016/j.brat.2005.12.009

Moore TJ, Glenmullen J, Mattison DR (2014) Reports of pathological gambling, hypersexuality, and compulsive shopping associated with dopamine receptor agonist drugs. JAMA Intern Med 174:1930–1933. https://doi.org/10.1001/jamainternmed.2014.5262

Mueller A, Mitchell JE, Crosby RD et al (2010) Estimated prevalence of compulsive buying in Germany and its association with sociodemographic characteristics and depressive symptoms. Psychiatry Res 180:137–142. https://doi.org/10.1016/j.psychres.2009.12.001

Mueller A, Mitchell JE, Peterson LA et al (2011) Depression, materialism, and excessive Internet use in relation to compulsive buying. Compr Psychiatry 52:420–424. https://doi.org/10.1016/j.comppsych.2010.09.001

Mueller A, Mueller U, Silberman A et al (2008) A randomized, controlled trial of group cognitive-behavioral therapy for compulsive buying disorder: posttreatment and 6-month follow-up results. J Clin Psychiatry 69:1131–1138. https://doi.org/10.4088/JCP.v69n0713

Müller A, Arikian A, de Zwaan M, Mitchell JE (2013) Cognitive–behavioural group therapy versus guided self-help for compulsive buying disorder: a preliminary study. Clin Psychol Psychother 20:28–35. https://doi.org/10.1002/cpp.773

Müller A, Brand M, Claes L, Demetrovics Z, De Zwaan M, Fernández-Aranda F, Frost RO, Jimenez-Murcia S, Lejoyeux M, Steins-Loeber S, Mitchell JE, Moulding R, Nedeljkovic M, Trotzke P, Weinstein A, Kyrios M (2019) Buying-shopping disorder—is there enough evidence to support its inclusion in ICD-11? CNS Spectr 24(4):374–379. https://doi.org/10.1017/S1092852918001323

Müller A, Laskowski NM, Trotzke P, Ali K, Fassnacht DB, de Zwaan M, Brand M, Häder M, Kyrios M (2020) Proposed diagnostic criteria for compulsive buying-shopping disorder: a Delphi expert consensus study. J Behav Addic 10(2):208–222. https://doi.org/10.1556/2006.2021.00013

Müller A, Mitchell JE, de Zwaan M (2015) Compulsive buying: compulsive buying. Am J Addict 24:132–137. https://doi.org/10.1111/ajad.12111

Neuner M, Raab G, Reisch LA (2005) Compulsive buying in maturing consumer societies: an empirical reinquiry. J Econ Psychol 26(4):509–522. https://doi.org/10.1016/j.joep.2004.08.002

Ninan PT, McElroy SL, Kane CP et al (2000) Placebo-controlled study of fluvoxamine in the treatment of patients with compulsive buying. J Clin Psychopharmacol 20:362–366

Odlaug BL, Grant JE (2010) Impulse-control disorders in a college sample: results from the self-administered Minnesota impulse disorders interview (MIDI). Prim Care Companion J Clin Psychiatry 12. https://doi.org/10.4088/PCC.09m00842whi

O'Guinn TC, Faber RJ (1989) Compulsive buying: a phenomenological exploration. J Consum Res 16:147. https://doi.org/10.1086/209204

Olsen CM (2011) Natural rewards, neuroplasticity, and non-drug addictions. Neuropharmacology 61(7):1109–1122. https://doi.org/10.1016/j.neuropharm.2011.03.010

Otero-López JM, Villardefrancos E (2013) Materialism and addictive buying in women: the mediating role of anxiety and depression. Psychol Rep 113:328–344. https://doi.org/10.2466/18.02.PR0.113x11z9

Pallanti S, Bernardi S, Quercioli L (2006) The shorter PROMIS questionnaire and the internet addiction scale in the assessment of multiple addictions in a high-school population: prevalence and related disability. CNS Spectr 11:966–974. https://doi.org/10.1017/S1092852900015157

Raab G, Elger CE, Neuner M, Weber B (2011) A neurological study of compulsive buying behaviour. J Consum Policy 34:401–413. https://doi.org/10.1007/s10603-011-9168-3

Ridgway NM, Kukar-Kinney M, Monroe KB (2008) An expanded conceptualization and a new measure of compulsive buying. J Consum Res 35:622–639. https://doi.org/10.1086/591108

Rook DW, Fisher RJ (1995) Normative influences on impulsive buying behavior. J Consum Res 22:305. https://doi.org/10.1086/209452

Rumpf HJ, Brandt D, Demetrovics Z, Billieux J, Carragher N, Brand M, Poznyak V et al (2019) Epidemiological challenges in the study of behavioral addictions: a call for high standard methodologies. Curr Addict Rep 6(3):331–337. https://doi.org/10.1556/2006.2020.00035

Schlosser S, Black DW, Repertinger S, Freet D (1994) Compulsive buying. Demography, phenomenology, and comorbidity in 46 subjects. Gen Hosp Psychiatry 16:205–212. https://doi.org/10.1016/0163-8343(94)90103-1

Schwartz HJ (1992) Psychoanalytic psychotherapy for a woman with diagnoses of kleptomania and bulimia. Psychiatr Serv 43:109–110. https://doi.org/10.1176/ps.43.2.109

Soares C, Fernandes N, Morgado P (2016) A review of pharmacologic treatment for compulsive buying disorder. CNS Drugs 30:281–291. https://doi.org/10.1007/s40263-016-0324-9

Solnais C, Andreu-Perez J, Sánchez-Fernández J, Andréu-Abela J (2013) The contribution of neuroscience to consumer research: a conceptual framework and empirical review. J Econ Psychol 36:68–81. https://doi.org/10.1016/j.joep.2013.02.011

Takeuchi DT, Chung RC, Lin KM et al (1998) Lifetime and twelve-month prevalence rates of major depressive episodes and dysthymia among Chinese Americans in Los Angeles. Am J Psychiatry 155:1407–1414. https://doi.org/10.1176/ajp.155.10.1407

Tavares H, Lobo DSS, Fuentes D, Black DW (2008) Compras compulsivas: uma revisão e um relato de caso. Braz J Psychiatry 30:S16–S23. https://doi.org/10.1590/S1516-44462008005000002

Trotzke P, Starcke K, Müller A, Brand M (2015) Pathological buying online as a specific form of internet addiction: a model-based experimental investigation. PLoS ONE 10:e0140296. https://doi.org/10.1371/journal.pone.0140296

Trotzke P, Starcke K, Pedersen A, Brand M (2014) Cue-induced craving in pathological buying: empirical evidence and clinical implications. Psychosom Med 76:694–700. https://doi.org/10.1097/PSY.0000000000000126

Weinstein A, Maraz A, Griffiths MD et al (2016) Chapter 98—compulsive buying—features and characteristics of addiction. In: Preedy VR (ed) Neuropathology of drug addictions and substance misuse. Academic Press, San Diego, pp 993–1007

Weinstein A, Mezig H, Mizrachi S, Lejoyeux M (2015) A study investigating the association between compulsive buying with measures of anxiety and obsessive-compulsive behavior among internet shoppers. Compr Psychiatry 57:46–50. https://doi.org/10.1016/j.comppsych.2014.11.003

Williams AD, Grisham JR (2012) Impulsivity, emotion regulation, and mindful attentional focus in compulsive buying. Cogn Ther Res 36:451–457. https://doi.org/10.1007/s10608-011-9384-9

Winestine MC (1985) Compulsive shopping as a derivative of a childhood deduction. Psychoanal Q 54:70–72. https://doi.org/10.1080/21674086.1985.11927095

Wise RA, Rompré P-P (1989) Brain dopamine and reward. Annu Rev Psychol 40:191–225. https://doi.org/10.1146/annurev.ps.40.020189.001203

Ye L, Kadia S, Lippmann S (2014) Topiramate and compulsive buying disorder. J Clin Psychopharmacol 34:174–175. https://doi.org/10.1097/JCP.0b013e3182aa0139

Yeh Y-W, Shiah I-S, Hu M-C et al (2008) Concurrence of compulsive buying and sexual addiction. Psychiatry Clin Neurosci 62:484. https://doi.org/10.1111/j.1440-1819.2008.01832.x

Yi S (2012) Shame-proneness as a risk factor of compulsive buying. J Consum Policy 35:393–410. https://doi.org/10.1007/s10603-012-9194-9

Yi S (2013) Heterogeneity of compulsive buyers based on impulsivity and compulsivity dimensions: a latent profile analytic approach. Psychiatry Res 208:174–182. https://doi.org/10.1016/j.psychres.2012.09.058

Zheng Y, Yang X, Liu Q et al (2020) Perceived stress and online compulsive buying among women: a moderated mediation model. Comput Hum Behav 103:13–20. https://doi.org/10.1016/j.chb.2019.09.012